THE VALUE
OF
SCIENCE

HENRI POINCARÉ

THE VALUE

OF

SCIENCE

ESSENTIAL WRITINGS OF HENRI POINCARÉ

STEPHEN JAY GOULD
SERIES EDITOR

THE MODERN LIBRARY

NEW YORK

2001 Modern Library Paperback Edition

Biographical note and compilation copyright © 2001 by Random House, Inc.
Series introduction copyright © 2001 by Stephen Jay Gould

This work is comprised of three seperate volumes by Henri Poincaré: *Science and
Hypothesis* (1905), *The Value of Science* (1913), and *Science and Method* (1914).

LIBRARY OF CONGRESS CATALOGING-IN-PUBLICATION DATA
Poincaré, Henri, 1854–1912.
[Selections. English]
The value of science : essential writings of Henri Poincaré /
Henri Poincaré.
p. cm. — (Modern Library science series)
Includes index.
ISBN 0-375-75848-8
1. Science—Philosophy. 2. Science—Methodology. 3. Mathematics—
Philosophy. I. Title. II. Modern Library science series (New York, N.Y.)
Q175.P7815213 2001
501—dc21 2001030834

Modern Library website address: www.modernlibrary.com

Printed in the United States of America

2 4 6 8 9 7 5 3 1

HENRI POINCARÉ

Jules Henri Poincaré, the illustrious French mathematician, theoretical astronomer, and philosopher, was born on April 29, 1854, in the city of Nancy in northeastern France. He was descended from a prominent family that had lived in the Lorraine region for several generations. His father was a physician and professor of medicine at the University of Nancy; his first cousin, Raymond Poincaré, later served as the president of France during World War I. Poincaré was educated at the local lycée, where a teacher deemed him a "monster of mathematics," and in 1873 he entered the École Polytéchnique at the top of his class. A genius who performed complex calculations in his head and committed them to paper only upon completion, he earned a doctorate from the École Nationale Supérieure des Mines in 1879 for a thesis on differential equations. Thereafter he taught briefly at the University of Caen before joining the faculty of the University of Paris in 1881.

Poincaré spent the rest of his life in Paris as the ruler of French mathematics. He achieved world renown before the age of thirty for developing the idea of "automorphic functions." Poincaré made substantial contributions to the theory of periodic orbits, and in 1889 he was awarded a prize by King Oscar II of Sweden for his work on the so-called "problem of *n* bodies." His breakthroughs in mathematical astronomy were outlined in a three-volume treatise, *Les méthodes nouvelles de la mécanique céleste* (1892–1899; translated as *New Methods of Celestial Mechanics* in 1993), the aim of which was, in Poincaré's own words, "to ascertain whether Newton's law of gravitation sufficed to explain all celestial phenomena." In 1906, in a paper on the dynamics of the electron, he obtained, quite separately from Albert Einstein, many of the results of the special theory of relativity. In addition he published numerous collections of lectures delivered at the Sorbonne, notably *Figures d'équilibre d'une masse fluide* (Figures of equilibrium of a fluid mass, 1902), *Leçons de mécanique céleste* (Lessons of celestial mechanics, 1905–1910), and *Leçons sur les hypothèses cosmogoniques* (Lessons on cosmological hypotheses, 1911).

"A scientist worthy of the name, above all a mathematician, experiences in his work the same impression as an artist," remarked Poincaré. "His pleasure is as great and of the same nature." Poincaré devoted much of the last decade of his life to sharing the meaning and human importance of science and mathematics with the general public. Ever absorbed by the philosophical implications of his subject, he gathered his views on the foundations of science in three lucid and compelling volumes. *La science et l'hypothèse* (1903; translated as *Science and Hypothesis* in 1952) presents Poincaré's famous discussion of creative psychology as it is revealed in the physical sciences. *La valeur de la science* (1905; translated as *The Value of Science* in 1958) deliberates the nature of scientific truth and considers whether order is innate in the universe or imposed upon it by man. *Science et méthode* (1908; translated as *Science and Method* in 1952) is concerned with the basic methodology and psychology of scientific discovery. All three books were bestsellers at the time of their initial publication and have since been widely translated. "[Poincaré's] essays on the foundations of science strike one as extemporaneous speeches rather than edited articles," observed one critic. "His prose is crisp, concise; it abounds in witty sentences, clever metaphors, and bold analogies." In 1908, in recognition of the literary excellence of his popular writings, Poincaré was elected to membership in the Académie Française, the highest honor accorded a French writer.

Henri Poincaré died suddenly from an embolism on July 17, 1912. *Dernières pensées*, a posthumous volume of his articles and lectures, was issued in 1913; it appeared in English as *Mathematics and Science: Last Essays* in 1963. An eleven-volume compilation of his lifework, *Oeuvres de Henri Poincaré* (*Works of Henri Poincaré*), was published in installments by the Académie des Sciences de Paris between 1916 and 1954. "Poincaré was the last man to take practically all mathematics, both pure and applied, as his province," said Eric T. Bell of the California Institute of Technology. "Few mathematicians have had the breadth of philosophical vision that Poincaré had, and none is his superior in the gift of clear exposition.... Poincaré spoke the universal languages of mathematics and science to all in accents which they recognized [and] was acknowledged as the foremost mathematician and leading popularizer of science of his time."

Introduction to the Modern Library

Science Series

THE NATURAL STATUS OF SCIENCE AS LITERATURE

Stephen Jay Gould

I have never quite figured out why standard college courses on Victorian literature invariably include Charles Dickens and George Eliot (as they should), but almost invariably exclude Thomas Henry Huxley and Charles Lyell, who wrote just as well, albeit in the different genre of science. Anyone who has accepted, passively and without personal examination, the conventional caricature of scientific writing as boring, inaccessible, illiterate, or unreadable might just consider Lyell's indictment, from the early 1830s, of catastrophism in geology, and his defense of the "uniformitarian" view that current causes, acting at their modest and observable rates throughout the immensity of geological time, can build the full panoply of earthly events, from Grand Canyons to Himalayan Mountains. (I don't even agree with Lyell's conclusion, for we now know that catastrophic impacts cause at least some mass extinctions, but I admire the literary quality of his largely correct assessment of old-style speculative catastrophism—even down to the slightly forced unsplit infinitive of the last line):

Never was there a dogma more calculated to foster indolence, and to blunt the keen edge of curiosity, than this assumption of the discor-

> dance between the former and existing causes of change.... The student was taught to despond from the first. Geology, it was affirmed, could never rise to the rank of an exact science....[With catastrophism] we see the ancient spirit of speculation revived, and a desire manifestly shown to cut, rather than patiently to untie, the Gordian Knot.

I would not claim, of course, that all science, or even science in general, appears in print as good writing. After all, literary quality does not rank high as a criterion (or even as a recognized property) for most editors of scientific journals, or for most practitioners themselves—and felicitous writers, as students, generally receive strong pushes, from their early teachers, towards the arts and humanities. Moreover, and for some perverse reason that I have never understood, editors of scientific journals have adopted several conventions that stifle good prose, albeit unintentionally—particularly the unrelenting passive voice required in descriptive sections, and often used throughout. The desired goals are, presumably, modesty, brevity, and objectivity; but why don't these editors understand that the passive voice, a pretty barbarous literary mode in most cases, but especially in this unrelenting and listlike form, offers no such guarantee? A person can be just as immodest thereby ("the discovery that was made will prove to be the greatest..."); moreover, the passive voice usually requires more words ("the work that was done showed...") than the far more eloquent direct statement ("I showed that...").

Nonetheless, and even within these strictures of Draconian editing and self-selection of people with little concern for literary quality into the profession, a remarkable amount of good writing does sneak into the pages of our technical journals. For example, the most famous short paper of modern science, Watson and Crick's announcement of the structure of DNA in 1953, begins with a crisp and elegant statement in the active voice (and certainly with requisite, if false, modesty): "We wish to suggest a structure for the salt of desoxyribose nucleic acid." The paper then ends with a lovely use of a classical literary device (understatement), invoked here to lay claim to a significant conclusion of equal importance to the discovery itself—an elucidation of the mechanism of replication, or self-copying, by DNA molecules—that the authors had inferred from the structure but had not yet proven: "It has not escaped our notice that the specific pairing mechanism we have postulated immediately suggests a possible copying mechanism for the genetic material."

In any case, we need hardly mine the technical literature to find works of sufficiently high literary quality and intellectual importance to merit inclusion within the Modern Library Science Series. After all, the parallel genre of "popular science writing" has developed, akin and apace, with technical publication. I do not speak of quick journalistic reads, hastily composed, consciously dumbeddown and hyped, and usually written by nonscientists without the requisite "feel" for the ethos of lab or field life in daily practice. Such works abound, and deserve their quick and permanent obsolescence. Rather, I speak of the fine literary skills possessed by several excellent scientists in each generation (not nearly a majority, of course, but we need representation, not saturation, and a little elitism is not a dangerous thing in this realm). These people write books of high literary quality and intellectual content—volumes accessible to all, and addressed equally to scientific peers and to a celebrated abstraction, doubted by some, but who really exist in large numbers, "the intelligent layperson" (again, far from a majority, but we Americans are 300 million strong, so even just a few percent of interested folks will buy more than enough books to justify the bottom line of the enterprise).

The books in this series represent excellent examples of this important genre, spanning literature and science, readable by all, and written by leading experts who did the original work in their technical fields. Thus, this series does not try to develop a compendium of the most influential works of science, in order of their putative importance—for some obvious contenders, Newton's *Principia,* for example, are truly inaccessible to general audiences. Similarly, we do not present a compendium of the best literature about scientific subjects, for some of these works display more style than content. Rather, we include in this series important books that stand both as landmarks of original scientific accomplishment and as highly readable works of substantial literary skill, explanatory depth and clarity.

The exclusion of some obvious candidates—Darwin's *Origin of Species* comes first to mind—only records their continued and ready availability in good and inexpensive editions by other publishers. (Over the years, many of my graduate students in evolutionary biology, after reading and enjoying Darwin's *Origin of Species,* then ask me, "Fine, I've now read the popular version, but where is the technical work that Darwin wrote for professionals before he watered down the *Origin* for lay readers?" I tell them that no such document exists, and that Darwin chose to present the greatest discovery in the history of biological thought as a volume for all intelligent readers. His achievement both validates the greatest traditions of humanism, and represents the common goal of the books reproduced in this series.)

We therefore pledge to give you, in this series, a set of the most readable and influential books in the history of public understanding of science, all written by primary doers, not just secondary interpreters, of this great enterprise. The last line of the preface written to introduce Copernicus's reconstruction of the universe in 1543 will therefore stand as the epitome of good advice to readers of this series: *"eme, lege, fruere,"* or "buy, read and enjoy!"

CONTENTS

THE VALUE OF SCIENCE
TRANSLATED BY GEORGE BRUCE HALSTED

SCIENCE AND METHOD
TRANSLATED BY FRANCIS MAITLAND

SCIENCE
AND
HYPOTHESIS

AUTHOR'S PREFACE

To the superficial observer scientific truth is unassailable, the logic of science is infallible; and if scientific men sometimes make mistakes, it is because they have not understood the rules of the game. Mathematical truths are derived from a few self-evident propositions, by a chain of flawless reasonings; they are imposed not only on us, but also on nature itself. By them the Creator is fettered, as it were, and His choice is limited to a relatively small number of solutions. A few experiments, therefore, will be sufficient to enable us to determine what choice He has made. From each experiment a number of consequences will follow by a series of mathematical deductions, and in this way each of them will reveal to us a corner of the universe. This, to the minds of most people, and to students who are getting their first ideas of physics, is the origin of certainty in science. This is what they take to be the role of experiment and mathematics. And thus, too, it was understood a hundred years ago by many men of science who dreamed of constructing the world with the aid of the smallest possible amount of material borrowed from experiment.

But upon more mature reflection the position held by hypothesis was seen; it was recognised that it is as necessary to the experimenter as it is to the mathematician. And then the doubt arose if all these constructions are built on solid foundations. The conclusion was drawn that a breath would bring them to the ground. This sceptical attitude does not escape the charge of superficiality. To doubt everything or to believe everything is two equally convenient solutions; both dispense with the necessity of reflection.

Instead of a summary condemnation we should examine with the utmost care the role of hypothesis; we shall then recognise not only that it is necessary, but that in most cases it is legitimate. We shall also see that there are several kinds of hypotheses; that some are verifiable, and when once confirmed by experiment become truths of great fertility; that others may be useful to us in fixing our ideas; and finally, that others are hypotheses only in appearance, and reduce to definitions or to conventions in disguise. The latter are to be met with especially in mathematics and in the sciences to which it is applied. From them, indeed, the sciences derive their rigour; such conventions are the result of the unrestricted activity of the mind, which in this domain recognises no obstacle. For here the mind may affirm because it lays down its own laws; but let us clearly understand that while these laws are imposed on *our* science, which otherwise could not exist, they are not imposed on nature. Are they then arbitrary? No; for if they were, they would not be fertile. Experience leaves us our freedom of choice, but it guides us by helping us to discern the most convenient path to follow. Our laws are therefore like those of an absolute monarch, who is wise and consults his council of state. Some people have been struck by this characteristic of free convention which may be recognised in certain fundamental principles of the sciences. Some have set no limits to their generalisations, and at the same time they have forgotten that there is a difference between liberty and the purely arbitrary. So that they are compelled to end in what is called *nominalism;* they have asked if the *savant* is not the dupe of his own definitions, and if the world he thinks he has discovered is not simply the creation of his own caprice.* Under these conditions science

---

* Cf. M. le Roy: "Science et Philosophie," *Revue de Métaphysique et de Morale,* 1901.

would retain its certainty, but would not attain its object, and would become powerless. Now, we daily see what science is doing for us. This could not be unless it taught us something about reality; the aim of science is not things themselves, as the dogmatists in their simplicity imagine, but the relations between things; outside those relations there is no reality knowable.

Such is the conclusion to which we are led; but to reach that conclusion we must pass in review the series of sciences from arithmetic and geometry to mechanics and experimental physics. What is the nature of mathematical reasoning? Is it really deductive, as is commonly supposed? Careful analysis shows us that it is nothing of the kind; that it participates to some extent in the nature of inductive reasoning, and for that reason it is fruitful. But nonetheless does it retain its character of absolute rigour; and this is what must first be shown.

When we know more of this instrument which is placed in the hands of the investigator by mathematics, we have then to analyse another fundamental idea, that of mathematical magnitude. Do we find it in nature, or have we ourselves introduced it? And if the latter be the case, are we not running a risk of coming to incorrect conclusions all round? Comparing the rough data of our senses with that extremely complex and subtle conception which mathematicians call magnitude, we are compelled to recognise a divergence. The framework into which we wish to make everything fit is one of our own construction; but we did not construct it at random, we constructed it by measurement so to speak; and that is why we can fit the facts into it without altering their essential qualities.

Space is another framework which we impose on the world. Whence are the first principles of geometry derived? Are they imposed on us by logic? Lobatschewsky, by inventing non-Euclidean geometries, has shown that this is not the case. Is space revealed to us by our senses? No; for the space revealed to us by our senses is absolutely different from the space of geometry. Is geometry derived from experience? Careful discussion will give the answer— no! We therefore conclude that the principles of geometry are only conventions; but these conventions are not arbitrary, and if transported into another world (which I shall call the non-Euclidean world, and which I shall endeavour to describe), we shall find ourselves compelled to adopt more of them.

In mechanics we shall be led to analogous conclusions, and we shall see that the principles of this science, although more directly based on experience, still share the conventional character of the geometrical postulates. So far, nominalism triumphs; but we now come to the physical sciences, properly so called, and here the scene changes. We meet with hypotheses of another kind, and we fully grasp how fruitful they are. No doubt at the outset theories seem unsound, and the history of science shows us how ephemeral they are; but they do not entirely perish, and of each of them some traces still remain. It is these traces which we must try to discover, because in them and in them alone is the true reality.

The method of the physical sciences is based upon the induction which leads us to expect the recurrence of a phenomenon when the circumstances which give rise to it are repeated. If all the circumstances could be simultaneously reproduced, this principle could be fearlessly applied; but this never happens; some of the circumstances will always be missing. Are we absolutely certain that they are unimportant? Evidently not! It may be probable, but it cannot be rigorously certain. Hence the importance of the role that is played in the physical sciences by the law of probability. The calculus of probabilities is therefore not merely a recreation, or a guide to the baccarat player; and we must thoroughly examine the principles on which it is based. In this connection I have but very incomplete results to lay before the reader, for the vague instinct which enables us to determine probability almost defies analysis. After a study of the conditions under which the work of the physicist is carried on, I have thought it best to show him at work. For this purpose I have taken instances from the history of optics and of electricity. We shall thus see how the ideas of Fresnel and Maxwell took their rise, and what unconscious hypotheses were made by Ampère and the other founders of electro-dynamics.

PART I

NUMBER

AND

MAGNITUDE

CHAPTER I

ON THE NATURE OF
MATHEMATICAL REASONING

I

The very possibility of mathematical science seems an insoluble contradiction. If this science is only deductive in appearance, from whence is derived that perfect rigour which is challenged by none? If, on the contrary, all the propositions which it enunciates may be derived in order by the rules of formal logic, how is it that mathematics is not reduced to a gigantic tautology? The syllogism can teach us nothing essentially new, and if everything must spring from the principle of identity, then everything should be capable of being reduced to that principle. Are we then to admit that the enunciations of all the theorems with which so many volumes are filled are only indirect ways of saying that A is A?

No doubt we may refer back to axioms which are at the source of all these reasonings. If it is felt that they cannot be reduced to the principle of contradiction, if we decline to see in them any more than experimental facts which have no part or lot in mathematical necessity, there is still one resource left to us: we may class them among *à priori* synthetic views. But this is no solution of the difficulty—it is merely giving it a name; and even if the nature of the synthetic views had no longer for us any mystery, the contradiction would not have disappeared; it would have only been shirked. Syllogistic reasoning remains incapable of adding anything to the data that are given it; the data are reduced to axioms, and that is all we should find in the conclusions.

No theorem can be new unless a new axiom intervenes in its demonstration; reasoning can only give us immediately evident truths borrowed from direct intuition; it would only be an intermediary parasite. Should we not therefore have reason for asking if the syllogistic apparatus serves only to disguise what we have borrowed?

The contradiction will strike us the more if we open any book on mathematics; on every page the author announces his intention of generalising some proposition already known. Does the mathematical method proceed from the particular to the general, and, if so, how can it be called deductive?

Finally, if the science of number were merely analytical, or could be analytically derived from a few synthetic intuitions, it seems that a sufficiently powerful mind could with a single glance perceive all its truths; nay, one might even hope that some day a language would be invented simple enough for these truths to be made evident to any person of ordinary intelligence.

Even if these consequences are challenged, it must be granted that mathematical reasoning has of itself a kind of creative virtue, and is therefore to be distinguished from the syllogism. The difference must be profound. We shall not, for instance, find the key to the mystery in the frequent use of the rule by which the same uniform operation applied to two equal numbers will give identical results. All these modes of reasoning, whether or not reducible to the syllogism, properly so called, retain the analytical character, and *ipso facto*, lose their power.

II

The argument is an old one. Let us see how Leibnitz tried to show that two and two make four. I assume the number one to be defined, and also the operation $x+1$—*i.e.,* the adding of unity to a given number x. These definitions, whatever they may be, do not enter into the subsequent reasoning. I next define the numbers 2, 3, 4 by the equalities:

(1) $1+1=2$; (2) $2+1=3$; (3) $3+1=4$, and in the same way I define the operation $x+2$ by the relation; (4) $x+2=(x+1)+1$.

Given this, we have:

$$2+2=(2+1)+1; \text{ (def. 4)}.$$
$$(2+1)+1=3+1 \quad \text{(def. 2)}.$$
$$3+1=4 \quad \text{(def. 3)}.$$
$$\text{whence } 2+2=4 \qquad \text{Q.E.D.}$$

It cannot be denied that this reasoning is purely analytical. But if we ask a mathematician, he will reply: "This is not a demonstration properly so called; it is a verification." We have confined ourselves to bringing together one or other of two purely conventional definitions, and we have verified their identity; nothing new has been learned. *Verification* differs from proof precisely because it is analytical, and because it leads to nothing. It leads to nothing because the conclusion is nothing but the premisses translated into another language. A real proof, on the other hand, is fruitful, because the conclusion is in a sense more general than the premisses. The equality $2+2=4$ can be verified because it is particular. Each individual enunciation in mathematics may be always verified in the same way. But if mathematics could be reduced to a series of such verifications it would not be a science. A chess-player, for instance, does not create a science by winning a piece. There is no science but the science of the general. It may even be said that the object of the exact sciences is to dispense with these direct verifications.

III

Let us now see the geometer at work, and try to surprise some of his methods. The task is not without difficulty; it is not enough to open a book at random and to analyse any proof we may come across. First of all, geometry must be excluded, or the question becomes complicated by difficult problems relating to the role of the postulates, the nature and the origin of the idea of space. For analogous reasons we cannot avail ourselves of the infinitesimal calculus. We must seek mathematical thought where it has remained pure—*i.e.*, in Arithmetic. But we still have to choose; in the higher parts of the theory of numbers the primitive mathematical ideas have already undergone so profound an elaboration that it becomes difficult to analyse them.

It is therefore at the beginning of Arithmetic that we must expect to find the explanation we seek; but it happens that it is precisely in the proofs of the most elementary theorems that the authors of classic treatises have displayed the least precision and rigour. We may not impute this to them as a crime; they have obeyed a necessity. Beginners are not prepared for real mathematical rigour; they would see in it nothing but empty, tedious subtleties. It would be a waste of time to try to make them more exacting; they have to pass rapidly and without stopping over the road which was trodden slowly by the founders of the science.

Why is so long a preparation necessary to habituate oneself to this perfect rigour, which it would seem should naturally be imposed on all minds? This is a logical and psychological problem which is well worthy of study. But we shall not dwell on it; it is foreign to our subject. All I wish to insist on is that we shall fail in our purpose unless we reconstruct the proofs of the elementary theorems, and give them, not the rough form in which they are left so as not to weary the beginner, but the form which will satisfy the skilled geometer.

DEFINITION OF ADDITION

I assume that the operation $x + 1$ has been defined; it consists in adding the number 1 to a given number x. Whatever may be said of this definition, it does not enter into the subsequent reasoning.

We now have to define the operation $x + a$, which consists in adding the number a to any given number x. Suppose that we have defined the operation $x + (a - 1)$; the operation $x + a$ will be defined by the equality: (1) $x + a = [x + (a - 1)] + 1$. We shall know what

$x+a$ is when we know what $x+(a\text{-}1)$ is, and as I have assumed that to start with we know what $x+1$ is, we can define successively and "by recurrence" the operations $x+2$, $x+3$, etc. This definition deserves a moment's attention; it is of a particular nature which distinguishes it even at this stage from the purely logical definition; the equality (1), in fact, contains an infinite number of distinct definitions, each having only one meaning when we know the meaning of its predecessor.

Properties of Addition

Associative. I say that $a+(b+c)=(a+b)+c$; in fact, the theorem is true for $c=1$. It may then be written $a+(b+1)=(a+b)+1$; which, remembering the difference of notation, is nothing but the equality (1) by which I have just defined addition. Assume the theorem true for $c=\gamma$, I say that it will be true for $c=\gamma+1$. Let $(a+b)+\gamma=a+(b+\gamma)$, it follows that $[(a+b)+\gamma]+1=[a+(b+\gamma)]+1$; or by def. (1)—$(a+b)+(\gamma+1)=a+(b+\gamma+1)=a+[b+(\gamma+1)]$, which shows by a series of purely analytical deductions that the theorem is true for $\gamma+1$. Being true for $c=1$, we see that it is successively true for $c=2$, $c=3$, etc.

Commutative. (1) I say that $a+1=1+a$. The theorem is evidently true for $a=1$; we can *verify* by purely analytical reasoning that if it is true for $a=\gamma$ it will be true for $a=\gamma+1$.* Now, it is true for $a=1$, and therefore is true for $a=2$, $a=3$, and so on. This is what is meant by saying that the proof is demonstrated "by recurrence."

(2) I say that $a+b=b+a$. The theorem has just been shown to hold good for $b=1$, and it may be verified analytically that if it is true for $b=\beta$, it will be true for $b=\beta+1$. The proposition is thus established by recurrence.

* For $(\gamma+1)+1=(1+\gamma)+1=1+(\gamma+1)$.—[Tr.]

DEFINITION OF MULTIPLICATION

We shall define multiplication by the equalities: (1) $a \times 1 = a$. (2) $a \times b = [a \times (b\text{-}1)] + a$. Both of these include an infinite number of definitions; having defined $a \times 1$, it enables us to define in succession $a \times 2$, $a \times 3$, and so on.

PROPERTIES OF MULTIPLICATION

Distributive. I say that $(a + b) \times c = (a \times c) + (b \times c)$. We can verify analytically that the theorem is true for $c = 1$; then if it is true for $c = \gamma$, it will be true for $c = \gamma + 1$. The proposition is then proved by recurrence.

Commutative. (1) I say that $a \times 1 = 1 \times a$. The theorem is obvious for $a = 1$. We can verify analytically that if it is true for $a = a$, it will be true for $a = a + 1$.

(2) I say that $a \times b = b \times a$. The theorem has just been proved for $b = 1$. We can verify analytically that if it be true for $b = \beta$ it will be true for $b = \beta + 1$.

IV

This monotonous series of reasonings may now be laid aside; but their very monotony brings vividly to light the process, which is uniform, and is met again at every step. The process is proof by recurrence. We first show that a theorem is true for $n = 1$; we then show that if it is true for $n - 1$ it is true for n, and we conclude that it is true for all integers. We have now seen how it may be used for the proof of the rules of addition and multiplication—that is to say, for the rules of the algebraic calculus. This calculus is an instrument of transformation which lends itself to many more different combinations than the simple syllogism; but it is still a purely analytical instrument, and is incapable of teaching us anything new. If mathematics had no other instrument, it would immediately be arrested in its development; but it has recourse anew to the same process—*i.e.*, to reasoning by recurrence, and it can continue its forward march. Then if we look carefully, we find this mode of reasoning at every step, either under the simple form which we have just given to it, or under a more or less modified form. It is therefore mathematical reasoning *par excellence*, and we must examine it closer.

V

The essential characteristic of reasoning by recurrence is that it contains, condensed, so to speak, in a single formula, an infinite number of syllogisms. We shall see this more clearly if we enunciate the syllogisms one after another. They follow one another, if one may use the expression, in a cascade. The following are the hypothetical syllogisms: The theorem is true of the number 1. Now, if it is true of 1, it is true of 2; therefore it is true of 2. Now, if it is true of 2, it is true of 3; hence it is true of 3, and so on. We see that the conclusion of each syllogism serves as the minor of its successor. Further, the majors of all our syllogisms may be reduced to a single form. If the theorem is true of $n-1$, it is true of n.

We see, then, that in reasoning by recurrence we confine ourselves to the enunciation of the minor of the first syllogism, and the general formula which contains as particular cases all the majors. This unending series of syllogisms is thus reduced to a phrase of a few lines.

It is now easy to understand why every particular consequence of a theorem may, as I have above explained, be verified by purely analytical processes. If, instead of proving that our theorem is true for all numbers, we only wish to show that it is true for the number 6 for instance, it will be enough to establish the first five syllogisms in our cascade. We shall require 9 if we wish to prove it for the number 10; for a greater number we shall require more still; but however great the number may be we shall always reach it, and the analytical verification will always be possible. But however far we went we should never reach the general theorem applicable to all numbers, which alone is the object of science. To reach it we should require an infinite number of syllogisms, and we should have to cross an abyss which the patience of the analyst, restricted to the resources of formal logic, will never succeed in crossing.

I asked at the outset why we cannot conceive of a mind powerful enough to see at a glance the whole body of mathematical truth. The answer is now easy. A chess-player can combine for four or five moves ahead; but, however extraordinary a player he may be, he cannot prepare for more than a finite number of moves. If he applies his faculties to Arithmetic, he cannot conceive its general truths by direct intuition alone; to prove even the smallest theorem he must use reasoning by recurrence, for that is the only instrument which enables us to pass from the finite to the infinite. This instrument is always useful, for it enables us to leap over as many stages as we wish; it frees us from the necessity of long, tedious, and monotonous verifications which would rapidly become impracticable. Then when we take in hand the general theorem it becomes indispensable, for otherwise we should ever be approaching the analytical verification without ever actually reaching it. In this domain of Arithmetic we may think ourselves very far from the infinitesimal analysis, but the idea of mathematical infinity is already playing a preponderating part, and without it there would be no science at all, because there would be nothing general.

VI

The views upon which reasoning by recurrence is based may be exhibited in other forms; we may say, for instance, that in any finite collection of different integers there is always one which is smaller than any other. We may readily pass from one enunciation to another, and thus give ourselves the illusion of having proved that reasoning by recurrence is legitimate. But we shall always be brought to a full stop—we shall always come to an indemonstrable axiom, which will at bottom be but the proposition we had to prove translated into another language. We cannot therefore escape the conclusion that the rule of reasoning by recurrence is irreducible to the principle of contradiction. Nor can the rule come to us from experiment. Experiment may teach us that the rule is true for the first ten or the first hundred numbers, for instance; it will not bring us to the indefinite series of numbers, but only to a more or less long, but always limited, portion of the series.

Now, if that were all that is in question, the principle of contradiction would be sufficient, it would always enable us to develop as many syllogisms as we wished. It is only when it is a question of a single formula to embrace an infinite number of syllogisms that this principle breaks down, and there, too, experiment is powerless to aid. This rule, inaccessible to analytical proof and to experiment, is the exact type of the *à priori* synthetic intuition. On the other hand, we cannot see in it a convention as in the case of the postulates of geometry.

Why then is this view imposed upon us with such an irresistible weight of evidence? It is because it is only the affirmation of the power of the mind which knows it can conceive of the indefinite repetition of the same act, when the act is once possible. The mind has a direct intuition of this power, and experiment can only be for it an opportunity of using it, and thereby of becoming conscious of it.

But it will be said, if the legitimacy of reasoning by recurrence cannot be established by experiment alone, is it so with experiment aided by induction? We see successively that a theorem is true of the number 1, of the number 2, of the number 3, and so on—the law is manifest, we say, and it is so on the same ground that every physical law is true which is based on a very large but limited number of observations.

It cannot escape our notice that here is a striking analogy with the usual processes of induction. But an essential difference exists. Induction applied to the physical sciences is always uncertain, because it is based on the belief in a general order of the universe, an order which is external to us. Mathematical induction—*i.e.,* proof by recurrence—is, on the contrary, necessarily imposed on us, because it is only the affirmation of a property of the mind itself.

VII

Mathematicians, as I have said before, always endeavour to generalise the propositions they have obtained. To seek no further example, we have just shown the equality, $a+1=1+a$, and we then used it to establish the equality, $a+b=b+a$, which is obviously more general. Mathematics may, therefore, like the other sciences, proceed from the particular to the general. This is a fact which might otherwise have appeared incomprehensible to us at the beginning of this study, but which has no longer anything mysterious about it, since we have ascertained the analogies between proof by recurrence and ordinary induction.

No doubt mathematical recurrent reasoning and physical inductive reasoning are based on different foundations, but they move in parallel lines and in the same direction—namely, from the particular to the general.

Let us examine the case a little more closely. To prove the equality $a+2=2+a\ldots\ldots(1)$, we need only apply the rule $a+1=1+a$, twice, and write $a+2=a+1+1=1+a+1=1+1+a=2+a\ldots\ldots(2)$.

The equality thus deduced by purely analytical means is not, however, a simple particular case. It is something quite different. We may not therefore even say in the really analytical and deductive part of mathematical reasoning that we proceed from the general to the particular in the ordinary sense of the words. The two sides of the equality (2) are merely more complicated combinations than the two sides of the equality (1), and analysis only serves to separate the elements which enter into these combinations and to study their relations.

Mathematicians therefore proceed "by construction," they "construct" more complicated combinations. When they analyse these combinations, these aggregates, so to speak, into their primitive elements, they see the relations of the elements and deduce the relations of the aggregates themselves. The process is purely analytical, but it is not a passing from the general to the particular, for the aggregates obviously cannot be regarded as more particular than their elements.

Great importance has been rightly attached to this process of "construction," and some claim to see in it the necessary and sufficient condition of the progress of the exact sciences. Necessary, no doubt, but not sufficient! For a construction to be useful and not mere waste of mental effort, for it to serve as a stepping-stone to higher things, it must first of all possess a kind of unity enabling us to see something more than the juxtaposition of its elements. Or more accurately, there must be some advantage in considering the construction rather than the elements themselves. What can this advantage be? Why reason on a polygon, for instance, which is always decomposable into triangles, and not on elementary triangles? It is because there are properties of polygons of any number of sides, and they can be immediately applied to any particular kind of polygon. In most cases it is only after long efforts that those properties can be discovered, by directly studying the relations of elementary triangles. If the quadrilateral is anything more than the juxtaposition of two triangles, it is because it is of the polygon type.

A construction only becomes interesting when it can be placed side by side with other analogous constructions for forming species of the same genus. To do this we must necessarily go back from the particular to the general, ascending one or more steps. The analytical process "by construction" does not compel us to descend, but it leaves us at the same level. We can only ascend by mathematical induction, for from it alone can we learn something new. Without the aid of this induction, which in certain respects differs from, but is as fruitful as, physical induction, construction would be powerless to create science.

Let me observe, in conclusion, that this induction is only possible if the same operation can be repeated indefinitely. That is why the theory of chess can never become a science, for the different moves of the same piece are limited and do not resemble each other.

CHAPTER II

MATHEMATICAL MAGNITUDE AND EXPERIMENT

If we want to know what the mathematicians mean by a continuum, it is useless to appeal to geometry. The geometer is always seeking, more or less, to represent to himself the figures he is studying, but his representations are only instruments to him; he uses space in his geometry just as he uses chalk; and further, too much importance must not be attached to accidents which are often nothing more than the whiteness of the chalk.

The pure analyst has not to dread this pitfall. He has disengaged mathematics from all extraneous elements, and he is in a position to answer our question: "Tell me exactly what this continuum is, about which mathematicians reason." Many analysts who reflect on their art have already done so—M. Tannery, for instance, in his *Introduction à la théorie des fonctions d'une variable*.

Let us start with the integers. Between any two consecutive sets, intercalate one or more intermediary sets, and then between these sets others again, and so on indefinitely. We thus get an unlimited number of terms, and these will be the numbers which we call fractional, rational, or commensurable. But this is not yet all; between these terms, which, be it marked, are already infinite in number, other terms are intercalated, and these are called irrational or incommensurable.

Before going any further, let me make a preliminary remark. The continuum thus conceived is no longer a collection of individuals arranged in a certain order, infinite in number, it is true, but external the one to the other. This is not the ordinary conception in which it is supposed that between the elements of the continuum exists an intimate connection making of it one whole, in which the point has no existence previous to the line, but the line does exist previous to the point. Multiplicity alone subsists, unity has disappeared—"the continuum is unity in multiplicity," according to the celebrated formula. The analysts have even less reason to define their continuum as they do, since it is always on this that they reason when they are particularly proud of their rigour. It is enough to warn the reader that the real mathematical continuum is quite different from that of the physicists and from that of the metaphysicians.

It may also be said, perhaps, that mathematicians who are contented with this definition are the dupes of words, that the nature of each of these sets should be precisely indicated, that it should be explained how they are to be intercalated, and that it should be shown how it is possible to do it. This, however, would be wrong; the only property of the sets which comes into the reasoning is that of preceding or succeeding these or those other sets; this alone should therefore intervene in the definition. So we need not concern ourselves with the manner in which the sets are intercalated, and no one will doubt the possibility of the operation if he only remembers that "possible" in the language of geometers simply means exempt from contradiction. But our definition is not yet complete, and we come back to it after this rather long digression.

Definition of Incommensurables. The mathematicians of the Berlin school, and Kronecker in particular, have devoted themselves to constructing this continuous scale of irrational and fractional numbers without using any other materials than the integer. The mathematical continuum from this point of view would be a pure creation of the mind in which experiment would have no part.

The idea of rational number not seeming to present to them any difficulty, they have confined their attention mainly to defining incommensurable numbers. But before reproducing their definition here, I must make an observation that will allay the astonishment which this will not fail to provoke in readers who are but little familiar with the habits of geometers.

Mathematicians do not study objects, but the relations between objects; to them it is a matter of indifference if these objects are replaced by others, provided that the relations do not change. Matter does not engage their attention, they are interested by form alone.

If we did not remember it, we could hardly understand that Kronecker gives the name of incommensurable number to a simple symbol—that is to say, something very different from the idea we think we ought to have of a quantity which should be measurable and almost tangible.

Let us see now what is Kronecker's definition. Commensurable numbers may be divided into classes in an infinite number of ways, subject to the condition that any number whatever of the first class is greater than any number of the second. It may happen that among the numbers of the first class there is one which is smaller than all the rest; if, for instance, we arrange in the first class all the numbers greater than 2, and 2 itself, and in the second class all the numbers smaller than 2, it is clear that 2 will be the smallest of all the numbers of the first class. The number 2 may therefore be chosen as the symbol of this division.

It may happen, on the contrary, that in the second class there is one which is greater than all the rest. This is what takes place, for example, if the first class comprises all the numbers greater than 2, and if, in the second, are all the numbers less than 2, and 2 itself. Here again the number 2 might be chosen as the symbol of this division.

But it may equally well happen that we can find neither in the first class a number smaller than all the rest, nor in the second class a number greater than all the rest. Suppose, for instance, we place in the first class all the numbers whose squares are greater than 2, and in the second all the numbers whose squares are smaller than 2. We know that in neither of them is a number whose square is equal to 2. Evidently there will be in the first class no number which is smaller than all the rest, for however near the square of a number may be to 2, we can always find a commensurable whose square is still nearer to 2. From Kronecker's point of view, the incommensurable number $\sqrt{2}$ is nothing but the symbol of this particular method of division of commensurable numbers; and to each mode of repartition corresponds in this way a number, commensurable or not, which serves as a symbol. But to be satisfied with this would be to forget the origin of these symbols; it remains to explain how we have been led to attribute to them a kind of concrete existence, and on the other hand, does not the difficulty begin with fractions? Should we have the notion of these numbers if we did not previously know a matter which we conceive as infinitely divisible—*i.e.*, as a continuum?

The Physical Continuum. We are next led to ask if the idea of the mathematical continuum is not simply drawn from experiment. If that be so, the rough data of experiment, which are our sensations, could be measured. We might, indeed, be tempted to believe that this is so, for in recent times there has been an attempt to measure them, and a law has even been formulated, known as Fechner's law, according to which sensation is proportional to the logarithm of the stimulus. But if we examine the experiments by which the endeavour has been made to establish this law, we shall be led to a diametrically opposite conclusion. It has, for instance, been observed that a weight A of 10 grammes and a weight B of 11 grammes produced identical sensations, that the weight B could no longer be distinguished from a weight C of 12 grammes, but that the weight A was readily distinguished from the weight C. Thus the rough results of the experiments may be expressed by the following relations: $A=B$, $B=C$, $A<C$, which may be regarded as the formula of the physical continuum. But here is an intolerable disagreement with the law of contradiction, and the necessity of banishing this

disagreement has compelled us to invent the mathematical continuum. We are therefore forced to conclude that this notion has been created entirely by the mind, but it is experiment that has provided the opportunity. We cannot believe that two quantities which are equal to a third are not equal to one another, and we are thus led to suppose that A is different from B, and B from C, and that if we have not been aware of this, it is due to the imperfections of our senses.

The Creation of the Mathematical Continuum: First Stage. So far it would suffice, in order to account for facts, to intercalate between A and B a small number of terms which would remain discrete. What happens now if we have recourse to some instrument to make up for the weakness of our senses? If, for example, we use a microscope? Such terms as A and B, which before were indistinguishable from one another, appear now to be distinct: but between A and B, which are distinct, is intercalated another new term D, which we can distinguish neither from A nor from B. Although we may use the most delicate methods, the rough results of our experiments will always present the characters of the physical continuum with the contradiction which is inherent in it. We only escape from it by incessantly intercalating new terms between the terms already distinguished, and this operation must be pursued indefinitely. We might conceive that it would be possible to stop if we could imagine an instrument powerful enough to decompose the physical continuum into discrete elements, just as the telescope resolves the Milky Way into stars. But this we cannot imagine; it is always with our senses that we use our instruments; it is with the eye that we observe the image magnified by the microscope, and this image must therefore always retain the characters of visual sensation, and therefore those of the physical continuum.

Nothing distinguishes a length directly observed from half that length doubled by the microscope. The whole is homogeneous to the part; and there is a fresh contradiction—or rather there would be one if the number of the terms was supposed to be finite; it is clear that the part containing fewer terms than the whole cannot be similar to the whole. The contradiction ceases as soon as the number of terms is regarded as infinite. There is nothing, for example, to prevent us from regarding the aggregate of integers as similar to the aggregate of even numbers, which is however only a part of it; in fact, to each integer corresponds another even number which is its double. But it is not only to escape this contradiction contained in the empiric data that the mind is led to create the concept of a continuum formed of an indefinite number of terms.

Here everything takes place just as in the series of the integers. We have the faculty of conceiving that a unit may be added to a collection of units. Thanks to experiment, we have had the opportunity of exercising this faculty and are conscious of it; but from this fact we feel that our power is unlimited, and that we can count indefinitely, although we have never had to count more than a finite number of objects. In the same way, as soon as we have intercalated terms between two consecutive terms of a series, we feel that this operation may be continued without limit, and that, so to speak, there is no intrinsic reason for stopping. As an abbreviation, I may give the name of a mathematical continuum of the first order to every aggregate of terms formed after the same law as the scale of commensurable numbers. If, then, we intercalate new sets according to the laws of incommensurable numbers, we obtain what may be called a continuum of the second order.

Second Stage. We have only taken our first step. We have explained the origin of continuums of the first order; we must now see why this is not sufficient, and why the incommensurable numbers had to be invented.

If we try to imagine a line, it must have the characters of the physical continuum—that is to say, our representation must have a certain breadth. Two lines will therefore appear to us under the form of two narrow bands, and if we are content with this rough image, it is clear that where two lines cross they must have some common part. But the pure geometer makes one further effort; without entirely renouncing the aid of his senses, he tries to imagine a line without breadth and a point without size. This he can do only by imagining a line as the limit towards which tends a band that is growing thinner and thinner, and the point as the limit towards which is tending an area that is growing smaller and smaller. Our two bands, however narrow they may be, will always have a common area; the smaller they are the smaller it will be, and its limit is what the geometer calls a point. This is why it is said that the two lines which cross must have a common point, and this truth seems intuitive.

But a contradiction would be implied if we conceived of lines as continuums of the first order—*i.e.,* the lines traced by the geometer should only give us points, the co-ordinates of which are rational numbers. The contradiction would be manifest if we were, for instance, to assert the existence of lines and circles. It is clear, in fact, that if the points whose co-ordinates are commensurable were alone regarded as real, the in-circle of a square and the diagonal of the square would not intersect, since the co-ordinates of the point of intersection are incommensurable.

Even then we should have only certain incommensurable numbers, and not all these numbers.

But let us imagine a line divided into two half-rays (*demi-droites*). Each of these half-rays will appear to our minds as a band of a certain breadth; these bands will fit close together, because there must be no interval between them. The common part will appear to us to be a point which will still remain as we imagine the bands to become thinner and thinner, so that we admit as an intuitive truth that if a line be divided into two half-rays the common frontier of these half-rays is a point. Here we recognise the conception of Kronecker, in which an incommensurable number was regarded as the common frontier of two classes of rational numbers. Such is the origin of the continuum of the second order, which is the mathematical continuum properly so called.

Summary. To sum up, the mind has the faculty of creating symbols, and it is thus that it has constructed the mathematical continuum, which is only a particular system of symbols. The only limit to its power is the necessity of avoiding all contradiction; but the mind only makes use of it when experiment gives a reason for it.

In the case with which we are concerned, the reason is given by the idea of the physical continuum, drawn from the rough data of the senses. But this idea leads to a series of contradictions from each of which in turn we must be freed. In this way we are forced to imagine a more and more complicated system of symbols. That on which we shall dwell is not merely exempt from internal contradiction—it was so already at all the steps we have taken—but it is no longer in contradiction with the various propositions which are called intuitive, and which are derived from more or less elaborate empirical notions.

Measurable Magnitude. So far we have not spoken of the *measure* of magnitudes; we can tell if any one of them is greater than any other, but we cannot say that it is two or three times as large.

So far, I have only considered the order in which the terms are arranged; but that is not sufficient for most applications. We must learn how to compare the interval which separates any two terms. On this condition alone will the continuum become measurable, and the operations of arithmetic be applicable. This can only be done by the aid of a new and special convention; and this convention is, that in such a case the interval between the terms A and B is equal to the interval which separates C and D. For instance, we started with the integers, and between two consecutive sets we intercalated n intermediary sets; by convention we now assume these new sets to be equidistant. This is one of the ways of defining the addition of two magnitudes; for if the interval AB is by definition equal to the interval CD, the interval AD will by definition be the sum of the intervals AB and AC. This definition is very largely, but not altogether, arbitrary. It must satisfy certain conditions—the commutative and associative laws of addition, for instance; but, provided the definition we choose satisfies these laws, the choice is indifferent, and we need not state it precisely.

Remarks. We are now in a position to discuss several important questions.

(1) Is the creative power of the mind exhausted by the creation of the mathematical continuum? The answer is in the negative, and this is shown in a very striking manner by the work of Du Bois Reymond.

We know that mathematicians distinguish between infinitesimals of different orders, and that infinitesimals of the second order are infinitely small, not only absolutely so, but also in relation to those of the first order. It is not difficult to imagine infinitesimals of fractional or even of irrational order, and here once more we find the mathematical continuum which has been dealt with in the preceding pages. Further, there are infinitesimals which are infinitely small with reference to those of the first order, and infinitely large with respect to the order $1 + \epsilon$, however small ϵ may be. Here, then, are new terms intercalated in our series; and if I may be permitted to revert to the terminology used in the preceding pages, a terminology which is very convenient, although it has not been consecrated by usage, I shall say that we have created a kind of continuum of the third order.

It is an easy matter to go further, but it is idle to do so, for we would only be imagining symbols without any possible application, and no one will dream of doing that. This continuum of the third order, to which we are led by the consideration of the different orders of infinitesimals, is in itself of but little use and hardly worth quoting. Geometers look on it as a mere curiosity. The mind only uses its creative faculty when experiment requires it.

(2) When we are once in possession of the conception of the mathematical continuum, are we protected from contradictions analogous to those which gave it birth? No, and the following is an instance:

He is a *savant* indeed who will not take it as evident that every curve has a tangent; and, in fact, if we think of a curve and a straight line as two narrow bands, we can always arrange them in such a way that they have a common part without intersecting. Suppose now that the breadth of the bands diminishes indefinitely: the common part will still remain, and in the limit, so to speak, the two lines will have a common point, although they do not intersect— *i.e.,* they will touch. The geometer who reasons in this way is only doing what we have done when we proved that two lines which intersect have a common point, and his intuition might also seem to be quite legitimate. But this is not the case. We can show that there are curves which have no tangent, if we define such a curve as an analytical continuum of the second order. No doubt some artifice analogous to those we have discussed above would enable us to get rid of this contradiction, but as the latter is only met with in very exceptional cases, we need not trouble to do so. Instead of endeavouring to reconcile intuition and analysis, we are content to sacrifice one of them, and as analysis must be flawless, intuition must go to the wall.

The Physical Continuum of Several Dimensions. We have discussed above the physical continuum as it is derived from the immediate evidence of our senses—or, if the reader prefers, from the rough results of Fechner's experiments; I have shown that these results are summed up in the contradictory formulæ: $A=B$, $B=C$, $A<C$.

Let us now see how this notion is generalised, and how from it may be derived the concept of continuums of several dimensions. Consider any two aggregates of sensations. We can either distinguish between them, or we cannot; just as in Fechner's experiments the weight of 10 grammes could be distinguished from the weight of 12 grammes, but not from the weight of 11 grammes. This is all that is required to construct the continuum of several dimensions.

Let us call one of these aggregates of sensations an *element.* It will be in a measure analogous to the *point* of the mathematicians, but will not be, however, the same thing. We cannot say that our element has no size, for we cannot distinguish it from its immediate neighbours, and it is thus surrounded by a kind of fog. If the astronomical comparison may be allowed, our "elements" would be like nebulæ, whereas the mathematical points would be like stars.

If this be granted, a system of elements will form a continuum, if we can pass from any one of them to any other by a series of consecutive elements such that each cannot be distinguished from its predecessor. This *linear* series is to the *line* of the mathematician what the isolated *element* was to the point.

Before going further, I must explain what is meant by a *cut*. Let us consider a continuum C, and remove from it certain of its elements, which for a moment we shall regard as no longer belonging to the continuum. We shall call the aggregate of elements thus removed a *cut*. By means of this cut, the continuum C will be *subdivided* into several distinct continuums; the aggregate of elements which remain will cease to form a single continuum. There will then be on C two elements, A and B, which we must look upon as belonging to two distinct continuums; and we see that this must be so, because it will be impossible to find a linear series of consecutive elements of C (each of the elements indistinguishable from the preceding, the first being A and the last B), *unless one of the elements of this series is indistinguishable from one of the elements of the cut.*

It may happen, on the contrary, that the cut may not be sufficient to subdivide the continuum C. To classify the physical continuums, we must first of all ascertain the nature of the cuts which must be made in order to subdivide them. If a physical continuum, C, may be subdivided by a cut reducing to a finite number of elements, all distinguishable the one from the other (and therefore forming neither one continuum nor several continuums), we shall call C a continuum *of one dimension.* If, on the contrary, C can only be subdivided by cuts which are themselves continuums, we shall say that C is of several dimensions; if the cuts are continuums of one dimension, then we shall say that C has two dimensions; if cuts of two dimensions are sufficient, we shall say that C is of three dimensions, and so on. Thus the notion of the physical continuum of several dimensions is defined, thanks to the very simple fact that two aggregates of sensations may be distinguishable or indistinguishable.

The Mathematical Continuum of Several Dimensions. The conception of the mathematical continuum of *n* dimensions may be led up to quite naturally by a process similar to that which we discussed at the beginning of this chapter. A point of such a continuum is defined by a system of *n* distinct magnitudes which we call its coordinates.

The magnitudes need not always be measurable; there is, for instance, one branch of geometry independent of the measure of magnitudes, in which we are only concerned with knowing, for example, if, on a curve ABC, the point B is between the points A and C, and in which it is immaterial whether the arc AB is equal to or twice the arc BC. This branch is called *Analysis Situs*. It contains quite a large body of doctrine which has attracted the attention of the greatest geometers, and from which are derived, one from another, a whole series of remarkable theorems. What distinguishes these theorems from those of ordinary geometry is that they are purely qualitative. They are still true if the figures are copied by an unskilful draughtsman, with the result that the proportions are distorted and the straight lines replaced by lines which are more or less curved.

As soon as measurement is introduced into the continuum we have just defined, the continuum becomes space, and geometry is born. But the discussion of this is reserved for Part II.

PART II

SPACE

CHAPTER III

NON–EUCLIDEAN GEOMETRIES

Every conclusion presumes premisses. These premisses are either self-evident and need no demonstration, or can be established only if based on other propositions; and, as we cannot go back in this way to infinity, every deductive science, and geometry in particular, must rest upon a certain number of indemonstrable axioms. All treatises of geometry begin therefore with the enunciation of these axioms. But there is a distinction to be drawn between them. Some of these, for example, "Things which are equal to the same thing are equal to one another," are not propositions in geometry but propositions in analysis. I look upon them as analytical *à priori* intuitions, and they concern me no further. But I must insist on other axioms which are special to geometry. Of these most treatises explicitly enunciate three: (1) Only one line can pass through two points; (2) a straight line is the shortest distance between two points; (3) through one point only one parallel can be drawn to a given straight line. Although we generally dispense with proving the second of these axioms, it would be possible to deduce it from the other two, and from those much more numerous axioms which are implicitly admitted without enunciation, as I shall explain further on. For a long time a proof of the third axiom known as Euclid's postulate was sought in vain. It is impossible to imagine the efforts that have been spent in pursuit of this chimera. Finally, at the beginning of the nineteenth century, and almost simultaneously, two scientists, a Russian and a Bulgarian, Lobatschewsky and Bolyai, showed irrefutably that this proof is impossible. They have nearly rid us of inventors of geometries without a postulate, and ever since the Académie des Sciences receives only about one or two new demonstrations a year. But the question was not exhausted, and it was not long before a great step was taken by the celebrated memoir of Riemann, entitled: *Ueber die Hypothesen welche der Geometrie zum Grunde liegen.* This little work has inspired most of the recent treatises to which I shall later on refer, and among which I may mention those of Beltrami and Helmholtz.

The Geometry of Lobatschewsky. If it were possible to deduce Euclid's postulate from the several axioms, it is evident that by rejecting the postulate and retaining the other axioms we should be led to contradictory consequences. It would be, therefore, impossible to found on those premisses a coherent geometry. Now, this is precisely what Lobatschewsky has done. He assumes at the outset that several parallels may be drawn through a point to a given straight line, and he retains all the other axioms of Euclid. From these hypotheses he deduces a series of theorems between which it is impossible to find any contradiction, and he constructs a geometry as impeccable in its logic as Euclidean geometry. The theorems are very different, however, from those to which we are accustomed, and at first will be found a little disconcerting. For instance, the sum of the angles of a triangle is always less than two right angles, and the difference between that sum and two right angles is proportional to the area of the triangle. It is impossible to construct a figure similar to a given figure but of different dimensions. If the circumference of a circle be divided into n equal parts, and tangents be drawn at the points of intersection, the n tangents will form a polygon if the radius of the circle is small enough, but if the radius is large enough they will never meet. We need not multiply these examples. Lobatschewsky's propositions have no relation to those of Euclid, but they are nonetheless logically interconnected.

Riemann's Geometry. Let us imagine to ourselves a world only peopled with beings of no thickness, and suppose these "infinitely flat" animals are all in one and the same plane, from which they cannot emerge. Let us further admit that this world is sufficiently distant from other worlds to be withdrawn from their influence, and while we are making these hypotheses it will not cost us much to endow these beings with reasoning power, and to believe them capable of making a geometry. In that case they will certainly attribute to space only two dimensions. But now suppose that these imaginary animals, while remaining without thickness, have the form of a spherical, and not of a plane figure, and are all on the same sphere, from which they cannot escape. What kind of a geometry will they construct? In the first place, it is clear that they will attribute to space only two dimensions. The straight line to them will be the shortest distance from one point on the sphere to another—that is to say, an arc of a great circle. In a word, their geometry will be spherical geometry. What they will call space will be the sphere on which they are confined, and on which take place all the phenomena with which they are acquainted. Their space will therefore be *unbounded,* since on a sphere one may always walk forward without ever being brought to a stop, and yet it will be *finite;* the end will never be found, but the complete tour can be made. Well, Riemann's geometry is spherical geometry extended to three dimensions. To construct it, the German mathematician had first of all to throw overboard, not only Euclid's postulate, but also the first axiom that *only one line can pass through two points.* On a sphere, through two given points, we can *in general* draw only one great circle which, as we have just seen, would be to our imaginary beings a straight line. But there was one exception. If the two given points are at the ends of a diameter, an infinite number of great circles can be drawn through them. In the same way, in Riemann's geometry—

at least in one of its forms—through two points only one straight line can in general be drawn, but there are exceptional cases in which through two points an infinite number of straight lines can be drawn. So there is a kind of opposition between the geometries of Riemann and Lobatschewsky. For instance, the sum of the angles of a triangle is equal to two right angles in Euclid's geometry, less than two right angles in that of Lobatschewsky, and greater than two right angles in that of Riemann. The number of parallel lines that can be drawn through a given point to a given line is one in Euclid's geometry, none in Riemann's, and an infinite number in the geometry of Lobatschewsky. Let us add that Riemann's space is finite, although unbounded in the sense which we have above attached to these words.

Surfaces with Constant Curvature. One objection, however, remains possible. There is no contradiction between the theorems of Lobatschewsky and Riemann; but however numerous are the other consequences that these geometers have deduced from their hypotheses, they had to arrest their course before they exhausted them all, for the number would be infinite; and who can say that if they had carried their deductions further they would not have eventually reached some contradiction? This difficulty does not exist for Riemann's geometry, provided it is limited to two dimensions. As we have seen, the two-dimensional geometry of Riemann, in fact, does not differ from spherical geometry, which is only a branch of ordinary geometry, and is therefore outside all contradiction. Beltrami, by showing that Lobatschewsky's two-dimensional geometry was only a branch of ordinary geometry, has equally refuted the objection as far as it is concerned. This is the course of his argument: Let us consider any figure whatever on a surface. Imagine this figure to be traced on a flexible and inextensible canvas applied to the surface, in such a way that when the canvas is displaced and deformed the different lines of the figure change their form without changing their length. As a rule, this flexible and inextensible figure cannot be displaced without leaving the surface. But there are certain surfaces for which such a movement would be possible. They are surfaces of constant curvature. If we resume the comparison that we made just now, and imagine beings without thickness living on one of these surfaces, they will regard as possible the motion of a figure all the lines of which remain of a constant length. Such a movement would appear absurd, on the other hand, to animals without thickness living on a surface of variable curvature. These surfaces of constant curvature are of two kinds. The curvature of some is *positive,* and they may be deformed so as to be applied to a sphere. The geometry of these surfaces is therefore reduced to spherical geometry—namely, Riemann's. The curvature of others is *negative.* Beltrami has shown that the geometry of these surfaces is identical with that of Lobatschewsky. Thus the two-dimensional geometries of Riemann and Lobatschewsky are connected with Euclidean geometry.

Interpretation of Non-Euclidean Geometries. Thus vanishes the objection so far as two-dimensional geometries are concerned. It would be easy to extend Beltrami's reasoning to three-dimensional geometries, and minds which do not recoil before space of four dimensions will see no difficulty in it; but such minds are few in number. I prefer, then, to proceed otherwise. Let us consider a certain plane, which I shall call the fundamental plane, and let us construct a kind of dictionary by making a double series of terms written in two columns, and corresponding each to each, just as in ordinary dictionaries the words in two languages which have the same signification correspond to one another:

Space	The portion of space situated above the fundamental plane.
Plane	Sphere cutting orthogonally the fundamental plane.
Line	Circle cutting orthogonally the fundamental plane.
Sphere	Sphere.
Circle	Circle.
Angle	Angle.

Distance between two points	... Logarithm of the anharmonic ratio of these two points and of the intersection of the fundamental plane with the circle passing through these two points and cutting it orthogonally.
Etc.	Etc.

Let us now take Lobatschewsky's theorems and translate them by the aid of this dictionary, as we would translate a German text with the aid of a German–French dictionary. *We shall then obtain the theorems of ordinary geometry.* For instance, Lobatschewsky's theorem: "The sum of the angles of a triangle is less than two right angles," may be translated thus: "If a curvilinear triangle has for its sides arcs of circles which if produced would cut orthogonally the fundamental plane, the sum of the angles of this curvilinear triangle will be less than two right angles." Thus, however far the consequences of Lobatschewsky's hypotheses are carried, they will never lead to a contradiction; in fact, if two of Lobatschewsky's theorems were contradictory, the translations of these two theorems made by the aid of our dictionary would be contradictory also. But these translations are theorems of ordinary geometry, and no one doubts that ordinary geometry is exempt from contradiction. Whence is the certainty derived, and how far is it justified? That is a question upon which I cannot enter here, but it is a very interesting question, and I think not insoluble. Nothing, therefore, is left of the objection I formulated above. But this is not all. Lobatschewsky's geometry being susceptible of a concrete interpretation ceases to be a useless logical exercise, and may be applied. I have no time here to deal with these applications, nor with what Herr Klein and I have done by using them in the integration of linear equations. Further, this interpretation is not unique, and several dictionaries may be constructed analogous to that above, which will enable us by a simple translation to convert Lobatschewsky's theorems into the theorems of ordinary geometry.

Implicit Axioms. Are the axioms implicitly enunciated in our text-books the only foundation of geometry? We may be assured of the contrary when we see that, when they are abandoned one after another, there are still left standing some propositions which are common to the geometries of Euclid, Lobatschewsky, and Riemann. These propositions must be based on premises that geometers admit without enunciation. It is interesting to try and extract them from the classical proofs.

John Stuart Mill asserted* that every definition contains an axiom, because by defining we implicitly affirm the existence of the object defined. That is going rather too far. It is but rarely in mathematics that a definition is given without following it up by the proof of the existence of the object defined, and when this is not done it is generally because the reader can easily supply it; and it must not be forgotten that the word "existence" has not the same meaning when it refers to a mathematical entity as when it refers to a material object.

A mathematical entity exists provided there is no contradiction implied in its definition, either in itself, or with the propositions previously admitted. But if the observation of John Stuart Mill cannot be applied to all definitions, it is nonetheless true for some of them. A plane is sometimes defined in the following manner: The plane is a surface such that the line which joins any two points upon it lies wholly on that surface. Now, there is obviously a new axiom concealed in this definition. It is true we might change it, and that would be preferable, but then we should have to enunciate the axiom explicitly. Other definitions may give rise to no less important reflections, such as, for example, that of the equality of two figures. Two figures are equal when they can be superposed. To superpose them, one of them must be displaced until it coincides with the other. But how must it be displaced? If we asked that question, no doubt we should be told that it ought to be done without deforming it, and as an invariable solid is displaced. The vicious

---

Logic, c. viii., cf. Definitions, § 5-6—Tr.

circle would then be evident. As a matter of fact, this definition defines nothing. It has no meaning to a being living in a world in which there are only fluids. If it seems clear to us, it is because we are accustomed to the properties of natural solids which do not much differ from those of the ideal solids, all of whose dimensions are invariable. However, imperfect as it may be, this definition implies an axiom. The possibility of the motion of an invariable figure is not a self-evident truth. At least it is only so in the application to Euclid's postulate, and not as an analytical *à priori* intuition would be. Moreover, when we study the definitions and the proofs of geometry, we see that we are compelled to admit without proof not only the possibility of this motion, but also some of its properties. This first arises in the definition of the straight line. Many defective definitions have been given, but the true one is that which is understood in all the proofs in which the straight line intervenes. "It may happen that the motion of an invariable figure may be such that all the points of a line belonging to the figure are motionless, while all the points situate outside that line are in motion. Such a line would be called a straight line." We have deliberately in this enunciation separated the definition from the axiom which it implies. Many proofs such as those of the cases of the equality of triangles, of the possibility of drawing a perpendicular from a point to a straight line, assume propositions the enunciations of which are dispensed with, for they necessarily imply that it is possible to move a figure in space in a certain way.

The Fourth Geometry. Among these explicit axioms there is one which seems to me to deserve some attention, because when we abandon it we can construct a fourth geometry as coherent as those of Euclid, Lobatschewsky, and Riemann. To prove that we can always draw a perpendicular at a point A to a straight line AB, we consider a straight line AC movable about the point A, and initially identical with the fixed straight line AB. We then can make it turn about the point A until it lies in AB produced. Thus we assume two propositions—first, that such a rotation is possible, and then that it may continue until the two lines lie the one in the other produced. If the first point is conceded and the second rejected, we are led to a series of theorems even stranger than those of Lobatschewsky and Riemann, but equally free from contradiction. I shall give only one of these theorems, and I shall not choose the least remarkable of them. *A real straight line may be perpendicular to itself.*

Lie's Theorem. The number of axioms implicitly introduced into classical proofs is greater than necessary, and it would be interesting to reduce them to a minimum. It may be asked, in the first place, if this reduction is possible—if the number of necessary axioms and that of imaginable geometries is not infinite? A theorem due to Sophus Lie is of weighty importance in this discussion. It may be enunciated in the following manner: Suppose the following premisses are admitted: (1) space has n dimensions; (2) the movement of an invariable figure is possible; (3) p conditions are necessary to determine the position of this figure in space.

The number of geometries compatible with these premisses will be limited. I may even add that if n is given, a superior limit can be assigned to p. If, therefore, the possibility of the movement is granted, we can only invent a finite and even a rather restricted number of three-dimensional geometries.

Riemann's Geometries. However, this result seems contradicted by Riemann, for that scientist constructs an infinite number of geometries, and that to which his name is usually attached is only a particular case of them. All depends, he says, on the manner in which the length of a curve is defined. Now, there is an infinite number of ways of defining this length, and each of them may be the starting-point of a new geometry. That is perfectly true, but most of these definitions are incompatible with the movement of a variable figure such as we assume to be possible in Lie's theorem. These geometries of Riemann, so interesting on various grounds, can never be, therefore, purely analytical, and would not lend themselves to proofs analogous to those of Euclid.

On the Nature of Axioms. Most mathematicians regard Lobat-schewsky's geometry as a mere logical curiosity. Some of them have, however, gone further. If several geometries are possible, they say, is it certain that our geometry is the one that is true? Experiment no doubt teaches us that the sum of the angles of a triangle is equal to two right angles, but this is because the triangles we deal with are too small. According to Lobatschewsky, the difference is proportional to the area of the triangle, and will not this become sensible when we operate on much larger triangles, and when our measurements become more accurate? Euclid's geometry would thus be a provisory geometry. Now, to discuss this view we must first of all ask ourselves, what is the nature of geometrical axioms? Are they synthetic *à priori* intuitions, as Kant affirmed? They would then be imposed upon us with such a force that we could not conceive of the contrary proposition, nor could we build upon it a theoretical edifice. There would be no non-Euclidean geometry. To convince ourselves of this, let us take a true synthetic *à priori* intuition—the following, for instance, which played an important part in the first chapter: If a theorem is true for the number 1, and if it has been proved that it is true of $n+1$, provided it is true of n, it will be true for all positive integers. Let us next try to get rid of this, and while rejecting this proposition let us construct a false arithmetic analogous to non-Euclidean geometry. We shall not be able to do it. We shall be even tempted at the outset to look upon these intuitions as analytical. Besides, to take up again our fiction of animals without thickness, we can scarcely admit that these beings, if their minds are like ours, would adopt the Euclidean geometry, which would be contradicted by all their experience. Ought we, then, to conclude that the axioms of geometry are experimental truths? But we do not make experiments on ideal lines or ideal circles; we can only make them on material objects. On what, therefore, would experiments serving as a foundation for geometry be based? The answer is easy. We have seen above that we constantly reason as if the geometrical figures behaved like solids. What geometry would borrow from experiment would therefore be the properties of these bodies. The properties of light and its propagation in a straight line have also given rise to some of the propositions of geometry, and in particular to those of projective geometry, so that from that point of view one would be tempted to say that metrical geometry is the study of solids, and projective geometry that of light. But a difficulty remains, and is unsurmountable. If geometry were an experimental science, it would not be an exact science. It would be subjected to continual revision. Nay, it would from that day forth be proved to be erroneous, for we know that no rigorously invariable solid exists. *The geometrical axioms are therefore neither synthetic à priori intuitions nor experimental facts.* They are conventions. Our choice among all possible conventions is *guided* by experimental facts; but it remains *free,* and is only limited by the necessity of avoiding every contradiction, and thus it is that postulates may remain rigorously true even when the experimental laws which have determined their adoption are only approximate. In other words, *the axioms of geometry* (I do not speak of those of arithmetic) *are only definitions in disguise.* What, then, are we to think of the question: Is Euclidean geometry true? It has no meaning. We might as well ask if the metric system is true, and if the old weights and measures are false; if Cartesian co-ordinates are true and polar co-ordinates false. One geometry cannot be more true than another; it can only be more convenient. Now, Euclidean geometry is, and will remain, the most convenient: first, because it is the simplest, and it is not so only because of our mental habits or because of the kind of direct intuition that we have of Euclidean space; it is the simplest in itself, just as a polynomial of the first degree is simpler than a polynomial of the second degree; second, because it sufficiently agrees with the properties of natural solids, those bodies which we can compare and measure by means of our senses.

CHAPTER IV

SPACE AND GEOMETRY

Let us begin with a little paradox. Beings whose minds were made as ours, and with senses like ours, but without any preliminary education, might receive from a suitably chosen external world impressions which would lead them to construct a geometry other than that of Euclid, and to localise the phenomena of this external world in a non-Euclidean space, or even in space of four dimensions. As for us, whose education has been made by our actual world, if we were suddenly transported into this new world, we should have no difficulty in referring phenomena to our Euclidean space. Perhaps somebody may appear on the scene some day who will devote his life to it, and be able to represent to himself the fourth dimension.

Geometrical Space and Representative Space. It is often said that the images we form of external objects are localised in space, and even that they can only be formed on this condition. It is also said that this space, which thus serves as a kind of framework ready prepared for our sensations and representations, is identical with the space of the geometers, having all the properties of that space. To all clearheaded men who think in this way, the preceding statement might well appear extraordinary; but it is as good to see if they are not the victims of some illusion which closer analysis may be able to dissipate. In the first place, what are the properties of space properly so called? I mean of that space which is the object of geometry, and which I shall call geometrical space. The following are some of the more essential:

First, it is continuous; second, it is infinite; third, it is of three dimensions; fourth, it is homogeneous—that is to say, all its points are identical one with another; fifth, it is isotropic. Compare this now with the framework of our representations and sensations, which I may call *representative space.*

Visual Space. First of all let us consider a purely visual impression, due to an image formed on the back of the retina. A cursory analysis shows us this image as continuous, but as possessing only two dimensions, which already distinguishes purely visual from what may be called geometrical space. On the other hand, the image is enclosed within a limited framework; and there is a no less important difference: *this pure visual space is not homogeneous.* All the points on the retina, apart from the images which may be formed, do not play the same role. The yellow spot can in no way be regarded as identical with a point on the edge of the retina. Not only does the same object produce on it much brighter impressions, but in the whole of the *limited* framework the point which occupies the centre will not appear identical with a point near one of the edges. Closer analysis no doubt would show us that this continuity of visual space and its two dimensions are but an illusion. It would make visual space even more different than before from geometrical space, but we may treat this remark as incidental.

However, sight enables us to appreciate distance, and therefore to perceive a third dimension. But everyone knows that this perception of the third dimension reduces to a sense of the effort of accommodation which must be made, and to a sense of the convergence of the two eyes, that must take place in order to perceive an object distinctly. These are muscular sensations quite different from the visual sensations which have given us the concept of the two first dimensions. The third dimension will therefore not appear to us as playing the same role as the two others. What may be called *complete visual space* is not therefore an isotropic space. It has, it is true, exactly three dimensions; which means that the elements of our visual sensations (those at least which concur in forming the concept of extension) will be completely defined if we know three of them; or, in mathematical language, they will be functions of three independent variables. But let us look at the matter a little closer. The third dimension is revealed to us in two different ways: by the effort of accommodation, and by the convergence of the eyes. No doubt these two indications are always in harmony; there is between them a constant relation; or, in mathematical language, the two variables which measure these two muscular sensations do not appear to us as independent. Or, again, to avoid an appeal to mathematical ideas which are already rather too refined, we may go back to the language of the preceding chapter and enunciate the same fact as follows: If two sensations of convergence A and B are indistinguishable, the two sensations of accommodation A' and B' which accompany them respectively will also be indistinguishable. But that is, so to speak, an experimental fact. Nothing prevents us *à priori* from assuming the contrary, and if the contrary takes place, if these two muscular sensations both vary independently, we must take into account one more independent variable, and complete visual space will appear to us as a physical continuum of four dimensions. And so in this there is also a fact of *external* experiment. Nothing prevents us from assuming that a being with a mind like ours, with the same sense-organs as ourselves, may be placed in a world in which light would only reach him after being passed through refracting media of complicated form. The two indications which enable us to appreciate distances would cease to be connected by a constant relation. A being educating his senses in such a world would no doubt attribute four dimensions to complete visual space.

Tactile and Motor Space. "Tactile space" is more complicated still than visual space, and differs even more widely from geometrical space. It is useless to repeat for the sense of touch my remarks on the sense of sight. But outside the data of sight and touch there are other sensations which contribute as much and more than they do to the genesis of the concept of space. They are those which everybody knows, which accompany all our movements, and which we usually call muscular sensations. The corresponding framework constitutes what may be called *motor space.* Each muscle gives rise to a special sensation which may be increased or diminished so that the aggregate of our muscular sensations will depend upon as many variables as we have muscles. From this point of view *motor space would have as many dimensions as we have muscles.* I know that it is said that if the muscular sensations contribute to form the concept of space, it is because we have the sense of the *direction* of each movement, and that this is an integral part of the sensation. If this were so, and if a muscular sense could not be aroused unless it were accompanied by this geometrical sense of direction, geometrical space would certainly be a form imposed upon our sensitiveness. But I do not see this at all when I analyse my sensations. What I do see is that the sensations which correspond to movements in the same direction are connected in my mind by a simple *association of ideas.* It is to this association that what we call the sense of direction is reduced. We cannot therefore discover this sense in a single sensation. This association is extremely complex, for the contraction of the same muscle may correspond, according to the position of the limbs, to very different movements of direction. Moreover, it is evidently acquired; it is like all associations of ideas, the result of a *habit.* This habit itself is the result of a very large number of *experiments,* and no doubt if the education of our senses had taken place in a different medium, where we would have been subjected to different impressions, then contrary habits would have been acquired, and our muscular sensations would have been associated according to other laws.

Characteristics of Representative Space. Thus representative space in its triple form—visual, tactile, and motor—differs essentially from geometrical space. It is neither homogeneous nor isotropic; we cannot even say that it is of three dimensions. It is often said that we "project" into geometrical space the objects of our external perception; that we "localise" them. Now, has that any meaning, and if so what is that meaning? Does it mean that we *represent* to ourselves external objects in geometrical space? Our representations are only the reproduction of our sensations; they cannot therefore be arranged in the same framework—that is to say, in representative space. It is also just as impossible for us to represent to ourselves external objects in geometrical space as it is impossible for a painter to paint on a flat surface objects with their three dimensions. Representative space is only an image of geometrical space, an image deformed by a kind of perspective, and we can only represent to ourselves objects by making them obey the laws of this perspective. Thus we do not *represent* to ourselves external bodies in geometrical space, but we *reason* about these bodies as if they were situated in geometrical space. When it is said, on the other hand, that we "localise" such an object in such a point of space, what does it mean? *It simply means that we represent to ourselves the movements that must take place to reach that object.* And it does not mean that to represent to ourselves these movements they must be projected into space, and that the concept of space must therefore pre-exist. When I say that we represent to ourselves these movements, I only mean that we represent to ourselves the muscular sensations which accompany them, and which have no geometrical character, and which therefore in no way imply the pre-existence of the concept of space.

Changes of State and Changes of Position. But, it may be said, if the concept of geometrical space is not imposed upon our minds, and if, on the other hand, none of our sensations can furnish us with that concept, how then did it ever come into existence? This is what we have now to examine, and it will take some time; but I can sum up in a few words the attempt at explanation which I am going to develop. *None of our sensations, if isolated, could have brought us to the concept of space; we are brought to it solely by studying the laws by which those sensations succeed one another.* We see at first that our impressions are subject to change; but among the changes that we ascertain, we are very soon led to make a distinction. Sometimes we say that the objects, the causes of these impressions, have changed their state, sometimes that they have changed their position, that they have only been displaced. Whether an object changes its state or only its position, this is always translated for us in the same manner, *by a modification in an aggregate of impressions.* How then have we been enabled to distinguish them? If there were only change of position, we could restore the primitive aggregate of impressions by making movements which would confront us with the movable object in the same *relative* situation. We thus *correct* the modification which was produced, and we reestablish the initial state by an inverse modification. If, for example, it were a question of the sight, and if an object be displaced before our eyes, we can "follow it with the eye," and retain its image on the same point of the retina by appropriate movements of the eyeball. These movements we are conscious of because they are voluntary, and because they are accompanied by muscular sensations. But that does not mean that we represent them to ourselves in geometrical space. So what characterises change of position, what distinguishes it from change of state, is that it can always be *corrected* by this means. It may therefore happen that we pass from the aggregate of impressions A to the aggregate B in two different ways. First, involuntarily and without experiencing muscular sensations—which happens when it is the object that is displaced; secondly, voluntarily, and with muscular sensation—which happens when the object is motionless, but when we displace ourselves in such a way that the object has relative motion with respect to us. If this be so, the translation of the aggregate A to the aggregate B is only a change of position. It follows that sight and touch could not have given us the idea of space without the help of the "muscular sense." Not only could this concept not be derived from a single sensation, or even from *a series of sensations;* but a *motionless* being could never have acquired it, because, not being able to correct by his movements the effects of the change of position of external objects, he would have had no reason to distinguish them from changes of state. Nor would he have been able to acquire it if his movements had not been voluntary, or if they were unaccompanied by any sensations whatever.

Conditions of Compensation. How is such a compensation possible in such a way that two changes, otherwise mutually independent, may be reciprocally corrected? A mind *already familiar with geometry* would reason as follows: If there is to be compensation, the different parts of the external object on the one hand, and the different organs of our senses on the other, must be in the same *relative* position after the double change. And for that to be the case, the different parts of the external body on the one hand, and the different organs of our senses on the other, must have the same relative position to each other after the double change; and so with the different parts of our body with respect to each other. In other words, the external object in the first change must be displaced as an invariable solid would be displaced, and it must also be so with the whole of our body in the second change, which is to correct the first. Under these conditions compensation may be produced. But we who as yet know nothing of geometry, whose ideas of space are not yet formed, we cannot reason in this way—we cannot predict *à priori* if compensation is possible. But experiment shows us that it sometimes does take place, and we start from this experimental fact in order to distinguish changes of state from changes of position.

Solid Bodies and Geometry. Among surrounding objects there are some which frequently experience displacements that may be thus corrected by a *correlative* movement of our own body—namely, *solid bodies.* The other objects, whose form is variable, only in exceptional circumstances undergo similar displacement (change of position without change of form). When the displacement of a body takes place with deformation, we can no longer by appropriate movements place the organs of our body in the same *relative* situation with respect to this body; we can no longer, therefore, reconstruct the primitive aggregate of impressions.

It is only later, and after a series of new experiments, that we learn how to decompose a body of variable form into smaller elements such that each is displaced approximately according to the same laws as solid bodies. We thus distinguish "deformations" from other changes of state. In these deformations each element undergoes a simple change of position which may be corrected; but the modification of the aggregate is more profound, and can no longer be corrected by a correlative movement. Such a concept is very complex even at this stage, and has been relatively slow in its appearance. It would not have been conceived at all had not the observation of solid bodies shown us beforehand how to distinguish changes of position.

If, then, there were no solid bodies in nature there would be no geometry.

Another remark deserves a moment's attention. Suppose a solid body to occupy successively the positions α and β; in the first position it will give us an aggregate of impressions A, and in the second position the aggregate of impressions B. Now let there be a second solid body, of qualities entirely different from the first—of different colour, for instance. Assume it to pass from the position α, where it gives us the aggregate of impressions A', to the position β, where it gives the aggregate of impressions B'. In general, the aggregate A will have nothing in common with the aggregate A', nor will the aggregate B have anything in common with the aggregate B'. The transition from the aggregate A to the aggregate B, and that of the aggregate A' to the aggregate B', are therefore two changes which *in themselves* have in general nothing in common. Yet we consider both these changes as displacements; and, further, we consider them the *same* displacement. How can this be? It is simply because they may be both corrected by the *same* correlative movement of our body. "Correlative movement," therefore, constitutes the *sole connection* between two phenomena which otherwise we should never have dreamed of connecting.

On the other hand, our body, thanks to the number of its articulations and muscles, may have a multitude of different movements, but all are not capable of "correcting" a modification of external objects; those alone are capable of it in which our whole body, or at least all those in which the organs of our senses enter into play are displaced *en bloc*—i.e., without any variation of their relative positions, as in the case of a solid body.

To sum up:

1. In the first place, we distinguish two categories of phenomena: The first involuntary, unaccompanied by muscular sensations, and attributed to external objects—they are external changes; the second, of opposite character and attributed to the movements of our own body, are internal changes.

2. We notice that certain changes of each in these categories may be corrected by a correlative change of the other category.

3. We distinguish among external changes those that have a correlative in the other category—which we call displacements; and in the same way we distinguish among the internal changes those which have a correlative in the first category.

Thus by means of this reciprocity is defined a particular class of phenomena called displacements. *The laws of these phenomena are the object of geometry.*

Law of Homogeneity. The first of these laws is the law of homogeneity. Suppose that by an external change we pass from the aggregate of impressions A to the aggregate B, and that then this change α is corrected by a correlative voluntary movement β, so that we are brought back to the aggregate A. Suppose now that another external change α' brings us again from the aggregate A to the aggregate B. Experiment then shows us that this change α', like the change α, may be corrected by a voluntary correlative movement β', and that this movement β' corresponds to the same muscular sensations as the movement β which corrected α.

This fact is usually enunciated as follows: *Space is homogeneous and isotropic.* We may also say that a movement which is once produced may be repeated a second and a third time, and so on, without any variation of its properties. In the first chapter, in which we discussed the nature of mathematical reasoning, we saw the importance that should be attached to the possibility of repeating the same operation indefinitely. The virtue of mathematical reasoning is due to this repetition; by means of the law of homogeneity geometrical facts are apprehended. To be complete, to the law of homogeneity must be added a multitude of other laws, into the details of which I do not propose to enter, but which mathematicians sum up by saying that these displacements form a "group."

The Non-Euclidean World. If geometrical space were a framework imposed on *each* of our representations considered individually, it would be impossible to represent to ourselves an image without this framework, and we should be quite unable to change our geometry. But this is not the case; geometry is only the summary of the laws by which these images succeed each other. There is nothing, therefore, to prevent us from imagining a series of representations, similar in every way to our ordinary representations, but succeeding one another according to laws which differ from those to which we are accustomed. We may thus conceive that beings whose education has taken place in a medium in which those laws would be so different, might have a very different geometry from ours.

Suppose, for example, a world enclosed in a large sphere and subject to the following laws: The temperature is not uniform; it is greatest at the centre, and gradually decreases as we move towards the circumference of the sphere, where it is absolute zero. The law of this temperature is as follows: If R be the radius of the sphere, and r the distance of the point considered from the centre, the absolute temperature will be proportional to R^2-r^2. Further, I shall suppose that in this world all bodies have the same co-efficient of dilatation, so that the linear dilatation of any body is proportional to its absolute temperature. Finally, I shall assume that a body transported from one point to another of different temperature is instantaneously in thermal equilibrium with its new environment. There is nothing in these hypotheses either contradictory or unimaginable. A moving object will become smaller and smaller as it approaches the circumference of the sphere. Let us observe, in the first place, that although from the point of view of our ordinary geometry this world is finite, to its inhabitants it will appear infinite. As they approach the surface of the sphere they become colder, and at the same time smaller and smaller. The steps they take are therefore also smaller and smaller, so that they can never reach the boundary of the sphere. If to us geometry is only the study of the laws according to which invariable solids move, to these imaginary beings it will be the study of the laws of motion of solids *deformed by the differences of temperature* alluded to.

No doubt, in our world, natural solids also experience variations of form and volume due to differences of temperature. But in laying the foundations of geometry we neglect these variations; for besides being but small they are irregular, and consequently appear to us to be accidental. In our hypothetical world this will no longer be the case, the variations will obey very simple and regular laws. On the other hand, the different solid parts of which the bodies of these inhabitants are composed will undergo the same variations of form and volume.

Let me make another hypothesis: suppose that light passes through media of different refractive indices, such that the index of refraction is inversely proportional to $R^2 - r^2$. Under these conditions it is clear that the rays of light will no longer be rectilinear but circular. To justify what has been said, we have to prove that certain changes in the position of external objects may be corrected by correlative movements of the beings which inhabit this imaginary world; and in such a way as to restore the primitive aggregate of the impressions experienced by these sentient beings. Suppose, for example, that an object is displaced and deformed, not like an invariable solid, but like a solid subjected to unequal dilatations in exact conformity with the law of temperature assumed above. To use an abbreviation, we shall call such a movement a non-Euclidean displacement.

If a sentient being be in the neighbourhood of such a displacement of the object, his impressions will be modified; but by moving in a suitable manner, he may reconstruct them. For this purpose, all that is required is that the aggregate of the sentient being and the object, considered as forming a single body, shall experience one of those special displacements which I have just called non-

Euclidean. This is possible if we suppose that the limbs of these beings dilate according to the same laws as the other bodies of the world they inhabit.

Although from the point of view of our ordinary geometry there is a deformation of the bodies in this displacement, and although their different parts are no longer in the same relative position, nevertheless we shall see that the impressions of the sentient being remain the same as before; in fact, though the mutual distances of the different parts have varied, yet the parts which at first were in contact are still in contact. It follows that tactile impressions will be unchanged. On the other hand, from the hypothesis as to refraction and the curvature of the rays of light, visual impressions will also be unchanged. These imaginary beings will therefore be led to classify the phenomena they observe, and to distinguish among them the "changes of position," which may be corrected by a voluntary correlative movement, just as we do.

If they construct a geometry, it will not be like ours, which is the study of the movements of our invariable solids; it will be the study of the changes of position which they will have thus distinguished, and will be "non-Euclidean displacements," and *this will be non-Euclidean geometry.* So that beings like ourselves, educated in such a world, will not have the same geometry as ours.

The World of Four Dimensions. Just as we have pictured to ourselves a non-Euclidean world, so we may picture a world of four dimensions.

The sense of light, even with one eye, together with the muscular sensations relative to the movements of the eyeball, will suffice to enable us to conceive of space of three dimensions. The images of external objects are painted on the retina, which is a plane of two dimensions; these are *perspectives.* But as eye and objects are movable, we see in succession different perspectives of the same body taken from different points of view. We find at the same time that the transition from one perspective to another is often accompanied by muscular sensations. If the transition from the perspective A to the perspective B, and that of the perspective A' to the perspective B′ are accompanied by the same muscular sensations, we connect them as we do other operations of the same nature. Then when we study the laws according to which these operations are combined, we see that they form a group, which has the same structure as that of the movements of invariable solids. Now, we have seen that it is from the properties of this group that we derive the idea of geometrical space and that of three dimensions. We thus understand how these perspectives gave rise to the conception of three dimensions, although each perspective is of only two dimensions—because *they succeed each other according to certain laws.* Well, in the same way that we draw the perspective of a three-dimensional figure on a plane, so we can draw that of a four-dimensional figure on a canvas of three (or two) dimensions. To a geometer this is but child's play. We can even draw several perspectives of the same figure from several different points of view. We can easily represent to ourselves these perspectives, since they are of only three dimensions. Imagine that the different perspectives of one and the same object to occur in succession, and that the transition from one to the other is accompanied by muscular sensations. It is understood that we shall consider two of these transitions as two operations of the same nature when they are associated with the same muscular sensations. There is nothing, then, to prevent us from imagining that these operations are combined according to any law we choose—for instance, by forming a group with the same structure as that of the movements of an invariable four-dimensional solid. In this there is nothing that we cannot represent to ourselves, and, moreover, these sensations are those which a being would experience who has a retina of two dimensions, and who may be displaced in space of four dimensions. In this sense we may say that we can represent to ourselves the fourth dimension.

Conclusions. It is seen that experiment plays a considerable role in the genesis of geometry; but it would be a mistake to conclude from that that geometry is, even in part, an experimental science. If it were experimental, it would only be approximative and provisory. And what a rough approximation it would be! Geometry would be only the study of the movements of solid bodies; but, in reality, it is not concerned with natural solids: its object is certain ideal solids, absolutely invariable, which are but a greatly simplified and very remote image of them. The concept of these ideal bodies is entirely mental, and experiment is but the opportunity which enables us to reach the idea. The object of geometry is the study of a particular "group"; but the general concept of group pre-exists in our minds, at least potentially. It is imposed on us not as a form of our sensitiveness, but as a form of our understanding; only, from among all possible groups, we must choose one that will be the *standard*, so to speak, to which we shall refer natural phenomena.

Experiment guides us in this choice, which it does not impose on us. It tells us not what is the truest, but what is the most convenient geometry. It will be noticed that my description of these fantastic worlds has required no language other than that of ordinary geometry. Then, were we transported to those worlds, there would be no need to change that language. Beings educated there would no doubt find it more convenient to create a geometry different from ours, and better adapted to their impressions; but as for us, in the presence of the same impressions, it is certain that we should not find it more convenient to make a change.

CHAPTER V

EXPERIMENT AND GEOMETRY

1. I have on several occasions in the preceding pages tried to show how the principles of geometry are not experimental facts, and that in particular Euclid's postulate cannot be proved by experiment. However convincing the reasons already given may appear to me, I feel I must dwell upon them, because there is a profoundly false conception deeply rooted in many minds.

2. Think of a material circle, measure its radius and circumference, and see if the ratio of the two lengths is equal to π. What have we done? We have made an experiment on the properties of the matter with which this *roundness* has been realised, and of which the measure we used is made.

3. *Geometry and Astronomy.* The same question may also be asked in another way. If Lobatschewsky's geometry is true, the parallax of a very distant star will be finite. If Riemann's is true, it will be negative. These are the results which seem within the reach of experiment, and it is hoped that astronomical observations may enable us to decide between the two geometries. But what we call a straight line in astronomy is simply the path of a ray of light. If, therefore, we were to discover negative parallaxes, or to prove that all parallaxes are higher than a certain limit, we should have a choice be-

tween two conclusions: we could give up Euclidean geometry, or modify the laws of optics, and suppose that light is not rigorously propagated in a straight line. It is needless to add that everyone would look upon this solution as the more advantageous. Euclidean geometry, therefore, has nothing to fear from fresh experiments.

4. Can we maintain that certain phenomena which are possible in Euclidean space would be impossible in non-Euclidean space, so that experiment in establishing these phenomena would directly contradict the non-Euclidean hypothesis? I think that such a question cannot be seriously asked. To me it is exactly equivalent to the following, the absurdity of which is obvious: There are lengths which can be expressed in metres and centimetres, but cannot be measured in toises, feet, and inches; so that experiment, by ascertaining the existence of these lengths, would directly contradict this hypothesis, that there are toises divided into six feet. Let us look at the question a little more closely. I assume that the straight line in Euclidean space possesses any two properties, which I shall call A and B; that in non-Euclidean space it still possesses the property A, but no longer possesses the property B; and, finally, I assume that in both Euclidean and non-Euclidean space the straight line is the only line that possesses the property A. If this were so, experiment would be able to decide between the hypotheses of Euclid and Lobatschewsky. It would be found that some concrete object, upon which we can experiment—for example, a pencil of rays of light—possesses the property A. We should conclude that it is rectilinear, and we should then endeavour to find out if it does, or does not, possess the property B. But *it is not so*. There exists no property which can, like this property A, be an absolute criterion enabling us to recognise the straight line, and to distinguish it from every other line. Shall we say, for instance, "This property will be the following: the straight line is a line such that a figure of which this line is a part can move without the mutual distances of its points varying, and in such a way that all the points in this straight line remain fixed"? Now, this is a property which in either Euclidean or non-Euclidean space belongs to the straight line, and belongs to it alone. But how can we ascertain by experiment if it belongs to any particular con-

crete object? Distances must be measured, and how shall we know that any concrete magnitude which I have measured with my material instrument really represents the abstract distance? We have only removed the difficulty a little further off. In reality, the property that I have just enunciated is not a property of the straight line alone; it is a property of the straight line and of distance. For it to serve as an absolute criterion, we must be able to show not only that it does not also belong to any other line than the straight line and to distance, but also that it does not belong to any other line than the straight line, and to any other magnitude than distance. Now, that is not true, and if we are not convinced by these considerations, I challenge anyone to give me a concrete experiment which can be interpreted in the Euclidean system, and which cannot be interpreted in the system of Lobatschewsky. As I am well aware that this challenge will never be accepted, I may conclude that no experiment will ever be in contradiction with Euclid's postulate; but, on the other hand, no experiment will ever be in contradiction with Lobatschewsky's postulate.

5. But it is not sufficient that the Euclidean (or non-Euclidean) geometry can ever be directly contradicted by experiment. Nor could it happen that it can only agree with experiment by a violation of the principle of sufficient reason, and of that of the relativity of space. Let me explain myself. Consider any material system whatever. We have to consider on the one hand the "state" of the various bodies of this system—for example, their temperature, their electric potential, etc.; and on the other hand their position in space. And among the data which enable us to define this position we distinguish the mutual distances of these bodies that define their relative positions, and the conditions which define the absolute position of the system and its absolute orientation in space. The law of the phenomena which will be produced in this system will depend on the state of these bodies, and on their mutual distances; but because of the relativity and the inertia of space, they will not depend on the absolute position and orientation of the system. In other words, the state of the bodies and their mutual distances at any moment will solely depend on the state of the same

bodies and on their mutual distances at the initial moment, but will in no way depend on the absolute initial position of the system and of its absolute initial orientation. This is what we shall call, for the sake of abbreviation, *the law of relativity.*

So far I have spoken as a Euclidean geometer. But I have said that an experiment, whatever it may be, requires an interpretation on the Euclidean hypothesis; it equally requires one on the non-Euclidean hypothesis. Well, we have made a series of experiments. We have interpreted them on the Euclidean hypothesis, and we have recognised that these experiments thus interpreted do not violate this "law of relativity." We now interpret them on the non-Euclidean hypothesis. This is always possible, only the non-Euclidean distances of our different bodies in this new interpretation will not generally be the same as the Euclidean distances in the primitive interpretation. Will our experiment interpreted in this new manner be still in agreement with our "law of relativity," and if this agreement had not taken place, would we not still have the right to say that experiment has proved the falsity of non-Euclidean geometry? It is easy to see that this is an idle fear. In fact, to apply the law of relativity in all its rigour, it must be applied to the entire universe; for if we were to consider only a part of the universe, and if the absolute position of this part were to vary, the distances of the other bodies of the universe would equally vary; their influence on the part of the universe considered might therefore increase or diminish, and this might modify the laws of the phenomena which take place in it. But if our system is the entire universe, experiment is powerless to give us any opinion on its position and its absolute orientation in space. All that our instruments, however perfect they may be, can let us know will be the state of the different parts of the universe, and their mutual distances. Hence, our law of relativity may be enunciated as follows: The readings that we can make with our instruments at any given moment will depend only on the readings that we were able to make on the same instruments at the initial moment. Now such an enunciation is independent of all interpretation by experiments. If the law is true in the Euclidean interpretation, it will be also true in the non-Euclidean interpretation. Allow me to make a short digression on this point. I have spoken above of the data which define the position of the different bodies of the system. I might also have spoken of those which define their velocities. I should then have to distinguish the velocity with which the mutual distances of the different bodies are changing, and on the other hand the velocities of translation and rotation of the system; that is to say, the velocities with which its absolute position and orientation are changing. For the mind to be fully satisfied, the law of relativity would have to be enunciated as follows: The state of bodies and their mutual distances at any given moment, as well as the velocities with which those distances are changing at that moment, will depend only on the state of those bodies, on their mutual distances at the initial moment, and on the velocities with which those distances were changing at the initial moment. But they will not depend on the absolute initial position of the system nor on its absolute orientation, nor on the velocities with which that absolute position and orientation were changing at the initial moment. Unfortunately, the law thus enunciated does not agree with experiments—at least, as they are ordinarily interpreted. Suppose a man were translated to a planet, the sky of which was constantly covered with a thick curtain of clouds, so that he could never see the other stars. On that planet he would live as if it were isolated in space. But he would notice that it revolves, either by measuring its ellipticity (which is ordinarily done by means of astronomical observations, but which could be done by purely geodesic means), or by repeating the experiment of Foucault's pendulum. The absolute rotation of this planet might be clearly shown in this way. Now, here is a fact which shocks the philosopher, but which the physicist is compelled to accept. We know that from this fact Newton concluded the existence of absolute space. I myself cannot accept this way of looking at it. I shall explain why in Part III, but for the moment it is not my intention to discuss this difficulty. I must therefore resign myself, in the enunciation of the law of relativity, to including velocities of every kind among the data which define the state of the bodies. However that may be, the difficulty is the same for both Euclid's geometry and for Lobatschewsky's. I need not therefore trouble about it further, and I have only mentioned it incidentally. To sum up, whichever way we look at it, it is impossible to discover in geometric empiricism a rational meaning.

6. Experiments only teach us the relations of bodies to one another. They do not and cannot give us the relations of bodies and space, nor the mutual relations of the different parts of space. "Yes!" you reply, "a single experiment is not enough, because it only gives us one equation with several unknowns; but when I have made enough experiments I shall have enough equations to calculate all my unknowns." If I know the height of the main-mast, that is not sufficient to enable me to calculate the age of the captain. When you have measured every fragment of wood in a ship you will have many equations, but you will be no nearer knowing the captain's age. All your measurements bearing on your fragments of wood can tell you only what concerns those fragments; and similarly, your experiments, however numerous they may be, referring only to the relations of bodies with one another, will tell you nothing about the mutual relations of the different parts of space.

7. Will you say that if the experiments have reference to the bodies, they at least have reference to the geometrical properties of the bodies. First, what do you understand by the geometrical properties of bodies? I assume that it is a question of the relations of the bodies to space. These properties therefore are not reached by experiments which only have reference to the relations of bodies to one another, and that is enough to show that it is not of those properties that there can be a question. Let us therefore begin by making ourselves clear as to the sense of the phrase: geometrical properties of bodies. When I say that a body is composed of several parts, I presume that I am thus enunciating a geometrical property, and that will be true even if I agree to give the improper name of points to the very small parts I am considering. When I say that this or that part of a certain body is in contact with this or that part of another body, I am enunciating a proposition which concerns the mutual relations of the two bodies, and not their relations with space. I assume that you will agree with me that these are not geo-

metrical properties. I am sure that at least you will grant that these properties are independent of all knowledge of metrical geometry. Admitting this, I suppose that we have a solid body formed of eight thin iron rods, *oa, ob, oc, od, oe, of, og, oh,* connected at one of their extremities, *o.* And let us take a second solid body—for example, a piece of wood, on which are marked three little spots of ink which I shall call α β γ. I now suppose that we find that we can bring into contact α β γ with *ago;* by that I mean α with *a,* and at the same time β with *g,* and γ with *o.* Then we can successively bring into contact αβγ with *bgo, cgo, dgo, ego, fgo,* then with *aho, bho, cho, dho, eho, fho;* and then αγ successively with *ab, bc, cd, de, ef, fa.* Now these are observations that can be made without having any idea beforehand as to the form or the metrical properties of space. They have no reference whatever to the "geometrical properties of bodies." These observations will not be possible if the bodies on which we experiment move in a group having the same structure as the Lobatschewskian group (I mean according to the same laws as solid bodies in Lobatschewsky's geometry). They therefore suffice to prove that these bodies move according to the Euclidean group; or at least that they do not move according to the Lobatschewskian group. That they may be compatible with the Euclidean group is easily seen; for we might make them so if the body αβγ were an invariable solid of our ordinary geometry in the shape of a right-angled triangle, and if the points *abcdefgh* were the vertices of a polyhedron formed of two regular hexagonal pyramids of our ordinary geometry having *abcdef* as their common base, and having the one *g* and the other *h* as their vertices. Suppose now, instead of the previous observations, we note that we can as before apply αβγ successively to *ago, bgo, cgo, dgo, ego, fgo, aho, bho, cho, dho, eho, fho,* and then that we can apply αβ (and no longer αγ) successively to *ab, bc, cd, de, ef,* and *fa.* These are observations that could be made if non-Euclidean geometry were true. If the bodies αβγ, *oabcdefgh* were invariable solids, if the former were a right-angled triangle, and the latter a double regular hexagonal pyramid of suitable dimensions. These new verifications are therefore impossible if the bodies move according to the Euclidean group; but they become possible if we suppose the bodies to move according to the Lobatschewskian group. They would therefore suffice to show, if we carried them out, that the bodies in question do not move according to the Euclidean group. And so, without making any hypothesis on the form and the nature of space, on the relations of the bodies and space, and without attributing to bodies any geometrical property, I have made observations which have enabled me to show in one case that the bodies experimented upon move according to a group, the structure of which is Euclidean, and in the other case, that they move in a group, the structure of which is Lobatschewskian. It cannot be said that all the first observations would constitute an experiment proving that space is Euclidean, and the second an experiment proving that space is non-Euclidean; in fact, it might be imagined (note that I use the word *imagined*) that there are bodies moving in such a manner as to render possible the second series of observations: and the proof is that the first mechanic who came our way could construct it if he would only take the trouble. But you must not conclude, however, that space is non-Euclidean. In the same way, just as ordinary solid bodies would continue to exist when the mechanic had constructed the strange bodies I have just mentioned, he would have to conclude that space is both Euclidean and non-Euclidean. Suppose, for instance, that we have a large sphere of radius R, and that its temperature decreases from the centre to the surface of the sphere according to the law of which I spoke when I was describing the non-Euclidean world. We might have bodies whose dilatation is negligible, and which would behave as ordinary invariable solids; and, on the other hand, we might have very dilatable bodies, which would behave as non-Euclidean solids. We might have two double pyramids $oabcdefgh$ and $o'a'b'c'd'e'f'g'h'$, and two triangles $\alpha\beta\gamma$ and $\alpha'\beta'\gamma'$. The first double pyramid would be rectilinear, and the second curvilinear. The triangle $\alpha\beta\gamma$ would consist of undilatable matter, and the other of very dilatable matter. We might therefore make our first observations with the double pyramid $o'a'h'$ and the triangle $\alpha'\beta'\gamma'$.

And then the experiment would seem to show—first, that Euclidean geometry is true, and then that it is false. Hence, *experiments have reference not to space but to bodies.*

Supplement

8. To round the matter off, I ought to speak of a very delicate question, which will require considerable development; but I shall confine myself to summing up what I have written in the *Revue de métaphysique et de morale* and in the *Monist*. When we say that space has three dimensions, what do we mean? We have seen the importance of these "internal changes" which are revealed to us by our muscular sensations. They may serve to characterise the different attitudes of our body. Let us take arbitrarily as our origin one of these attitudes, A. When we pass from this initial attitude to another attitude B we experience a series of muscular sensations, and this series S of muscular sensations will define B. Observe, however, that we shall often look upon two series S and S′ as defining the same attitude B (since the initial and final attitudes A and B remaining the same, the intermediary attitudes of the corresponding sensations may differ). How then can we recognise the equivalence of these two series? Because they may serve to compensate for the same external change, or more generally, because, when it is a question of compensation for an external change, one of the series may be replaced by the other. Among these series we have distinguished those which can alone compensate for an external change, and which we have called "displacements." As we cannot distinguish two displacements which are very close together, the aggregate of these displacements presents the characteristics of a physical continuum. Experience teaches us that they are the characteristics of a physical continuum of six dimensions; but we do not know as yet how many dimensions space itself possesses, so we must first of all answer another question. What is a point in space? Everyone thinks he knows, but that is an illusion. What we see when we try to represent to ourselves a point in space is a black spot on white paper, a spot of chalk on a blackboard, always an object. The question should therefore be understood as follows: What do I mean when I

say the object B is at the point which a moment before was occupied by the object A? Again, what criterion will enable me to recognise it? I mean that *although I have not moved* (my muscular sense tells me this), my finger, which just now touched the object A, is now touching the object B. I might have used other criteria—for instance, another finger or the sense of sight—but the first criterion is sufficient. I know that if it answers in the affirmative all other criteria will give the same answer. I know it from experiment. I cannot know it *à priori*. For the same reason I say that touch cannot be exercised at a distance; that is another way of enunciating the same experimental fact. If I say, on the contrary, that sight is exercised at a distance, it means that the criterion furnished by sight may give an affirmative answer while the others reply in the negative.

To sum up. For each attitude of my body my finger determines a point, and it is that and that only which defines a point in space. To each attitude corresponds in this way a point. But it often happens that the same point corresponds to several different attitudes (in this case we say that our finger has not moved, but the rest of our body has). We distinguish, therefore, among changes of attitude those in which the finger does not move. How are we led to this? It is because we often remark that in these changes the object which is in touch with the finger remains in contact with it. Let us arrange then in the same class all the attitudes which are deduced one from the other by one of the changes that we have thus distinguished. To all these attitudes of the same class will correspond the same point in space. Then to each class will correspond a point, and to each point a class. Yet it may be said that what we get from this experiment is not the point, but the class of changes, or, better still, the corresponding class of muscular sensations. Thus, when we say that space has three dimensions, we merely mean that the aggregate of these classes appears to us with the characteristics of a physical continuum of three dimensions. Then if, instead of defining the points in space with the aid of the first finger, I use, for example, another finger, would the results be the same? That is by no means *à priori* evident. But, as we have seen, experiment has shown us that all our criteria are in agreement, and this enables us to answer in the affirmative. If we recur to what we have called displacements, the aggregate of which forms, as we have seen, a group, we shall be brought to distinguish those in which a finger does not move; and by what has preceded, those are the displacements which characterise a point in space, and their aggregate will form a sub-group of our group. To each sub-group of this kind, then, will correspond a point in space. We might be tempted to conclude that experiment has taught us the number of dimensions of space; but in reality our experiments have referred not to space, but to our body and its relations with neighbouring objects. What is more, our experiments are exceedingly crude. In our mind the latent idea of a certain number of groups pre-existed; these are the groups with which Lie's theory is concerned. Which shall we choose to form a kind of standard by which to compare natural phenomena? And when this group is chosen, which of the sub-groups shall we take to characterise a point in space? Experiment has guided us by showing us what choice adapts itself best to the properties of our body; but there its role ends.

FORCE

CHAPTER VI

THE CLASSICAL MECHANICS

The English teach mechanics as an experimental science; on the Continent it is taught always more or less as a deductive and *à priori* science. The English are right, no doubt. How is it that the other method has been persisted in for so long; how is it that Continental scientists who have tried to escape from the practice of their predecessors have in most cases been unsuccessful? On the other hand, if the principles of mechanics are only of experimental origin, are they not merely approximate and provisory? May we not be some day compelled by new experiments to modify or even to abandon them? These are the questions which naturally arise, and the difficulty of solution is largely due to the fact that treatises on mechanics do not clearly distinguish between what is experiment, what is mathematical reasoning, what is convention, and what is hypothesis. This is not all.

1. There is no absolute space, and we only conceive of relative motion; and yet in most cases mechanical facts are enunciated as if there were an absolute space to which they can be referred.

2. There is no absolute time. When we say that two periods are equal, the statement has no meaning, and can only acquire a meaning by a convention.

3. Not only have we no direct intuition of the equality of two periods, but we have not even direct intuition of the simultaneity of two events occurring in two different places. I have explained this in an article entitled "Mesure du Temps."*

4. Finally, is not our Euclidean geometry in itself only a kind of convention of language? Mechanical facts might be enunciated with reference to a non-Euclidean space which would be less convenient but quite as legitimate as our ordinary space; the enunciation would become more complicated, but it still would be possible.

Thus, absolute space, absolute time, and even geometry are not conditions which are imposed on mechanics. All these things no more existed before mechanics than the French language can be logically said to have existed before the truths which are expressed in French. We might endeavour to enunciate the fundamental law of mechanics in a language independent of all these conventions; and no doubt we should in this way get a clearer idea of those laws in themselves. This is what M. Andrade has tried to do, to some extent at any rate, in his *Leçons de mécanique physique*. Of course the enunciation of these laws would become much more complicated, because all these conventions have been adopted for the very purpose of abbreviating and simplifying the enunciation. As far as we are concerned, I shall ignore all these difficulties; not because I disregard them, far from it; but because they have received sufficient attention in the first two parts of the book. Provisionally, then, we shall admit absolute time and Euclidean geometry.

The Principle of Inertia. A body under the action of no force can only move uniformly in a straight line. Is this a truth imposed on the mind *à priori?* If this be so, how is it that the Greeks ignored it? How could they have believed that motion ceases with the cause of motion? or, again, that every body, if there is nothing to prevent it, will move in a circle, the noblest of all forms of motion?

If it be said that the velocity of a body cannot change, if there is no reason for it to change, may we not just as legitimately maintain

---

* *Revue de métaphysique et de morale*, t. vi., pp. 1–13, January, 1898.

that the position of a body cannot change, or that the curvature of its path cannot change, without the agency of an external cause? Is, then, the principle of inertia, which is not an *à priori* truth, an experimental fact? Have there ever been experiments on bodies acted on by no forces? and, if so, how did we know that no forces were acting? The usual instance is that of a ball rolling for a very long time on a marble table; but why do we say it is under the action of no force? Is it because it is too remote from all other bodies to experience any sensible action? It is not farther from the earth than if it were thrown freely into the air; and we all know that in that case it would be subject to the attraction of the earth. Teachers of mechanics usually pass rapidly over the example of the ball, but they add that the principle of inertia is verified indirectly by its consequences. This is very badly expressed; they evidently mean that various consequences may be verified by a more general principle, of which the principle of inertia is only a particular case. I shall propose for this general principle the following enunciation: The acceleration of a body depends only on its position and that of neighbouring bodies, and on their velocities. Mathematicians would say that the movements of all the material molecules of the universe depend on differential equations of the second order. To make it clear that this is really a generalisation of the law of inertia we may again have recourse to our imagination. The law of inertia, as I have said above, is not imposed on us *à priori;* other laws would be just as compatible with the principle of sufficient reason. If a body is not acted upon by a force, instead of supposing that its velocity is unchanged we may suppose that its position or its acceleration is unchanged.

Let us for a moment suppose that one of these two laws is a law of nature, and substitute it for the law of inertia: what will be the natural generalisation? A moment's reflection will show us. In the first case, we may suppose that the velocity of a body depends only on its position and that of neighbouring bodies; in the second case, that the variation of the acceleration of a body depends only on the position of the body and of neighbouring bodies, on their velocities and accelerations; or, in mathematical terms, the differential equations of the motion would be of the first order in the first case and of the third order in the second.

Let us now modify our supposition a little. Suppose a world analogous to our solar system, but one in which by a singular chance the orbits of all the planets have neither eccentricity nor inclination; and further, I suppose that the masses of the planets are too small for their mutual perturbations to be sensible. Astronomers living in one of these planets would not hesitate to conclude that the orbit of a star can only be circular and parallel to a certain plane; the position of a star at a given moment would then be sufficient to determine its velocity and path. The law of inertia which they would adopt would be the former of the two hypothetical laws I have mentioned.

Now, imagine this system to be some day crossed by a body of vast mass and immense velocity coming from distant constellations. All the orbits would be profoundly disturbed. Our astronomers would not be greatly astonished. They would guess that this new star is in itself quite capable of doing all the mischief; but, they would say, as soon as it has passed by, order will again be established. No doubt the distances of the planets from the sun will not be the same as before the cataclysm, but the orbits will become circular again as soon as the disturbing cause has disappeared. It would be only when the perturbing body is remote, and when the orbits, instead of being circular are found to be elliptical, that the astronomers would find out their mistake, and discover the necessity of reconstructing their mechanics.

I have dwelt on these hypotheses, for it seems to me that we can clearly understand our generalised law of inertia only by opposing it to a contrary hypothesis.

Has this generalised law of inertia been verified by experiment, and can it be so verified? When Newton wrote *Principia,* he certainly regarded this truth as experimentally acquired and demonstrated. It was so in his eyes, not only from the anthropomorphic conception to which I shall later refer, but also because of the work of Galileo. It was so proved by the laws of Kepler. According to those laws, in fact, the path of a planet is entirely determined by its initial position and initial velocity; this, indeed, is what our generalised law of inertia requires.

For this principle to be only true in appearance—lest we should fear that some day it must be replaced by one of the analogous principles which I opposed to it just now—we must have been led astray by some amazing chance such as that which had led into error our imaginary astronomers. Such an hypothesis is so unlikely that it need not delay us. No one will believe that there can be such chances; no doubt the probability that two eccentricities are both exactly zero is not smaller than the probability that one is 0.1 and the other 0.2. The probability of a simple event is not smaller than that of a complex one. If, however, the former does occur, we shall not attribute its occurrence to chance; we shall not be inclined to believe that nature has done it deliberately to deceive us. The hypothesis of an error of this kind being discarded, we may admit that so far as astronomy is concerned our law has been verified by experiment.

But Astronomy is not the whole of Physics. May we not fear that some day a new experiment will falsify the law in some domain of physics? An experimental law is always subject to revision; we may always expect to see it replaced by some other and more exact law. But no one seriously thinks that the law of which we speak will ever be abandoned or amended. Why? Precisely because it will never be submitted to a decisive test.

In the first place, for this test to be complete, all the bodies of the universe must return with their initial velocities to their initial positions after a certain time. We ought then to find that they would resume their original paths. But this test is impossible; it can be only partially applied, and even when it is applied there will still be some bodies which will not return to their original positions. Thus there will be a ready explanation of any breaking down of the law.

Yet this is not all. In Astronomy we *see* the bodies whose motion we are studying, and in most cases we grant that they are not subject to the action of other invisible bodies. Under these conditions, our law must certainly be either verified or not. But it is not so in Physics. If physical phenomena are due to motion, it is to the motion of molecules which we cannot see. If, then, the acceleration of bodies we cannot see depends on something else than the positions or velocities of other visible bodies or of invisible molecules, the existence of which we have been led previously to admit, there is nothing to prevent us from supposing that this something else is the position or velocity of other molecules of which we have not so far suspected the existence. The law will be safeguarded. Let me express the same thought in another form in mathematical language. Suppose we are observing n molecules, and find that their $3n$ co-ordinates satisfy a system of $3n$ differential equations of the fourth order (and not of the second, as required by the law of inertia). We know that by introducing $3n$ variable auxiliaries, a system of $3n$ equations of the fourth order may be reduced to a system of $6n$ equations of the second order. If, then, we suppose that the $3n$ auxiliary variables represent the co-ordinates of n invisible molecules, the result is again conformable to the law of inertia. To sum up, this law, verified experimentally in some particular cases, may be extended fearlessly to the most general cases; for we know that in these general cases it can neither be confirmed nor contradicted by experiment.

The Law of Acceleration. The acceleration of a body is equal to the force which acts on it divided by its mass.

Can this law be verified by experiment? If so, we have to measure the three magnitudes mentioned in the enunciation: acceleration, force, and mass. I admit that acceleration may be measured, because I pass over the difficulty arising from the measurement of time. But how are we to measure force and mass? We do not even know what they are. What is mass? Newton replies: "The product of the volume and the density." "It were better to say," answer Thomson and Tait, "that density is the quotient of the mass by the volume." What is force? "It is," replies Lagrange, "that which moves or tends to move a body." "It is," according to Kirchoff, "the product of the mass and the acceleration." Then why not say that mass is the quotient of the force by the acceleration? These difficulties are insurmountable.

When we say force is the cause of motion, we are talking metaphysics; and this definition, if we had to be content with it, would be absolutely fruitless, would lead to absolutely nothing. For a definition to be of any use it must tell us how to measure force; and that is quite sufficient, for it is by no means necessary to tell what force is in itself, nor whether it is the cause or the effect of motion. We must therefore first define what is meant by the equality of two forces. When are two forces equal? We are told that it is when they give the same acceleration to the same mass, or when acting in opposite directions they are in equilibrium. This definition is a sham. A force applied to a body cannot be uncoupled and applied to another body as an engine is uncoupled from one train and coupled to another. It is therefore impossible to say what acceleration such a force, applied to such a body, would give to another body if it were applied to it. It is impossible to tell how two forces which are not acting in exactly opposite directions would behave if they were acting in opposite directions. It is this definition which we try to materialise, as it were, when we measure a force with a dynamometer or with a balance. Two forces, F and F′, which I suppose, for simplicity, to be acting vertically upwards, are respectively applied to two bodies, C and C′. I attach a body weighing P first to C and then to C′; if there is equilibrium in both cases I conclude that the two forces F and F′ are equal, for they are both equal to the weight of the body P. But am I certain that the body P has kept its weight when I transferred it from the first body to the second? Far from it. I am certain of the contrary. I know that the magnitude of the weight varies from one point to another, and that it is greater, for instance, at the pole than at the equator. No doubt the difference is very small, and we neglect it in practice; but a definition must have mathematical rigour; this rigour does not exist. What I say of weight would apply equally to the force of the spring of a dynamometer, which would vary according to temperature and many other circumstances. Nor is this all. We cannot say that the weight of the body P is applied to the body C and keeps in equilibrium the force F. What is applied to the body C is the action of the body P on the body C. On the other hand, the body P is acted on by its weight, and by the reaction R of the body C on P the forces F and A are equal, because they are in equilibrium; the forces A and R are equal by virtue of the principle of action and reaction; and finally, the force R and the weight P are equal because they are in equilibrium. From these three equalities we deduce the equality of the weight P and the force F.

Thus we are compelled to bring into our definition of the equality of two forces the principle of the equality of action and reaction; *hence this principle can no longer be regarded as an experimental law but only as a definition.*

To recognise the equality of two forces we are then in possession of two rules: the equality of two forces in equilibrium and the equality of action and reaction. But, as we have seen, these are not sufficient, and we are compelled to have recourse to a third rule, and to admit that certain forces—the weight of a body, for instance—are constant in magnitude and direction. But this third rule is an experimental law. It is only approximately true: *it is a bad definition.* We are therefore reduced to Kirchoff's definition: force is the product of the mass and the acceleration. This law of Newton in its turn ceases to be regarded as an experimental law, it is now only a definition. But as a definition it is insufficient, for we do not know what mass is. It enables us, no doubt, to calculate the ratio of two forces applied at different times to the same body, but it tells us nothing about the ratio of two forces applied to two different bodies. To fill up the gap we must have recourse to Newton's third law, the equality of action and reaction, still regarded not as an experimental law but as a definition. Two bodies, A and B, act on each other; the acceleration of A, multiplied by the mass of A, is equal to the action of B on A; in the same way the acceleration of B, multiplied by the mass of B, is equal to the reaction of A on B. As, by definition, the action and the reaction are equal, the masses of A and B are respectively in the inverse ratio of their masses. Thus is the ratio of the two masses defined, and it is for experiment to verify that the ratio is constant.

This would do very well if the two bodies were alone and could be abstracted from the action of the rest of the world; but this is by no means the case. The acceleration of A is not solely due to the action of B, but to that of a multitude of other bodies, C, D, ... To apply the preceding rule we must decompose the acceleration of A into many components, and find out which of these components is due to the action of B. The decomposition would still be possible if we suppose that the action of C on A is simply added to that of B on A, and that the presence of the body C does not in any way modify the action of B on A, or that the presence of B does not modify the action of C on A; that is, if we admit that any two bodies attract each other, that their mutual action is along their join, and is only dependent on their distance apart; if, in a word, we admit the *hypothesis of central forces*.

We know that to determine the masses of the heavenly bodies we adopt quite a different principle. The law of gravitation teaches us that the attraction of two bodies is proportional to their masses; if r is their distance apart, m and m' their masses, k a constant, then their attraction will be kmm'/r^2. What we are measuring is therefore not mass, the ratio of the force to the acceleration, but the attracting mass; not the inertia of the body, but its attracting power. It is an indirect process, the use of which is not indispensable theoretically. We might have said that the attraction is inversely proportional to the square of the distance, without being proportional to the product of the masses, that it is equal to f/r^2 and not to kmm'. If it were so, we should nevertheless, by observing the *relative* motion of the celestial bodies, be able to calculate the masses of these bodies.

But have we any right to admit the hypothesis of central forces? Is this hypothesis rigorously accurate? Is it certain that it will never be falsified by experiment? Who will venture to make such an assertion? And if we must abandon this hypothesis, the building which has been so laboriously erected must fall to the ground.

We have no longer any right to speak of the component of the acceleration of A which is due to the action of B. We have no means of distinguishing it from that which is due to the action of C or of any other body. The rule becomes inapplicable in the measurement of masses. What then is left of the principle of the equality of action and reaction? If we reject the hypothesis of central forces this principle must go too; the geometrical resultant of all the forces applied to the different bodies of a system abstracted from all external action will be zero. In other words, *the motion of the centre of gravity of this system will be uniform and in a straight line*. Here would seem to be a means of defining mass. The position of the centre of gravity evidently depends on the values given to the masses; we must select these values so that the motion of the centre of gravity is uniform and rectilinear. This will always be possible if Newton's third law holds good, and it will be in general possible only in one way. But no system exists which is abstracted from all external action; every part of the universe is subject, more or less, to the action of the other parts. *The law of the motion of the centre of gravity is only rigorously true when applied to the whole universe.*

But then, to obtain the values of the masses we must find the motion of the centre of gravity of the universe. The absurdity of this conclusion is obvious; the motion of the centre of gravity of the universe will be forever to us unknown. Nothing, therefore, is left, and our efforts are fruitless. There is no escape from the following definition, which is only a confession of failure: *Masses are coefficients which it is found convenient to introduce into calculations.*

We could reconstruct our mechanics by giving to our masses different values. The new mechanics would be in contradiction neither with experiment nor with the general principles of dynamics (the principle of inertia, proportionality of masses and accelerations, equality of action and reaction, uniform motion of the centre of gravity in a straight line, and areas). But the equations of this mechanics *would not be so simple*. Let us clearly understand this. It would be only the first terms which would be less simple—*i.e.,* those we already know through experiment; perhaps the small masses could be slightly altered without the *complete* equations gaining or losing in simplicity.

Hertz has inquired if the principles of mechanics are rigorously true. "In the opinion of many physicists it seems inconceivable that experiment will ever alter the impregnable principles of mechanics; and yet, what is due to experiment may always be rectified by experiment." From what we have just seen these fears would appear to be groundless. The principles of dynamics appeared to us first as experimental truths, but we have been compelled to use them as definitions. It is *by definition* that force is equal to the product of the mass and the acceleration; this is a principle which is henceforth beyond the reach of any future experiment. Thus it is by definition that action and reaction are equal and opposite. But then it will be said, these unverifiable principles are absolutely devoid of any significance. They cannot be disproved by experiment, but we can learn from them nothing of any use to us; what then is the use of studying dynamics? This somewhat rapid condemnation would be rather unfair. There is not in Nature any system *perfectly* isolated, perfectly abstracted from all external action; but there are systems which are *nearly* isolated. If we observe such a system, we can study not only the relative motion of its different parts with respect to each other, but the motion of its centre of gravity with respect to the other parts of the universe. We then find that the motion of its centre of gravity is *nearly* uniform and rectilinear in conformity with Newton's third law. This is an experimental fact, which cannot be invalidated by a more accurate experiment. What, in fact, would a more accurate experiment teach us? It would teach us that the law is only approximately true, and we know that already. *Thus is explained how experiment may serve as a basis for the principles of mechanics, and yet will never invalidate them.*

Anthropomorphic Mechanics. It will be said that Kirchoff has only followed the general tendency of mathematicians towards nominalism; from this his skill as a physicist has not saved him. He wanted a definition of a force, and he took the first that came handy; but we do not require a definition of force; the idea of force is primitive, irreducible, indefinable; we all know what it is; of it we have direct intuition. This direct intuition arises from the idea of effort which is familiar to us from childhood. But in the first place, even if this direct intuition made known to us the real nature of force in itself, it would prove to be an insufficient basis for mechanics; it would, moreover, be quite useless. The important thing is not to know what force is, but how to measure it. Everything which does not teach us how to measure it is as useless to the mechanician as, for instance, the subjective idea of heat and cold to the student of heat. This subjective idea cannot be translated into numbers, and is therefore useless; a scientist whose skin is an absolutely bad conductor of heat, and who, therefore, has never felt the sensation of heat or cold, would read a thermometer in just the same way as anyone else, and would have enough material to construct the whole of the theory of heat.

Now this immediate notion of effort is of no use to us in the measurement of force. It is clear, for example, that I shall experience more fatigue in lifting a weight of 100 lb. than a man who is accustomed to lifting heavy burdens. But there is more than this. This notion of effort does not teach us the nature of force; it is definitively reduced to a recollection of muscular sensations, and no one will maintain that the sun experiences a muscular sensation when it attracts the earth. All that we can expect to find from it is a symbol, less precise and less convenient than the arrows (to denote direction) used by geometers, and quite as remote from reality.

Anthropomorphism plays a considerable historic role in the genesis of mechanics; perhaps it may yet furnish us with a symbol which some minds may find convenient; but it can be the foundation of nothing of a really scientific or philosophical character.

The Thread School. M. Andrade, in his *Leçons de mécanique physique,* has modernised anthropomorphic mechanics. To the school of mechanics with which Kirchoff is identified, he opposes a school which is quaintly called the "Thread School."

This school tries to reduce everything to the consideration of certain material systems of negligible mass, regarded in a state of tension and capable of transmitting considerable effort to distant bodies—systems of which the ideal type is the fine string, wire, or *thread.* A thread which transmits any force is slightly lengthened in the direction of that force; the direction of the thread tells us the direction of the force, and the magnitude of the force is measured by the lengthening of the thread.

We may imagine such an experiment as the following: A body *A* is attached to a thread; at the other extremity of the thread acts a force which is made to vary until the length of the thread is increased by α, and the acceleration of the body A is recorded. A is then detached, and a body B is attached to the same thread, and the same or another force is made to act until the increment of length again is α, and the acceleration of B is noted. The experiment is then renewed with both A and B until the increment of length is β. The four accelerations observed should be proportional. Here we have an experimental verification of the law of acceleration enunciated above. Again, we may consider a body under the action of several threads in equal tension, and by experiment we determine the direction of those threads when the body is in equilibrium. This is an experimental verification of the law of the composition of forces. But, as a matter of fact, what have we done? We have defined the force acting on the string by the deformation of the thread, which is reasonable enough; we have then assumed that if a body is attached to this thread, the effort which is transmitted to it by the thread is equal to the action exercised by the body on the thread; in fact, we have used the principle of action and reaction by considering it, not as an experimental truth, but as the very definition of force. This definition is quite as conventional as that of Kirchoff, but it is much less general.

All the forces are not transmitted by the thread (and to compare them they would all have to be transmitted by identical threads). If we even admitted that the earth is attached to the sun by an invisible thread, at any rate it will be agreed that we have no means of measuring the increment of the thread. Nine times out of ten, in consequence, our definition will be in default; no sense of any kind can be attached to it, and we must fall back on that of Kirchoff. Why then go on in this roundabout way? You admit a certain definition of force which has a meaning only in certain particular cases. In those cases you verify by experiment that it leads to the law of acceleration. On the strength of these experiments you then take the law of acceleration as a definition of force in all the other cases.

Would it not be simpler to consider the law of acceleration as a definition in all cases, and to regard the experiments in question, not as verifications of that law, but as verifications of the principle of action and reaction, or as proving the deformations of an elastic body depend only on the forces acting on that body? Without taking into account the fact that the conditions in which your definition could be accepted can only be very imperfectly fulfilled, that a thread is never without mass, that it is never isolated from all other forces than the reaction of the bodies attached to its extremities.

The ideas expounded by M. Andrade are none the less very interesting. If they do not satisfy our logical requirements, they give us a better view of the historical genesis of the fundamental ideas of mechanics. The reflections they suggest show us how the human mind passed from a naïve anthropomorphism to the present conception of science.

We see that we end with an experiment which is very particular, and as a matter of fact very crude, and we start with a perfectly general law, perfectly precise, the truth of which we regard as absolute. We have, so to speak, freely conferred this certainty on it by looking upon it as a convention.

Are the laws of acceleration and of the composition of forces only arbitrary conventions? Conventions, yes; arbitrary, no—they would be so if we lost sight of the experiments which led the founders of the science to adopt them, and which, imperfect as they were, were sufficient to justify their adoption. It is good from time to time to let our attention dwell on the experimental origin of these conventions.

CHAPTER VII

RELATIVE AND ABSOLUTE MOTION

The Principle of Relative Motion. Sometimes endeavours have been made to connect the law of acceleration with a more general principle. The movement of any system whatever ought to obey the same laws, whether it is referred to fixed axes or to the movable axes which are implied in uniform motion in a straight line. This is the principle of relative motion; it is imposed upon us for two reasons: the commonest experiment confirms it; the consideration of the contrary hypothesis is singularly repugnant to the mind.

Let us admit it then, and consider a body under the action of a force. The relative motion of this body with respect to an observer moving with a uniform velocity equal to the initial velocity of the body should be identical with what would be its absolute motion if it started from rest. We conclude that its acceleration must not depend upon its absolute velocity, and from that we attempt to deduce the complete law of acceleration.

For a long time there have been traces of this proof in the regulations for the degree of B. ès Sc. It is clear that the attempt has failed. The obstacle which prevented us from proving the law of acceleration is that we have no definition of force. This obstacle subsists in its entirety, since the principle invoked has not furnished us with the missing definition. The principle of relative motion is nonetheless very interesting, and deserves to be considered for its own sake. Let us try to enunciate it in an accurate manner. We have said above that the accelerations of the different bodies which form part of an isolated system only depend on their velocities and their relative positions, and not on their velocities and their absolute positions, provided that the movable axes to which the relative motion is referred move uniformly in a straight line; or, if it is preferred, their accelerations depend only on the differences of their velocities and the differences of their co-ordinates, and not on the absolute values of these velocities and co-ordinates. If this principle is true for relative accelerations, or rather for differences of acceleration, by combining it with the law of reaction we shall deduce that it is true for absolute accelerations. It remains to be seen how we can prove that differences of acceleration depend only on differences of velocities and co-ordinates; or, to speak in mathematical language, that these differences of co-ordinates satisfy differential equations of the second order. Can this proof be deduced from experiment or from *à priori* conditions? Remembering what we have said before, the reader will give his own answer. Thus enunciated, in fact, the principle of relative motion curiously resembles what I called above the generalised principle of inertia; it is not quite the same thing, since it is a question of differences of co-ordinates, and not of the co-ordinates themselves. The new principle teaches us something more than the old, but the same discussion applies to it, and would lead to the same conclusions. We need not recur to it.

Newton's Argument. Here we find a very important and even slightly disturbing question. I have said that the principle of relative motion was not for us simply a result of experiment; and that *à priori* every contrary hypothesis would be repugnant to the mind. But, then, why is the principle only true if the motion of the movable axes is uniform and in a straight line? It seems that it should be imposed upon us with the same force if the motion is accelerated, or at any rate if it reduces to a uniform rotation. In these two cases, in fact, the principle is not true. I need not dwell on the case in which the motion of the axes is in a straight line and not uniform. The paradox does not bear a moment's examination. If I am in a railway carriage, and if the train, striking against any obstacle whatever, is suddenly stopped, I shall be projected on to the opposite side, although I have not been directly acted upon by any force. There is nothing mysterious in that, and if I have not been subject to the action of any external force, the train has experienced an external impact. There can be nothing paradoxical in the relative motion of two bodies being disturbed when the motion of one or the other is modified by an external cause. Nor need I dwell on the case of relative motion referring to axes which rotate uniformly. If the sky were forever covered with clouds, and if we had no means of observing the stars, we might, nevertheless, conclude that the earth turns round. We should be warned of this fact by the flattening at the poles, or by the experiment of Foucault's pendulum. And yet, would there in this case be any meaning in saying that the earth turns round? If there is no absolute space, can a thing turn without turning with respect to something; and, on the other hand, how can we admit Newton's conclusion and believe in absolute space? But it is not sufficient to state that all possible solutions are equally unpleasant to us. We must analyse in each case the reason of our dislike, in order to make our choice with the knowledge of the cause. The long discussion which follows must, therefore, be excused.

Let us resume our imaginary story. Thick clouds hide the stars from men who cannot observe them, and even are ignorant of their existence. How will those men know that the earth turns round? No doubt, for a longer period than did our ancestors, they will regard the soil on which they stand as fixed and immovable! They will wait a much longer time than we did for the coming of a Copernicus; but this Copernicus will come at last. How will he come? In the first place, the mechanical school of this world would not run their heads against an absolute contradiction. In the theory of relative motion we observe, besides real forces, two imaginary forces, which we call ordinary centrifugal force and compounded centrifugal force. Our imaginary scientists can thus explain everything by looking upon these two forces as real, and they would not see in this a contradiction of the generalised principle of inertia, for these forces would depend, the one on the relative positions of the different parts of the system, such as real attractions, and the other on their relative velocities, as in the case of real frictions. Many difficulties, however, would before long awaken their attention. If they succeeded in realising an isolated system, the centre of gravity of this system would not have an approximately rectilinear path. They could invoke, to explain this fact, the centrifugal forces which they would regard as real, and which, no doubt, they would attribute to the mutual actions of the bodies—only they would not see these forces vanish at great distances—that is to say, in proportion as the isolation is better realised. Far from it. Centrifugal force increases indefinitely with distance. Already this difficulty would seem to them sufficiently serious, but it would not detain them for long. They would soon imagine some very subtle medium analogous to our ether, in which all bodies would be bathed, and which would exercise on them a repulsive action. But that is not all. Space is symmetrical—yet the laws of motion would present no symmetry. They should be able to distinguish between right and left. They would see, for instance, that cyclones always turn in the same direction, while for reasons of symmetry they should turn indifferently in any direction. If our scientists were able by dint of much hard work to make their universe perfectly symmetrical, this symmetry would not subsist, although there is no apparent reason why it should be disturbed in one direction more than in another. They would extract this from the situation no doubt—they would invent something which would not be more extraordinary than the glass spheres of Ptolemy, and would thus go on accumulating complications until the long-expected Copernicus would sweep them all away with a single blow, saying it is much more simple to admit that the earth turns round. Just as our Copernicus said to us: "It is more convenient to suppose that the earth turns round, because the laws of astronomy are thus expressed in a more simple language," so he would say to them: "It is more convenient to suppose that the earth turns round, because the laws of mechanics are thus expressed in much more simple language. That does not prevent absolute space—that is to say, the point to which we must refer the earth to know if it really does turn round—from having no objective existence. And hence this affirmation: "the earth turns round," has no meaning, since it cannot be verified by experiment; since such an experiment not only cannot be realised or even dreamed of by the most daring Jules Verne, but cannot even be conceived of without contradiction; or, in other words, these two propositions, "the earth turns round," and "it is more convenient to suppose that the earth turns round," have one and the same meaning. There is nothing more in one than in the other. Perhaps they will not be content with this, and may find it surprising that among all the hypotheses, or rather all the conventions, that can be made on this subject there is one which is more convenient than the rest? But if we have admitted it without difficulty when it is a question of the laws of astronomy, why should we object when it is a question of the laws of mechanics? We have seen that the co-ordinates of bodies are determined by differential equations of the second order, and that so are the differences of these co-ordinates. This is what we have called the generalised principle of inertia, and the principle of relative motion. If the distances of these bodies were determined in the same way by equations of the second order, it seems that the mind should be entirely satisfied. How far does the mind receive this satisfaction, and why is it not content with it? To explain this we had better take a simple example. I assume a system analogous to our solar system, but in which fixed stars foreign to this system cannot be perceived, so that astronomers can only observe the mutual distances of planets and the sun, and not the absolute longitudes of the planets. If we deduce directly from Newton's law the differential equations which define the variation of these distances, these equations will not be of the second order. I mean that if, outside Newton's law, we knew the initial values of these distances and of their derivatives with respect to time—that would not be sufficient to determine the values of these same distances at an ulterior moment. A datum would be still lacking, and this datum might be, for example, what astronomers call the area-constant. But here we may look at it from two different points of view. We may consider two kinds of constants. In the eyes of the physicist the world reduces to a series of phenomena depending, on the one hand, solely on initial phenomena, and, on the other hand, on the laws connecting consequence and antecedent. If observation then teaches us that a certain quantity is a constant, we shall have a choice of two ways of looking at it. So let us admit that there is a law which requires that this quantity shall not vary, but that by chance it has been found to have had in the beginning of time this value rather than that, a value that it has kept ever since. This quantity might then be called an *accidental* constant. Or again, let us admit on the contrary that there is a law of nature which imposes on this quantity this value and not that. We shall then have what may be called an *essential* constant. For example, in virtue of the laws of Newton the duration of the revolution of the earth must be constant. But if it is 366 and something sidereal days, and not 300 or 400, it is because of some initial chance or other. It is an *accidental* constant. If, on the other hand, the exponent of the distance which figures in the expression of the attractive force is equal to -2 and not to -3, it is not by chance, but because it is required by Newton's law. It is an *essential* constant. I do not know if this manner of giving to chance its share is legitimate in itself, and if there is not some artificiality about this distinction; but it is certain at least that in proportion as nature has secrets, she will be strictly arbitrary and always uncertain in their application. As far as the area-constant is concerned, we are accustomed to look upon it as accidental. Is it certain that our imaginary astronomers would do the same? If they were able to compare two different solar systems, they would get the idea that this constant may assume several different values. But I supposed at the outset, as I was entitled to do, that their system would appear isolated, and that they would see no star which was foreign to their system. Under these conditions they could only detect a single constant, which would have an absolutely invariable, unique value. They would be led no doubt to look upon it as an essential constant.

One word in passing to forestall an objection. The inhabitants of this imaginary world could neither observe nor define the area-constant as we do, because absolute longitudes escape their notice; but that would not prevent them from being rapidly led to remark a certain constant which would be naturally introduced into their equations, and which would be nothing but what we call the area-constant. But then what would happen? If the area-constant is regarded as essential, as dependent upon a law of nature, then in order to calculate the distances of the planets at any given moment it would be sufficient to know the initial values of these distances and those of their first derivatives. From this new point of view, distances will be determined by differential equations of the second order. Would this completely satisfy the minds of these astronomers? I think not. In the first place, they would very soon see that in differentiating their equations so as to raise them to a higher order, these equations would become much more simple, and they would be especially struck by the difficulty which arises from symmetry. They would have to admit different laws, according as the aggregate of the planets presented the figure of a certain polyhedron or rather of a regular polyhedron, and these consequences can only be escaped by regarding the area-constant as accidental. I have taken this particular example, because I have imagined astronomers who would not be in the least concerned with terrestrial mechanics and whose vision would be bounded by the solar system. But our conclusions apply in all cases. Our universe is more extended than theirs, since we have fixed stars; but it, too, is very limited, so we might reason on the whole of our universe just as these astronomers do on their solar system. We thus see that we should be definitively led to conclude that the equations which define distances are of an order higher than the second. Why should this alarm us—why do we find it perfectly natural that the sequence of phenomena depends on initial values of the first derivatives of these distances, while we hesitate to admit that they may depend on the initial values of the second derivatives? It can only be because of mental habits created in us by the constant study of the generalised principle of inertia and of its consequences. The values of the distances at any given moment depend upon their initial values, on that of their first derivatives, and something else. What is that *something else?* If we do not want it to be merely one of the second derivatives, we have only the choice of hypotheses. Suppose, as is usually done, that this something else is the absolute orientation of the universe in space, or the rapidity with which this orientation varies; this may be, it certainly is, the most convenient solution for the geometer. But it is not the most satisfactory for the philosopher, because this orientation does not exist. We may assume that this something else is the position or the velocity of some invisible body, and this is what is done by certain persons, who have even called the body Alpha, although we are destined to never know anything about this body except its name. This is an artifice entirely analogous to that of which I spoke at the end of the paragraph containing my reflections on the principle of inertia. But as a matter of fact the difficulty is artificial. Provided that the future indications of our instruments can only depend on the indications which they have given us, or that they might have formerly given us, such is all we want, and with these conditions we may rest satisfied.

CHAPTER VIII

ENERGY AND THERMO-DYNAMICS

Energetics. The difficulties raised by the classical mechanics have led certain minds to prefer a new system which they call Energetics. Energetics took its rise in consequence of the discovery of the principle of the conservation of energy. Helmholtz gave it its definite form. We begin by defining two quantities which play a fundamental part in this theory. They are *kinetic energy,* or *vis viva,* and *potential energy.* Every change that the bodies of nature can undergo is regulated by two experimental laws. First, the sum of the kinetic and potential energies is constant. This is the principle of the conservation of energy. Second, if a system of bodies is at A at the time t_0, and at B at the time t_1, it always passes from the first position to the second by such a path that the *mean* value of the difference between the two kinds of energy in the interval of time which separates the two epochs t_0 and t_1 is a minimum. This is Hamilton's principle, and is one of the forms of the principle of least action. The energetic theory has the following advantages over the classical. First, it is less incomplete—that is to say, the principles of the conservation of energy and of Hamilton teach us more than the fundamental principles of the classical theory, and exclude certain motions which do not occur in nature and which would be compatible with the classical theory. Second, it frees us from the hypothesis of atoms, which it was almost impossible to avoid with the classical theory. But in its turn it raises fresh difficulties. The definitions of the two kinds of energy would raise difficulties almost as great as those of force and mass in the first system. However, we can get out of these difficulties more easily, at any rate in the simplest cases. Assume an isolated system formed of a certain number of material points. Assume that these points are acted upon by forces depending only on their relative position and their distances apart, and independent of their velocities. In virtue of the principle of the conservation of energy there must be a function of forces. In this simple case the enunciation of the principle of the conservation of energy is of extreme simplicity. A certain quantity, which may be determined by experiment, must remain constant. This quantity is the sum of two terms. The first depends only on the position of the material points, and is independent of their velocities; the second is proportional to the squares of these velocities. This decomposition can only take place in one way. The first of these terms, which I shall call U, will be potential energy; the second, which I shall call T, will be kinetic energy. It is true that if $T+U$ is constant, so is any function of $T+U$, $\phi(T+U)$. But this function $\phi(T+U)$ will not be the sum of two terms, the one independent of the velocities, and the other proportional to the square of the velocities. Among the functions which remain constant there is only one which enjoys this property. It is $T+U$ (or a linear function of $T+U$), it matters not which, since this linear function may always be reduced to $T+U$ by a change of unit and of origin. This, then, is what we call energy. The first term we shall call potential energy, and the second kinetic energy. The definition of the two kinds of energy may therefore be carried through without any ambiguity.

So it is with the definition of mass. Kinetic energy, or *vis viva*, is expressed very simply by the aid of the masses, and of the relative velocities of all the material points with reference to one of them. These relative velocities may be observed, and when we have the expression of the kinetic energy as a function of these relative velocities, the co-efficients of this expression will give us the masses.

So in this simple case the fundamental ideas can be defined without difficulty. But the difficulties reappear in the more complicated cases if the forces, instead of depending solely on the distances, depend also on the velocities. For example, Weber supposes the mutual action of two electric molecules to depend not only on their distance but on their velocity and on their acceleration. If material points attracted each other according to an analogous law, U would depend on the velocity, and it might contain a term proportional to the square of the velocity. How can we detect among such terms those that arise from T or U? and how, therefore, can we distinguish the two parts of the energy? But there is more than this. How can we define energy itself? We have no more reason to take as our definition T+U rather than any other function of T+U, when the property which characterised T+U has disappeared—namely, that of being the sum of two terms of a particular form. But that is not all. We must take account not only of mechanical energy properly so called, but of the other forms of energy—heat, chemical energy, electrical energy, etc. The principle of the conservation of energy must be written T+U+Q= a constant, where T is the sensible kinetic energy, U the potential energy of position, depending only on the position of the bodies, Q the internal molecular energy under the thermal, chemical, or electrical form. This would be all right if the three terms were absolutely distinct; if T were proportional to the square of the velocities, U independent of these velocities and of the state of the bodies, Q independent of the velocities and of the positions of the bodies, and depending only on their internal state. The expression for the energy could be decomposed in one way only into three terms of this form. But this is not the case. Let us consider electrified bodies. The electrostatic energy due to their mutual action will evidently depend on their charge—*i.e.,* on their state; but it will equally depend on their position. If these bodies are in motion, they will act electrodynamically on one another, and the electro-dynamic energy will depend not only on their state and their position but on their velocities. We have therefore no means of making the selection of the terms which should form part of T, and U, and Q, and of separating the three parts of the energy. If T + U + Q is constant, the same is true of any function whatever, ϕ (T + U + Q).

If T + U + Q were of the particular form that I have suggested above, no ambiguity would ensue. Among the functions ϕ (T + U + Q) which remain constant, there is only one that would be of this particular form, namely the one which I would agree to call energy. But I have said this is not rigorously the case. Among the functions that remain constant there is not one which can rigorously be placed in this particular form. How then can we choose from among them that which should be called energy? We have no longer any guide in our choice.

Of the principle of the conservation of energy there is nothing left then but an enunciation: *there is something which remains constant*. In this form it, in its turn, is outside the bounds of experiment and reduced to a kind of tautology. It is clear that if the world is governed by laws there will be quantities which remain constant. Like Newton's laws, and for an analogous reason, the principle of the conservation of energy being based on experiment, can no longer be invalidated by it.

This discussion shows that, in passing from the classical system to the energetic, an advance has been made; but it shows, at the same time, that we have not advanced far enough.

Another objection seems to be still more serious. The principle of least action is applicable to reversible phenomena, but it is by no means satisfactory as far as irreversible phenomena are concerned. Helmholtz attempted to extend it to this class of phenomena, but he did not and could not succeed. So far as this is concerned all has yet to be done. The very enunciation of the principle of least action is objectionable. To move from one point to another, a material molecule, acted upon by no force, but compelled to move on a surface, will take as its path the geodesic line—*i.e.,* the shortest path. This molecule seems to know the point to which we want to take it, to foresee the time that it will take it to reach it by such a path, and then to know how to choose the most convenient path. The enunciation of the principle presents it to us, so to speak, as a living and free entity. It is clear that it would be better to replace it by a less objectionable enunciation, one in which, as philosophers would say, final effects do not seem to be substituted for acting causes.

Thermo-dynamics. The role of the two fundamental principles of thermo-dynamics becomes daily more important in all branches of natural philosophy. Abandoning the ambitious theories of forty years ago, encumbered as they were with molecular hypotheses, we now try to rest on thermo-dynamics alone the entire edifice of mathematical physics. Will the two principles of Mayer and of Clausius assure to it foundations solid enough to last for some time? We all feel it, but whence does our confidence arise? An eminent physicist said to me one day, *àpropos* of the law of errors: everyone stoutly believes it, because mathematicians imagine that it is an effect of observation, and observers imagine that it is a mathematical theorem. And this was for a long time the case with the principle of the conservation of energy. It is no longer the same now. There is no one who does not know that it is an experimental fact. But then who gives us the right of attributing to the principle itself more generality and more precision than to the experiments which have served to demonstrate it? This is asking, if it is legitimate to generalise, as we do every day, empiric data, and I shall not be so foolhardy as to discuss this question, after so many philosophers have vainly tried to solve it. One thing alone is certain. If this permission were refused to us, science could not exist; or at least would be reduced to a kind of inventory, to the ascertaining of isolated facts. It would no longer be to us of any value, since it could not satisfy our need of order and harmony, and because it would be at the same time incapable of prediction. As the circumstances which have preceded any fact whatever will never again, in all probability, be simultaneously reproduced, we already require a first generalisation to predict whether the fact will be renewed as soon as the least of these circumstances is changed. But every proposition may be generalised in an infinite number of ways. Among all possible generalisations we must choose, and we cannot but choose the simplest. We are therefore led to adopt the same course as if a simple law were, other things being equal, more probable than a complex law. A century ago it was frankly confessed and proclaimed abroad that nature loves simplicity; but nature has proved the contrary since then on more than one occasion. We no longer confess this tendency, and we only keep of it what is indispensable, so that science may not become impossible. In formulating a general, simple, and formal law, based on a comparatively small number of not altogether consistent experiments, we have only obeyed a necessity from which the human mind cannot free itself. But there is something more, and that is why I dwell on this topic. No one doubts that Mayer's principle is not called upon to survive all the particular laws from which it was deduced, in the same way that Newton's law has survived the laws of Kepler from which it was derived, and which are no longer anything but approximations, if we take perturbations into account. Now why does this principle thus occupy a kind of privileged position among physical laws? There are many reasons for that. At the outset we think that we cannot reject it, or even doubt its absolute rigour, without admitting the possibility of perpetual motion; we certainly feel distrust at such a prospect, and we believe ourselves less rash in affirming it than in denying it. That perhaps is not quite accurate. The impossibility of perpetual motion only implies the conservation of energy for reversible phenomena. The imposing simplicity of Mayer's principle equally contributes to strengthen our faith. In a law immediately deduced from experiments, such as Mariotte's law, this simplicity would rather appear to us a reason for distrust; but here this is no longer the case. We take elements which at the first glance are unconnected; these arrange themselves in an unexpected order, and form a harmonious whole. We cannot believe that this unexpected harmony is a mere result of chance. Our conquest appears to be valuable to us in proportion to the efforts it has cost, and we feel the more certain of having snatched its true secret from nature in proportion as nature has appeared more jealous of our attempts to discover it. But these are only small reasons. Before we raise Mayer's law to the dignity of an absolute principle, a deeper discussion is necessary. But if we embark on this discussion we see that this absolute principle is not even easy to enunciate. In every particular case we clearly see what energy is, and we can give it at least a provisory definition; but it is impossible to find a general definition of it. If we wish to enunciate the principle in all its generality and apply it to the universe, we see it vanish, so to speak, and nothing is left but this—*there is something which remains constant.* But has this a meaning? In the determinist hypothesis the state of the universe is determined by an extremely large number n of parameters, which I shall call $x_1, x_2, x_3 \ldots x_n$. As soon as we know at a given moment the values of these n parameters, we also know their derivatives with respect to time, and we can therefore calculate the values of these same parameters at an anterior or ulterior moment. In other words, these n parameters specify n differential equations of the first order. These equations have $n-1$ integrals, and therefore there are $n-1$ functions of $x_1, x_2, x_3 \ldots x_n$, which remain constant. If we say then, *there is something which remains constant,* we are only enunciating a tautology. We would be even embarrassed to decide which among all our integrals is that which should retain the name of energy. Besides, it is not in this sense that Mayer's principle is understood when it is applied to a limited system. We admit, then, that p of our n parameters vary independently so that we have only $n-p$ relations, generally linear, between our n parameters and their derivatives. Suppose, for the sake of simplicity, that the sum of the work done by the external forces is zero, as well as that of all the quantities of heat given off from the interior: what will then be the meaning of our principle? *There is a combination of these $n-p$ relations, of which the first member is an exact differential;* and then this differential vanishing in virtue of our $n-p$ relations, its integral is a constant, and it is this integral which we call energy. But how can it be that there are several parameters whose variations are independent? That can only take place in the case of external forces (although we have supposed, for the sake of simplicity, that the algebraical sum of all the work done by these forces has vanished). If, in fact, the system were completely isolated from all external action, the values of our n parameters at a given moment would suffice to determine the state of the system at any ulterior moment whatever, provided that we still clung to the determinist hypothesis. We should therefore fall back on the same difficulty as before. If the future state of the system is not entirely determined by its present state, it is because it further depends on the state of bodies external to the system. But then, is it likely that there exist among the parameters x which define the state of the system of equations independent of this state of the external bodies? and if in certain cases we think we can find them, is it not only because of our ignorance, and because the influence of these bodies is too weak for our experiment to be able to detect it? If the system is not regarded as completely isolated, it is probable that the rigorously exact expression of its internal energy will depend upon the state of the external bodies. Again, I have supposed above that the sum of all the external work is zero, and if we wish to be free from this rather artificial restriction the enunciation becomes still more difficult. To formulate Mayer's principle by giving it an absolute meaning, we must extend it to the whole universe, and then we find ourselves face to face with the very difficulty we have endeavoured to avoid. To sum up, and to use ordinary language, the law of the conservation of energy can have only one significance, because there is in it a property common to all possible properties; but in the determinist hypothesis there is only one possible, and then the law has no meaning. In the indeterminist hypothesis, on the other hand, it would have a meaning even if we wished to regard it in an absolute sense. It would appear as a limitation imposed on freedom.

But this word warns me that I am wandering from the subject, and that I am leaving the domain of mathematics and physics. I check myself, therefore, and I wish to retain only one impression of the whole of this discussion, and that is, that Mayer's law is a form subtle enough for us to be able to put into it almost anything we like. I do not mean by that that it corresponds to no objective reality, nor that it is reduced to mere tautology; since, in each particular case, and provided we do not wish to extend it to the absolute, it has a perfectly clear meaning. This subtlety is a reason for believing that it will last long; and as, on the other hand, it will only disappear to be blended in a higher harmony, we may work with confidence and utilise it, certain beforehand that our work will not be lost.

Almost everything that I have just said applies to the principle of Clausius. What distinguishes it is, that it is expressed by an inequality. It will be said perhaps that it is the same with all physical laws, since their precision is always limited by errors of observation. But they at least claim to be first approximations, and we hope to replace them little by little by more exact laws. If, on the other hand, the principle of Clausius reduces to an inequality, this is not caused by the imperfection of our means of observation, but by the very nature of the question.

General Conclusions on Part III. The principles of mechanics are therefore presented to us under two different aspects. On the one hand, there are truths founded on experiment, and verified approximately as far as almost isolated systems are concerned; on the other hand, there are postulates applicable to the whole of the universe and regarded as rigorously true. If these postulates possess a generality and a certainty which falsify the experimental truths from which they were deduced, it is because they reduce in final analysis to a simple convention that we have a right to make, because we are certain beforehand that no experiment can contradict it. This convention, however, is not absolutely arbitrary; it is not the child of our caprice. We admit it because certain experiments have shown us that it will be convenient, and thus is explained how experiment has built up the principles of mechanics, and why, moreover, it cannot reverse them. Take a comparison with geometry. The fundamental propositions of geometry, for instance, Euclid's postulate, are only conventions, and it is quite as unreasonable to ask if they are true or false as to ask if the metric system is true or false. Only, these conventions are convenient, and there are certain experiments which prove it to us. At the first glance, the analogy is complete, the role of experiment seems the same. We shall therefore be tempted to say, either mechanics must be looked upon as experimental science and then it should be the same with geometry; or, on the contrary, geometry is a deductive science, and then we can say the same of mechanics. Such a conclusion would be illegitimate. The experiments which have led us to adopt as more convenient the fundamental conventions of geometry refer to bodies which have nothing in common with those that are studied by geometry. They refer to the properties of solid bodies and to the propagation of light in a straight line. These are mechanical, optical experiments. In no way can they be regarded as geometrical experiments. And even the probable reason why our geometry seems convenient to us is that our bodies, our hands, and our limbs enjoy the properties of solid bodies. Our fundamental experiments are pre-eminently physiological experiments which refer, not to the space which is the object that geometry must study, but to our body—that is to say, to the instrument which we use for that study. On the other hand, the fundamental conventions of mechanics and the experiments which prove to us that they are convenient, certainly refer to the same objects or to analogous objects. Conventional and general principles are the natural and direct generalisations of experimental and particular principles. Let it not be said that I am thus tracing artificial frontiers between the sciences; that I am separating by a barrier geometry properly so called from the study of solid bodies. I might just as well raise a barrier between experimental mechanics and the conventional mechanics of general principles. Who does not see, in fact, that by separating these two sciences we mutilate both, and that what will remain of the conventional mechanics when it is isolated will be but very little, and can in no way be compared with that grand body of doctrine which is called geometry.

We now understand why the teaching of mechanics should remain experimental. Thus only can we be made to understand the genesis of the science, and that is indispensable for a complete knowledge of the science itself. Besides, if we study mechanics, it is in order to apply it; and we can only apply it if it remains objective. Now, as we have seen, when principles gain in generality and certainty they lose in objectivity. It is therefore especially with the objective side of principles that we must be early familiarised, and this can only be by passing from the particular to the general, instead of from the general to the particular.

Principles are conventions and definitions in disguise. They are, however, deduced from experimental laws, and these laws have, so to speak, been erected into principles to which our mind attributes an absolute value. Some philosophers have generalised far too much. They have thought that the principles were the whole of science, and therefore that the whole of science was conventional. This paradoxical doctrine, which is called Nominalism, cannot stand examination. How can a law become a principle? It expressed a relation between two real terms, A and B; but it was not rigorously true, it was only approximate. We introduce arbitrarily an intermediate term, C, more or less imaginary, and C is *by definition* that which has with A *exactly* the relation expressed by the law. So our law is decomposed into an absolute and rigorous principle which expresses the relation of A to C, and an approximate experimental and revisable law which expresses the relation of C to B. But it is clear that however far this decomposition may be carried, laws will always remain. We shall now enter into the domain of laws properly so called.

PART IV

NATURE

CHAPTER IX

HYPOTHESES IN PHYSICS

The Role of Experiment and Generalisation. Experiment is the sole source of truth. It alone can teach us something new; it alone can give us certainty. These are two points that cannot be questioned. But then, if experiment is everything, what place is left for mathematical physics? What can experimental physics do with such an auxiliary—an auxiliary, moreover, which seems useless, and even may be dangerous?

However, mathematical physics exists. It has rendered undeniable service, and that is a fact which has to be explained. It is not sufficient merely to observe; we must use our observations, and for that purpose we must generalise. This is what has always been done, only as the recollection of past errors has made man more and more circumspect, he has observed more and more and generalised less and less. Every age has scoffed at its predecessor, accusing it of having generalised too boldly and too naïvely. Descartes used to commiserate the Ionians. Descartes in his turn makes us smile, and no doubt some day our children will laugh at us. Is there no way of getting at once to the gist of the matter, and thereby escaping the raillery which we foresee? Cannot we be content with experiment alone? No, that is impossible; that would be a complete misunderstanding of the true character of science. The man of science must work with method. Science is built up of facts, as a house is built of stones; but an accumulation of facts is no more a science than a heap of stones is a house. Most important of all, the man of science must exhibit foresight. Carlyle has written somewhere something after this fashion. "Nothing but facts are of importance. John Lackland passed by here. Here is something that is admirable. Here is a reality for which I would give all the theories in the world."* Carlyle was a compatriot of Bacon, and, like him, he wished to proclaim his worship of *the God of Things as they are*.

But Bacon would not have said that. That is the language of the historian. The physicist would most likely have said: "John Lackland passed by here. It is all the same to me, for he will not pass this way again."

We all know that there are good and bad experiments. The latter accumulate in vain. Whether there are a hundred or a thousand, one single piece of work by a real master—by a Pasteur, for example—will be sufficient to sweep them into oblivion. Bacon would have thoroughly understood that, for he invented the phrase *experimentum crucis;* but Carlyle would not have understood it. A fact is a fact. A student has read such and such a number on his thermometer. He has taken no precautions. It does not matter; he has read it, and if it is only the fact which counts, this is a reality that is as much entitled to be called a reality as the peregrinations of King John Lackland. What, then, is a good experiment? It is that which teaches us something more than an isolated fact. It is that which enables us to predict, and to generalise. Without generalisation, prediction is impossible. The circumstances under which one has operated will never again be reproduced simultaneously. The fact observed will never be repeated. All that can be affirmed is that under analogous circumstances an analogous fact will be produced. To predict it, we must therefore invoke the aid of analogy— that is to say, even at this stage, we must generalise. However timid we may be, there must be interpolation. Experiment only gives us a

* *Past and Present,* end of Chapter I, Book II.—[Tr.]

certain number of isolated points. They must be connected by a continuous line, and this is a true generalisation. But more is done. The curve thus traced will pass between and near the points observed; it will not pass through the points themselves. Thus we are not restricted to generalising our experiment, we correct it; and the physicist who would abstain from these corrections, and really content himself with experiment pure and simple, would be compelled to enunciate very extraordinary laws indeed. Detached facts cannot therefore satisfy us, and that is why our science must be ordered, or, better still, generalised.

It is often said that experiments should be made without preconceived ideas. That is impossible. Not only would it make every experiment fruitless, but even if we wished to do so, it could not be done. Every man has his own conception of the world, and this he cannot so easily lay aside. We must, for example, use language, and our language is necessarily steeped in preconceived ideas. Only they are unconscious preconceived ideas, which are a thousand times the most dangerous of all. Shall we say, that if we cause others to intervene of which we are fully conscious, that we shall only aggravate the evil? I do not think so. I am inclined to think that they will serve as ample counterpoises—I was almost going to say antidotes. They will generally disagree, they will enter into conflict one with another, and *ipso facto,* they will force us to look at things under different aspects. This is enough to free us. He is no longer a slave who can choose his master.

Thus, by generalisation, every fact observed enables us to predict a large number of others; only, we ought not to forget that the first alone is certain, and that all the others are merely probable. However solidly founded a prediction may appear to us, we are never *absolutely* sure that experiment will not prove it to be baseless if we set to work to verify it. But the probability of its accuracy is often so great that practically we may be content with it. It is far better to predict without certainty than never to have predicted at all. We should never, therefore, disdain to verify when the opportunity presents itself. But every experiment is long and difficult, and the labourers are few, and the number of facts which we require to predict is enormous; and besides this mass, the number of direct verifications that we can make will never be more than a negligible quantity. Of this little that we can directly attain we must choose the best. Every experiment must enable us to make a maximum number of predictions having the highest possible degree of probability. The problem is, so to speak, to increase the output of the scientific machine. I may be permitted to compare science to a library which must go on increasing indefinitely; the librarian has limited funds for his purchases, and he must, therefore, strain every nerve not to waste them. Experimental physics has to make the purchases, and experimental physics alone can enrich the library. As for mathematical physics, her duty is to draw up the catalogue. If the catalogue is well done the library is none the richer for it; but the reader will be enabled to utilise its riches; and also by showing the librarian the gaps in his collection, it will help him to make a judicious use of his funds, which is all the more important, inasmuch as those funds are entirely inadequate. That is the role of mathematical physics. It must direct generalisation so as to increase what I called just now the output of science. By what means it does this, and how it may do it without danger, is what we have now to examine.

The Unity of Nature. Let us first of all observe that every generalisation supposes in a certain measure a belief in the unity and simplicity of nature. As far as the unity is concerned, there can be no difficulty. If the different parts of the universe were not as the organs of the same body, they would not re-act one upon the other; they would mutually ignore each other, and we in particular should only know one part. We need not, therefore, ask if nature is one, but how she is one.

As for the second point, that is not so clear. It is not certain that nature is simple. Can we without danger act as if she were?

There was a time when the simplicity of Mariotte's law was an argument in favour of its accuracy: when Fresnel himself, after having said in a conversation with Laplace that nature cares naught for analytical difficulties, was compelled to explain his words so as not to give offence to current opinion. Nowadays, ideas have changed considerably; but those who do not believe that natural laws must be simple are still often obliged to act as if they did believe it. They cannot entirely dispense with this necessity without making all generalisation, and therefore all science, impossible. It is clear that any fact can be generalised in an infinite number of ways, and it is a question of choice. The choice can only be guided by considerations of simplicity. Let us take the most ordinary case, that of interpolation. We draw a continuous line as regularly as possible between the points given by observation. Why do we avoid angular points and inflexions that are too sharp? Why do we not make our curve describe the most capricious zigzags? It is because we know beforehand, or think we know, that the law we have to express cannot be so complicated as all that. The mass of Jupiter may be deduced either from the movements of his satellites, or from the perturbations of the major planets, or from those of the minor planets. If we take the mean of the determinations obtained by these three methods, we find three numbers very close together, but not quite identical. This result might be interpreted by supposing that the gravitation constant is not the same in the three cases; the observations would be certainly much better represented. Why do we reject this interpretation? Not because it is absurd, but because it is uselessly complicated. We shall only accept it when we are forced to, and it is not imposed upon us yet. To sum up, in most cases every law is held to be simple until the contrary is proved.

This custom is imposed upon physicists by the reasons that I have indicated, but how can it be justified in the presence of discoveries which daily show us fresh details, richer and more complex? How can we even reconcile it with the unity of nature? For if all things are interdependent, the relations in which so many different objects intervene can no longer be simple.

If we study the history of science we see produced two phenomena which are, so to speak, each the inverse of the other. Sometimes it is simplicity which is hidden under what is apparently complex; sometimes, on the contrary, it is simplicity which is apparent, and which conceals extremely complex realities. What is there more complicated than the disturbed motions of the planets, and what more simple than Newton's law? There, as Fresnel said, Nature playing with analytical difficulties only uses simple means, and creates by their combination I know not what tangled skein. Here it is the hidden simplicity which must be disentangled. Examples to the contrary abound. In the kinetic theory of gases, molecules of tremendous velocity are discussed, whose paths, deformed by incessant impacts, have the most capricious shapes, and plough their way through space in every direction. The result observable is Mariotte's simple law. Each individual fact was complicated. The law of great numbers has re-established simplicity in the mean. Here the simplicity is only apparent, and the coarseness of our senses alone prevents us from seeing the complexity.

Many phenomena obey a law of proportionality. But why? Because in these phenomena there is something which is very small. The simple law observed is only the translation of the general analytical rule by which the infinitely small increment of a function is proportional to the increment of the variable. As in reality our increments are not infinitely small, but only very small, the law of proportionality is only approximate, and simplicity is only apparent. What I have just said applies to the law of the superposition of small movements, which is so fruitful in its applications and which is the foundation of optics.

And Newton's law itself? Its simplicity, so long undetected, is perhaps only apparent. Who knows if it be not due to some complicated mechanism, to the impact of some subtle matter animated by irregular movements, and if it has not become simple merely through the play of averages and large numbers? In any case, it is difficult not to suppose that the true law contains complementary terms which may become sensible at small distances. If in astronomy they are negligible, and if the law thus regains its simplicity, it is solely on account of the enormous distances of the celestial bodies. No doubt, if our means of investigation became more and more penetrating, we should discover the simple beneath the complex, and then the complex from the simple, and then again the simple beneath the complex, and so on, without ever being able to predict what the last term will be. We must stop somewhere, and for science to be possible we must stop where we have found simplicity. That is the only ground on which we can erect the edifice of our generalisations. But, this simplicity being only apparent, will the ground be solid enough? That is what we have now to discover.

For this purpose let us see what part is played in our generalisations by the belief in simplicity. We have verified a simple law in a considerable number of particular cases. We refuse to admit that this coincidence, so often repeated, is a result of mere chance, and we conclude that the law must be true in the general case.

Kepler remarks that the positions of a planet observed by Tycho are all on the same ellipse. Not for one moment does he think that, by a singular freak of chance, Tycho had never looked at the heavens except at the very moment when the path of the planet happened to cut that ellipse. What does it matter then if the simplicity be real or if it hide a complex truth? Whether it be due to the influence of great numbers which reduces individual differences to a level, or to the greatness or the smallness of certain quantities which allow of certain terms to be neglected—in no case is it due to chance. This simplicity, real or apparent, has always a cause. We shall therefore always be able to reason in the same fashion, and if a simple law has been observed in several particular cases, we may legitimately suppose that it still will be true in analogous cases. To refuse to admit this would be to attribute an inadmissible role to chance. However, there is a difference. If the simplicity were real and profound it would bear the test of the increasing precision of our methods of measurement. If, then, we believe Nature to be profoundly simple, we must conclude that it is an approximate and not a rigorous simplicity. This is what was formerly done, but it is what we have no longer the right to do. The simplicity of Kepler's laws, for instance, is only apparent; but that does not prevent them from being applied to almost all systems analogous to the solar system, though that prevents them from being rigorously exact.

Role of Hypothesis. Every generalisation is a hypothesis. Hypothesis therefore plays a necessary role, which no one has ever contested. Only, it should always be as soon as possible submitted to verification. It goes without saying that, if it cannot stand this test, it must be abandoned without any hesitation. This is, indeed, what is generally done; but sometimes with a certain impatience. Ah well! this impatience is not justified. The physicist who has just given up one of his hypotheses should, on the contrary, rejoice, for he found an unexpected opportunity of discovery. His hypothesis, I imagine, had not been lightly adopted. It took into account all the known factors which seem capable of intervention in the phenomenon. If it is not verified, it is because there is something unexpected and extraordinary about it, because we are on the point of finding something unknown and new. Has the hypothesis thus rejected been sterile? Far from it. It may be even said that it has rendered more service than a true hypothesis. Not only has it been the occasion of a decisive experiment, but if this experiment had been made by chance, without the hypothesis, no conclusion could have been drawn; nothing extraordinary would have been seen; and only one fact the more would have been catalogued, without deducing from it the remotest consequence.

Now, under what conditions is the use of hypothesis without danger? The proposal to submit all to experiment is not sufficient. Some hypotheses are dangerous—first and foremost those which are tacit and unconscious. And since we make them without knowing them, we cannot get rid of them. Here again, there is a service that mathematical physics may render us. By the precision which is its characteristic, we are compelled to formulate all the hypotheses that we would unhesitatingly make without its aid. Let us also notice that it is important not to multiply hypotheses indefinitely. If we construct a theory based upon multiple hypotheses, and if experiment condemns it, which of the premises must be changed? It is impossible to tell. Conversely, if the experiment succeeds, must we suppose that it has verified all these hypotheses at once? Can several unknowns be determined from a single equation?

We must also take care to distinguish between the different kinds of hypotheses. First of all, there are those which are quite natural and necessary. It is difficult not to suppose that the influence of very distant bodies is quite negligible, that small movements obey a linear law, and that effect is a continuous function of its cause. I will say as much for the conditions imposed by symmetry. All these hypotheses affirm, so to speak, the common basis of all the theories of mathematical physics. They are the last that should be abandoned. There is a second category of hypotheses which I shall qualify as indifferent. In most questions the analyst assumes, at the beginning of his calculations, either that matter is continuous, or the reverse, that it is formed of atoms. In either case, his results would have been the same. On the atomic supposition he has a little more difficulty in obtaining them—that is all. If, then, experiment confirms his conclusions, will he suppose that he has proved, for example, the real existence of atoms?

In optical theories two vectors are introduced, one of which we consider as a velocity and the other as a vortex. This again is an indifferent hypothesis, since we should have arrived at the same conclusions by assuming the former to be a vortex and the latter to be a velocity. The success of the experiment cannot prove, therefore, that the first vector is really a velocity. It only proves one thing— namely, that it is a vector; and that is the only hypothesis that has really been introduced into the premises. To give it the concrete appearance that the fallibility of our minds demands, it was necessary to consider it either as a velocity or as a vortex. In the same way, it was necessary to represent it by an x or a y, but the result will not prove that we were right or wrong in regarding it as a velocity; nor will it prove we are right or wrong in calling it x and not y.

These indifferent hypotheses are never dangerous provided their characters are not misunderstood. They may be useful either as artifices for calculation, or to assist our understanding by concrete images, to fix the ideas, as we say. They need not therefore be rejected. The hypotheses of the third category are real general isations. They must be confirmed or invalidated by experiment. Whether verified or condemned, they will always be fruitful; but, for the reasons I have given, they will only be so if they are not too numerous.

Origin of Mathematical Physics. Let us go further and study more closely the conditions which have assisted the development of mathematical physics. We recognise at the outset that the efforts of men of science have always tended to resolve the complex phenomenon given directly by experiment into a very large number of elementary phenomena, and that in three different ways.

First, with respect to time. Instead of embracing in its entirety the progressive development of a phenomenon, we simply try to connect each moment with the one immediately preceding. We admit that the present state of the world only depends on the immediate past, without being directly influenced, so to speak, by the recollection of a more distant past. Thanks to this postulate, instead of studying directly the whole succession of phenomena, we may confine ourselves to writing down its *differential equation;* for the laws of Kepler we substitute the law of Newton.

Next, we try to decompose the phenomena in space. What experiment gives us is a confused aggregate of facts spread over a scene of considerable extent. We must try to deduce the elementary phenomenon, which will still be localised in a very small region of space.

A few examples perhaps will make my meaning clearer. If we wished to study in all its complexity the distribution of temperature in a cooling solid, we could never do so. This is simply because, if we only reflect that a point in the solid can directly impart some of its heat to a neighbouring point, it will immediately impart that heat only to the nearest points, and it is but gradually that the flow of heat will reach other portions of the solid. The elementary phenomenon is the interchange of heat between two contiguous points. It is strictly localised and relatively simple if, as is natural, we admit that it is not influenced by the temperature of the molecules whose distance apart is small.

I bend a rod: it takes a very complicated form, the direct investigation of which would be impossible. But I can attack the problem, however, if I notice that its flexure is only the resultant of the deformations of the very small elements of the rod, and that the deformation of each of these elements only depends on the forces which are directly applied to it, and not in the least on those which may be acting on the other elements.

In all these examples, which may be increased without difficulty, it is admitted that there is no action at a distance or at great distances. That is an hypothesis. It is not always true, as the law of gravitation proves. It must therefore be verified. If it is confirmed, even approximately, it is valuable, for it helps us to use mathematical physics, at any rate by successive approximations. If it does not stand the test, we must seek something else that is analogous, for there are other means of arriving at the elementary phenomenon. If several bodies act simultaneously, it may happen that their actions are independent, and may be added one to the other, either as vectors or as scalar quantities. The elementary phenomenon is then the action of an isolated body. Or suppose, again, it is a question of small movements, or more generally of small variations which obey the well-known law of mutual or relative independence. The movement observed will then be decomposed into simple movements— for example, sound into its harmonics, and white light into its monochromatic components. When we have discovered in which direction to seek for the elementary phenomenon, by what means may we reach it? First, it will often happen that in order to predict it, or rather in order to predict what is useful to us, it will not be necessary to know its mechanism. The law of great numbers will suffice. Take for example the propagation of heat. Each molecule radiates towards its neighbour—we need not inquire according to what law; and if we make any supposition in this respect, it will be an indifferent hypothesis, and therefore useless and unverifiable. In fact, by the action of averages and thanks to the symmetry of the medium, all differences are levelled, and, whatever the hypothesis may be, the result is always the same.

The same feature is presented in the theory of elasticity, and in that of capillarity. The neighbouring molecules attract and repel each other, we need not inquire by what law. It is enough for us that this attraction is sensible at small distances only, and that the molecules are very numerous, that the medium is symmetrical, and we have only to let the law of great numbers come into play.

Here again the simplicity of the elementary phenomenon is hidden beneath the complexity of the observable resultant phenomenon; but in its turn this simplicity was only apparent and disguised a very complex mechanism. Evidently the best means of reaching the elementary phenomenon would be experiment. It would be necessary by experimental artifices to dissociate the complex system which nature offers for our investigations and carefully to study the elements as dissociated as possible; for example, natural white light would be decomposed into monochromatic lights by the aid of the prism, and into polarised lights by the aid of the polariser. Unfortunately, that is neither always possible nor always sufficient, and sometimes the mind must run ahead of experiment. I shall only give one example which has always struck me rather forcibly. If I decompose white light, I shall be able to isolate a portion of the spectrum, but however small it may be, it will always be a certain width. In the same way the natural lights which are called *monochromatic* give us a very fine array, but a ray which is not, however, infinitely fine. It might be supposed that in the experimental study of the properties of these natural lights, by operating with finer and finer rays, and passing on at last to the limit, so to speak, we should eventually obtain the properties of a rigorously monochromatic light. That would not be accurate. I assume that two rays emanate from the same source, that they are first polarised in planes at right angles, that they are then brought back again to the same plane of polarisation, and that we try to obtain interference. If the light were *rigorously* monochromatic, there would be interference; but with our nearly monochromatic lights, there will be no interference, and that, however narrow the ray may be. For it to be otherwise, the ray would have to be several million times finer than the finest known rays.

Here then we should be led astray by proceeding to the limit. The mind has to run ahead of the experiment, and if it has done so with success, it is because it has allowed itself to be guided by the instinct of simplicity. The knowledge of the elementary fact enables us to state the problem in the form of an equation. It only remains to deduce from it by combination the observable and verifiable complex fact. That is what we call *integration,* and it is the province of the mathematician. It might be asked, why in physical science generalisation so readily takes the mathematical form. The reason is now easy to see. It is not only because we have to express numerical laws; it is because the observable phenomenon is due to the superposition of a large number of elementary phenomena which are *all similar to each other;* and in this way differential equations are quite naturally introduced. It is not enough that each elementary phenomenon should obey simple laws: all those that we have to combine must obey the same law; then only is the intervention of mathematics of any use. Mathematics teaches us, in fact, to combine like with like. Its object is to divine the result of a combination without having to reconstruct that combination element by element. If we have to repeat the same operation several times, mathematics enables us to avoid this repetition by telling the result beforehand by a kind of induction. This I have explained before in the chapter on mathematical reasoning. But for that purpose all these operations must be similar; in the contrary case we must evidently make up our minds to working them out in full one after the other, and mathematics will be useless. It is therefore, thanks to the approximate homogeneity of the matter studied by physicists, that mathematical physics came into existence. In the natural sciences the following conditions are no longer to be found: homogeneity, relative independence of remote parts, simplicity of the elementary fact; and that is why the student of natural science is compelled to have recourse to other modes of generalisation.

CHAPTER X

THE THEORIES OF MODERN PHYSICS

Significance of Physical Theories. The ephemeral nature of scientific theories takes by surprise the man of the world. Their brief period of prosperity ended, he sees them abandoned one after another; he sees ruins piled upon ruins; he predicts that the theories in fashion today will in a short time succumb in their turn, and he concludes that they are absolutely in vain. This is what he calls the *bankruptcy of science.*

His skepticism is superficial; he does not take into account the object of scientific theories and the part they play, or he would understand that the ruins may be still good for something. No theory seemed established on firmer ground than Fresnel's, which attributed light to the movements of the ether. Then if Maxwell's theory is today preferred, does that mean that Fresnel's work was in vain? No; for Fresnel's object was not to know whether there really is an ether, if it is or is not formed of atoms, if these atoms really move in this way or that; his object was to predict optical phenomena.

This Fresnel's theory enables us to do today as well as it did before Maxwell's time. The differential equations are always true, they may be always integrated by the same methods, and the results of this integration still preserve their value. It cannot be said that this is reducing physical theories to simple practical recipes; these equations express relations, and if the equations remain true, it is because the relations preserve their reality. They teach us now, as they did then, that there is such and such a relation between this thing and that; only, the something which we then called *motion,* we now call *electric current.* But these are merely names of the images we substituted for the real objects which Nature will hide forever from our eyes. The true relations between these real objects are the only reality we can attain, and the sole condition is that the same relations shall exist between these objects as between the images we are forced to put in their place. If the relations are known to us, what does it matter if we think it convenient to replace one image by another?

That a given periodic phenomenon (an electric oscillation, for instance) is really due to the vibration of a given atom, which, behaving like a pendulum, is really displaced in this manner or that, all this is neither certain nor essential. But that there is between the electric oscillation, the movement of the pendulum, and all periodic phenomena an intimate relationship which corresponds to a profound reality; that this relationship, this similarity, or rather this parallelism, is continued in the details; that it is a consequence of more general principles such as that of the conservation of energy, and that of least action; this we may affirm; this is the truth which will ever remain the same in whatever garb we may see fit to clothe it.

Many theories of dispersion have been proposed. The first were imperfect, and contained but little truth. Then came that of Helmholtz, and this in its turn was modified in different ways; its author himself conceived another theory, founded on Maxwell's principles. But the remarkable thing is, that all the scientists who followed Helmholtz obtain the same equations, although their starting-points were to all appearance widely separated. I venture to say that these theories are all simultaneously true; not merely because they express a true relation—that between absorption and abnormal dispersion. In the premisses of these theories the part that is true is the part common to all: it is the affirmation of this or that relation between certain things, which some call by one name and some by another.

The kinetic theory of gases has given rise to many objections, to which it would be difficult to find an answer were it claimed that the theory is absolutely true. But all these objections do not alter the fact that it has been useful, particularly in revealing to us one true relation which would otherwise have remained profoundly hidden—the relation between gaseous and osmotic pressures. In this sense, then, it may be said to be true.

When a physicist finds a contradiction between two theories which are equally dear to him, he sometimes says: "Let us not be troubled, but let us hold fast to the two ends of the chain, lest we lose the intermediate links." This argument of the embarrassed theologian would be ridiculous if we were to attribute to physical theories the interpretation given them by the man of the world. In case of contradiction one of them at least should be considered false. But this is no longer the case if we only seek in them what should be sought. It is quite possible that they both express true relations, and that the contradictions only exist in the images we have formed to ourselves of reality. To those who feel that we are going too far in our limitations of the domain accessible to the scientist, I reply: these questions which we forbid you to investigate, and which you so regret, are not only insoluble, they are illusory and devoid of meaning.

Such a philosopher claims that all physics can be explained by the mutual impact of atoms. If he simply means that the same relations obtain between physical phenomena as between the mutual impact of a large number of billiard balls—well and good! this is verifiable, and perhaps is true. But he means something more, and we think we understand him, because we think we know what an impact is. Why? Simply because we have often watched a game of billiards. Are we to understand that God experiences the same sensations in the contemplation of His work that we do in watching a game of billiards? If it is not our intention to give his assertion this fantastic meaning, and if we do not wish to give it the more restricted meaning I have already mentioned, which is the sound meaning, then it has no meaning at all. Hypotheses of this kind have therefore only a metaphorical sense. The scientist should no more banish them than a poet banishes metaphor; but he ought to know what they are worth. They may be useful to give satisfaction to the mind, and they will do no harm as long as they are only indifferent hypotheses.

These considerations explain to us why certain theories, that were thought to be abandoned and definitively condemned by experiment, are suddenly revived from their ashes and begin a new life. It is because they expressed true relations, and had not ceased to do so when for some reason or other we felt it necessary to enunciate the same relations in another language. Their life had been latent, as it were.

Barely fifteen years ago, was there anything more ridiculous, more quaintly old-fashioned, than the fluids of Coulomb? And yet, here they are re-appearing under the name of *electrons*. In what do these permanently electrified molecules differ from the electric molecules of Coulomb? It is true that in the electrons the electricity is supported by a little, a very little matter; in other words, they have mass. Yet Coulomb did not deny mass to his fluids, or if he did, it was with reluctance. It would be rash to affirm that the belief in electrons will not also undergo an eclipse, but it was nonetheless curious to note this unexpected renaissance.

But the most striking example is Carnot's principle. Carnot established it, starting from false hypotheses. When it was found that heat was indestructible, and may be converted into work, his ideas were completely abandoned; later, Clausius returned to them, and to him is due their definitive triumph. In its primitive form, Carnot's theory expressed in addition to true relations, other inexact relations, the *débris* of old ideas; but the presence of the latter did not alter the reality of the others. Clausius had only to separate them, just as one lops off dead branches.

The result was the second fundamental law of thermo-dynamics. The relations were always the same, although they did not hold, at least to all appearance, between the same objects. This was sufficient for the principle to retain its value. Nor have the reasonings of Carnot perished on this account; they were applied to an imperfect conception of matter, but their form—*i.e.,* the essential part of them, remained correct. What I have just said throws some light at the same time on the role of general principles, such as those of the principle of least action or of the conservation of energy. These principles are of very great value. They were obtained in the search for what there was in common in the enunciation of numerous physical laws; they thus represent the quintessence of innumerable observations. However, from their very generality results a consequence to which I have called attention in Chapter VIII—namely, that they are no longer capable of verification. As we cannot give a general definition of energy, the principle of the conservation of energy simply signifies that there is a *something* which remains constant. Whatever fresh notions of the world may be given us by future experiments, we are certain beforehand that there is something which remains constant, and which may be called *energy.* Does this mean that the principle has no meaning and vanishes into a tautology? Not at all. It means that the different things to which we give the name of *energy* are connected by a true relationship; it affirms between them a real relation. But then, if this principle has a meaning, it may be false; it may be that we have no right to extend indefinitely its applications, and yet it is certain beforehand to be verified in the strict sense of the word. How, then, shall we know when it has been extended as far as is legitimate? Simply when it ceases to be useful to us—*i.e.,* when we can no longer use it to predict correctly new phenomena. We shall be certain in such a case that the relation affirmed is no longer real, for otherwise it would be fruitful; experiment without directly contradicting a new extension of the principle will nevertheless have condemned it.

Physics and Mechanism. Most theorists have a constant predilection for explanations borrowed from physics, mechanics, or dynamics. Some would be satisfied if they could account for all phenomena by the motion of molecules attracting one another according to certain laws. Others are more exact; they would suppress attractions acting at a distance; their molecules would follow rectilinear paths, from which they would only be deviated by impacts. Others again, such as Hertz, suppress the forces as well, but suppose their molecules subjected to geometrical connections analogous, for instance, to those of articulated systems; thus, they wish to reduce dynamics to a kind of kinematics. In a word, they all wish to bend nature into a certain form, and unless they can do this they cannot be satisfied. Is Nature flexible enough for this?

We shall examine this question in Chapter XII, *àpropos* of Maxwell's theory. Every time that the principles of least action and energy are satisfied, we shall see that not only is there always a mechanical explanation possible, but that there is an unlimited number of such explanations. By means of a well-known theorem due to Königs, it may be shown that we can explain everything in an unlimited number of ways, by connections after the manner of Hertz, or, again, by central forces. No doubt it may be just as easily demonstrated that everything may be explained by simple impacts. For this, let us bear in mind that it is not enough to be content with the ordinary matter of which we are aware by means of our senses, and the movements of which we observe directly. We may conceive of ordinary matter as either composed of atoms, whose internal movements escape us, our senses being able to estimate only the displacement of the whole; or we may imagine one of those subtle fluids, which under the name of *ether* or other names, have from all time played so important a role in physical theories. Often we go further, and regard the ether as the only primitive, or even as the only true matter. The more moderate consider ordinary matter to be condensed ether, and there is nothing startling in this conception; but others only reduce its importance still further, and see in matter nothing more than the geometrical locus of singularities in the ether. Lord Kelvin, for instance, holds what we call matter to be only the locus of those points at which the ether is animated by vortex motions. Riemann believes it to be locus of those points at which ether is constantly destroyed; to Wiechert or Larmor, it is the locus of the points at which the ether has undergone a kind of torsion of a very particular kind. Taking any one of these points of view, I ask by what right do we apply to the ether the mechanical properties observed in ordinary matter, which is but false matter? The ancient fluids, caloric, electricity, etc., were abandoned when it was seen that heat is not indestructible. But they were also laid aside for another reason. In materialising them, their individuality was, so to speak, emphasised—gaps were opened between them; and these gaps had to be filled in when the sentiment of the unity of Nature became stronger, and when the intimate relations which connect all the parts were perceived. In multiplying the fluids, not only did the ancient physicists create unnecessary entities, but they destroyed real ties. It is not enough for a theory not to affirm false relations; it must not conceal true relations.

Does our ether actually exist? We know the origin of our belief in the ether. If light takes several years to reach us from a distant star, it is no longer on the star, nor is it on the earth. It must be somewhere, and supported, so to speak, by some material agency.

The same idea may be expressed in a more mathematical and more abstract form. What we note are the changes undergone by the material molecules. We see, for instance, that the photographic plate experiences the consequences of a phenomenon of which the incandescent mass of a star was the scene several years before. Now, in ordinary mechanics, the state of the system under consideration depends only on its state at the moment immediately preceding; the system therefore satisfies certain differential equations. On the other hand, if we did not believe in the ether, the state of the material universe would depend not only on the state immediately preceding, but also on much older states; the system would satisfy equations of finite differences. The ether was invented to escape this breaking down of the laws of general mechanics.

Still, this would only compel us to fill the interplanetary space with ether, but not to make it penetrate into the midst of the material media. Fizeau's experiment goes further. By the interference of rays which have passed through the air or water in motion, it seems to show us two different media penetrating each other, and yet being displaced with respect to each other. The ether is all but in our grasp. Experiments can be conceived in which we come closer still to it. Assume that Newton's principle of the equality of action and re-action is not true if applied to matter *alone*, and that this can be proved. The geometrical sum of all the forces applied to all the molecules would no longer be zero. If we did not wish to change the whole of the science of mechanics, we should have to introduce the ether, in order that the action which matter apparently undergoes should be counterbalanced by the re-action of matter on something.

Or again, suppose we discover that optical and electrical phenomena are influenced by the motion of the earth. It would follow that those phenomena might reveal to us not only the relative motion of material bodies, but also what would seem to be their absolute motion. Again, it would be necessary to have an ether in order that these so-called absolute movements should not be their displacements with respect to empty space, but with respect to something concrete.

Will this ever be accomplished? I do not think so, and I shall explain why; and yet, it is not absurd, for others have entertained this view. For instance, if the theory of Lorentz, of which I shall speak in more detail in Chapter XIII, were true, Newton's principle would not apply to matter *alone*, and the difference would not be very far from being within reach of experiment. On the other hand, many experiments have been made on the influence of the motion of the earth. The results have always been negative. But if these experiments have been undertaken, it is because we have not been certain beforehand; and indeed, according to current theories, the compensation would be only approximate, and we might expect to find accurate methods giving positive results. I think that such a hope is illusory; it was nonetheless interesting to show that a success of this kind would, in a certain sense, open to us a new world.

And now allow me to make a digression; I must explain why I do not believe, in spite of Lorentz, that more exact observations will ever make evident anything else but the relative displacements of material bodies. Experiments have been made that should have disclosed the terms of the first order; the results were nugatory. Could that have been by chance? No one has admitted this; a general explanation was sought, and Lorentz found it. He showed that the terms of the first order should cancel each other, but not the terms of the second order. Then more exact experiments were made, which were also negative; neither could this be the result of chance. An explanation was necessary, and was forthcoming; they always are; hypotheses are what we lack the least. But this is not enough. Who is there who does not think that this leaves to chance far too important a role? Would it not also be a chance that this singular concurrence should cause a certain circumstance to destroy the terms of the first order, and that a totally different but very opportune circumstance should cause those of the second order to vanish? No; the same explanation must be found for the two cases, and everything tends to show that this explanation would serve equally well for the terms of the higher order, and that the mutual destruction of these terms will be rigorous and absolute.

The Present State of Physics. Two opposite tendencies may be distinguished in the history of the development of physics. On the one hand, new relations are continually being discovered between objects which seemed destined to remain forever unconnected; scattered facts cease to be strangers to each other and tend to be marshalled into an imposing synthesis. The march of science is towards unity and simplicity.

On the other hand, new phenomena are continually being revealed; it will be long before they can be assigned their place— sometimes it may happen that to find them a place a corner of the edifice must be demolished. In the same way, we are continually perceiving details ever more varied in the phenomena we know, where our crude senses used to be unable to detect any lack of unity. What we thought to be simple becomes complex, and the march of science seems to be towards diversity and complication.

Here, then, are two opposing tendencies, each of which seems to triumph in turn. Which will win? If the first wins, science is possible; but nothing proves this *à priori*, and it may be that after unsuccessful efforts to bend Nature to our ideal of unity in spite of herself, we shall be submerged by the ever-rising flood of our new riches and compelled to renounce all idea of classification—to abandon our ideal, and to reduce science to the mere recording of innumerable recipes.

In fact, we can give this question no answer. All that we can do is to observe the science of today, and compare it with that of yesterday. No doubt after this examination we shall be in a position to offer a few conjectures.

Half a century ago hopes ran high indeed. The unity of force had just been revealed to us by the discovery of the conservation of energy and of its transformation. This discovery also showed that the phenomena of heat could be explained by molecular movements. Although the nature of these movements was not exactly known, no one doubted but that they would be ascertained before long. As for light, the work seemed entirely completed. So far as electricity was concerned, there was not so great an advance. Electricity had just annexed magnetism. This was a considerable and a definitive step towards unity. But how was electricity in its turn to be brought into the general unity, and how was it to be included in the general universal mechanism? No one had the slightest idea. As to the possibility of the inclusion, all were agreed; they had faith. Finally, as far as the molecular properties of material bodies are concerned, the inclusion seemed easier, but the details were very hazy. In a word, hopes were vast and strong, but vague.

Today, what do we see? In the first place, a step in advance— immense progress. The relations between light and electricity are now known; the three domains of light, electricity, and magnetism, formerly separated, are now one; and this annexation seems definitive.

Nevertheless the conquest has caused us some sacrifices. Optical phenomena become particular cases in electric phenomena; as long as the former remained isolated, it was easy to explain them by movements which were thought to be known in all their details. That was easy enough; but any explanation to be accepted must now cover the whole domain of electricity. This cannot be done without difficulty.

The most satisfactory theory is that of Lorentz; it is unquestionably the theory that best explains the known facts, the one that throws into relief the greatest number of known relations, the one in which we find most traces of definitive construction. That it still possesses a serious fault I have shown above. It is in contradiction with Newton's law that action and re-action are equal and opposite—or rather, this principle according to Lorentz cannot be applicable to matter alone; if it be true, it must take into account the action of the ether on matter, and the re-action of the matter on the ether. Now, in the new order, it is very likely that things do not happen in this way.

However this may be, it is due to Lorentz that the results of Fizeau on the optics of moving bodies, the laws of normal and abnormal dispersion and of absorption are connected with each other and with the other properties of the ether, by bonds which no doubt will not be readily severed. Look at the ease with which the new Zeeman phenomenon found its place, and even aided the classification of Faraday's magnetic rotation, which had defied all Maxwell's efforts. This facility proves that Lorentz's theory is not a mere artificial combination which must eventually find its solvent. It will probably have to be modified, but not destroyed.

The only object of Lorentz was to include in a single whole all the optics and electro-dynamics of moving bodies; he did not claim to give a mechanical explanation. Larmor goes further; keeping the essential part of Lorentz's theory, he grafts upon it, so to speak, MacCullagh's ideas on the direction of the movement of the ether. MacCullagh held that the velocity of the ether is the same in magnitude and direction as the magnetic force. Ingenious as is this attempt, the fault in Lorentz's theory remains, and is even aggravated. According to Lorentz, we do not know what the movements of the ether are; and because we do not know this, we may suppose them to be movements compensating those of matter, and re-affirming that action and re-action are equal and opposite. According to Larmor we know the movements of the ether, and we can prove that the compensation does not take place.

If Larmor has failed, as in my opinion he has, does it necessarily follow that a mechanical explanation is impossible? Far from it. I said above that as long as a phenomenon obeys the two principles of energy and least action, so long it allows of an unlimited number of mechanical explanations. And so with the phenomena of optics and electricity.

But this is not enough. For a mechanical explanation to be good it must be simple; to choose it from among all the explanations that are possible there must be other reasons than the necessity of making a choice. Well, we have no theory as yet which will satisfy this condition and consequently be of any use. Are we then to complain? That would be to forget the end we seek, which is not the mechanism; the true and only aim is unity.

We ought therefore to set some limits to our ambition. Let us not seek to formulate a mechanical explanation; let us be content to show that we can always find one if we wish. In this we have succeeded. The principle of the conservation of energy has always been confirmed, and now it has a fellow in the principle of least action, stated in the form appropriate to physics. This has also been verified, at least as far as concerns the reversible phenomena which obey Lagrange's equations—in other words, which obey the most general laws of physics. The irreversible phenomena are much more difficult to bring into line; but they, too, are being coordinated and tend to come into the unity. The light which illuminates them comes from Carnot's principle. For a long time thermo-dynamics was confined to the study of the dilatations of bodies and of their change of state. For some time past it has been growing bolder, and has considerably extended its domain. We owe to it the theories of the voltaic cell and of their thermo-electric phenomena; there is not a corner in physics which it has not explored, and it has even attacked chemistry itself. The same laws hold good; everywhere, disguised in some form or other, we find Carnot's principle; everywhere also appears that eminently abstract concept of entropy which is as universal as the concept of energy, and like it, seems to conceal a reality. It seemed that radiant heat must escape, but recently that, too, has been brought under the same laws.

In this way fresh analogies are revealed which may be often pursued in detail; electric resistance resembles the viscosity of fluids; hysteresis would rather be like the friction of solids. In all cases friction appears to be the type most imitated by the most diverse irreversible phenomena, and this relationship is real and profound.

A strictly mechanical explanation of these phenomena has also been sought, but, owing to their nature, it is hardly likely that it will be found. To find it, it has been necessary to suppose that the irreversibility is but apparent, that the elementary phenomena are reversible and obey the known laws of dynamics. But the elements are extremely numerous, and become blended more and more, so that to our crude sight all appears to tend towards uniformity—*i.e.,* all seems to progress in the same direction, and that without hope of return. The apparent irreversibility is therefore but an effect of the law of great numbers. Only a being of infinitely subtle senses, such as Maxwell's demon, could unravel this tangled skein and turn back the course of the universe.

This conception, which is connected with the kinetic theory of gases, has cost great effort and has not, on the whole, been fruitful; it may become so. This is not the place to examine if it leads to contradictions, and if it is in conformity with the true nature of things.

Let us notice, however, the original ideas of M. Gouy on the Brownian movement. According to this scientist, this singular movement does not obey Carnot's principle. The particles which it sets moving would be smaller than the meshes of that tightly drawn net; they would thus be ready to separate them, and thereby to set back the course of the universe. One can almost see Maxwell's demon at work.*

To resume, phenomena long known are gradually being better classified, but new phenomena come to claim their place, and most of them, like the Zeeman effect, find it at once. Then we have the cathode rays, the X-rays, uranium and radium rays; in fact, a whole world of which none had suspected the existence. How many un-

---

* Clerk-Maxwell imagined some supernatural agency at work, sorting molecules in a gas of uniform temperature into (*a*) those possessing kinetic energy above the average, (*b*) those possessing kinetic energy below the average.—[Tr.]

expected guests to find a place for! No one can yet predict the place they will occupy, but I do not believe they will destroy the general unity; I think that they will rather complete it. On the one hand, indeed, the new radiations seem to be connected with the phenomena of luminosity; not only do they excite fluorescence, but they sometimes come into existence under the same conditions as that property; neither are they unrelated to the cause which produces the electric spark under the action of ultra-violet light. Finally, and most important of all, it is believed that in all these phenomena there exist ions, animated, it is true, with velocities far greater than those of electrolytes. All this is very vague, but it will all become clearer.

Phosphorescence and the action of light on the spark were regions rather isolated, and consequently somewhat neglected by investigators. It is to be hoped that a new path will now be made which will facilitate their communications with the rest of science. Not only do we discover new phenomena, but those we think we know are revealed in unlooked-for aspects. In the free ether the laws preserve their majestic simplicity, but matter properly so called seems more and more complex; all we can say of it is but approximate, and our formulæ are constantly requiring new terms.

But the ranks are unbroken, the relations that we have discovered between objects we thought simple still hold good between the same objects when their complexity is recognised, and that alone is the important thing. Our equations become, it is true, more and more complicated, so as to embrace more closely the complexity of nature; but nothing is changed in the relations which enable these equations to be derived from each other. In a word, the form of these equations persists. Take for instance the laws of reflection. Fresnel established them by a simple and attractive theory which experiment seemed to confirm. Subsequently, more accurate researches have shown that this verification was but approximate; traces of elliptic polarisation were detected everywhere. But it is owing to the first approximation that the cause of these anomalies was found in the existence of a transition layer, and all the essentials of Fresnel's theory have remained. We cannot help reflecting that all these relations would never have been noted if there had been doubt in the first place as to the complexity of the objects they connect. Long ago it was said: If Tycho had had instruments ten times as precise, we would never have had a Kepler, or a Newton, or Astronomy. It is a misfortune for a science to be born too late, when the means of observation have become too perfect. That is what is happening at this moment with respect to physical chemistry; the founders are hampered in their general grasp by third and fourth decimal places; happily they are men of robust faith. As we get to know the properties of matter better we see that continuity reigns. From the work of Andrews and Van der Waals, we see how the transition from the liquid to the gaseous state is made, and that it is not abrupt. Similarly, there is no gap between the liquid and solid states, and in the proceedings of a recent Congress we see memoirs on the rigidity of liquids side by side with papers on the flow of solids.

With this tendency there is no doubt a loss of simplicity. Such and such an effect was represented by straight lines; it is now necessary to connect these lines by more or less complicated curves. On the other hand, unity is gained. Separate categories quieted but did not satisfy the mind.

Finally, a new domain, that of chemistry, has been invaded by the method of physics, and we see the birth of physical chemistry. It is still quite young, but already it has enabled us to connect such phenomena as electrolysis, osmosis, and the movements of ions.

From this cursory exposition what can we conclude? Taking all things into account, we have approached the realisation of unity. This has not been done as quickly as was hoped fifty years ago, and the path predicted has not always been followed; but, on the whole, much ground has been gained.

CHAPTER XI

THE CALCULUS OF PROBABILITIES

No doubt the reader will be astonished to find reflections on the calculus of probabilities in such a volume as this. What has that calculus to do with physical science? The questions I shall raise—without, however, giving them a solution—are naturally raised by the philosopher who is examining the problems of physics. So far this is the case, that in the two preceding chapters I have several times used the words "probability" and "chance." "Predicted facts," as I said above, "can only be probable." However solidly founded a prediction may appear to be, we are never absolutely certain that experiment will not prove it false; but the probability is often so great that practically it may be accepted. And a little farther on I added: "See what a part the belief in simplicity plays in our gener-alisations. We have verified a simple law in a large number of par-ticular cases, and we refuse to admit that this so-often-repeated coincidence is a mere effect of chance." Thus, in a multitude of circumstances the physicist is often in the same position as the gambler who reckons up his chances. Every time that he reasons by induction, he more or less consciously requires the calculus of probabilities, and that is why I am obliged to open this chapter parenthetically, and to interrupt our discussion of method in the physical sciences in order to examine a little closer what this calculus is worth, and what dependence we may place upon it. The very name of the calculus of probabilities is a paradox. Probability as opposed to certainty is what one does not know, and how can we calculate the unknown? Yet many eminent scientists have devoted themselves to this calculus, and it cannot be denied that science has drawn therefrom no small advantage. How can we explain this apparent contradiction? Has probability been defined? Can it even be defined? And if it cannot, how can we venture to reason upon it? The definition, it will be said, is very simple. The probability of an event is the ratio of the number of cases favourable to the event to the total number of possible cases. A simple example will show how incomplete this definition is: I throw two dice. What is the probability that one of the two at least turns up a 6? Each can turn up in six different ways; the number of possible cases is $6 \times 6 = 36$. The number of favourable cases is 11; the probability is $\frac{11}{36}$. That is the correct solution. But why cannot we just as well proceed as follows?—The points which turn up on the two dice form $\frac{6 \times 7}{2} = 21$ different combinations. Among these combinations, six are favourable; the probability is $\frac{6}{21}$. Now why is the first method of calculating the number of possible cases more legitimate than the second? In any case it is not the definition that tells us. We are therefore bound to complete the definition by saying, "… to the total number of possible cases, provided the cases are equally probable." So we are compelled to define the probable by the probable. How can we know that two possible cases are equally probable? Will it be by a convention? If we insert at the beginning of every problem an explicit convention, well and good! We then have nothing to do but to apply the rules of arithmetic and algebra, and we complete our calculation, when our result cannot be called in question. But if we wish to make the slightest application of this result, we must prove that our convention is legitimate, and we shall find ourselves in the presence of the very difficulty we thought we had avoided. It may be said that common-sense is enough to show us the convention that should be adopted. Alas! M. Bertrand has amused himself by discussing the following simple problem: "What is the probability that a chord of a circle may be greater than the side of the inscribed equilateral triangle?" The illustrious geometer successively adopted two conventions which seemed to be equally imperative in the eyes of common-sense, and with one convention he finds $\frac{1}{2}$, and with the other $\frac{1}{3}$. The conclusion which seems to follow from this is that the calculus of probabilities is a useless science, that the obscure instinct which we call common-sense, and to which we appeal for the legitimisation of our conventions, must be distrusted. But to this conclusion we can no longer subscribe. We cannot do without that obscure instinct. Without it, science would be impossible, and without it we could neither discover nor apply a law. Have we any right, for instance, to enunciate Newton's law? No doubt numerous observations are in agreement with it, but is not that a simple fact of chance? and how do we know, besides, that this law which has been true for so many generations will not be untrue in the next? To this objection the only answer you can give is: it is very improbable. But grant the law. By means of it I can calculate the position of Jupiter in a year from now. Yet have I any right to say this? Who can tell if a gigantic mass of enormous velocity is not going to pass near the solar system and produce unforeseen perturbations? Here again the only answer is: it is very improbable. From this point of view all the sciences would only be unconscious applications of the calculus of probabilities. And if this calculus be condemned, then the whole of the sciences must also be condemned. I shall not dwell at length on scientific problems in which the intervention of the calculus of probabilities is more evident. In the forefront of these is the problem of interpolation, in which, knowing a certain number of values of a function, we try to discover the intermediary values. I may also mention the celebrated theory of errors of observation, to which I shall return later; the kinetic theory of gases, a well-known hypothesis wherein each gaseous molecule is supposed to describe an extremely complicated path, but in which, through the effect of great numbers, the mean phenomena which are all we observe obey the simple laws of Mariotte and Gay-Lussac. All these theories are based upon the laws of great numbers, and the calculus of probabilities would evidently involve them in its ruin. It is true that they have only a particular interest, and that, save as far as interpolation is concerned, they are sacrifices to which we might readily be resigned. But I have said above, it would not be these partial sacrifices that would be in question; it would be the legitimacy of the whole of science that would be challenged. I quite see that it might be said: we do not know, and yet we must act. As for action, we have not time to devote ourselves to an inquiry that will suffice to dispel our ignorance. Besides, such an inquiry would demand unlimited time. We must therefore make up our minds without knowing. This must be often done whatever may happen, and we must follow the rules although we may have but little confidence in them. What I know is, not that such a thing is true, but that the best course for me is to act as if it were true. The calculus of probabilities, and therefore science itself, would be no longer of any practical value.

Unfortunately the difficulty does not thus disappear. A gambler wants to try a *coup*, and he asks my advice. If I give it to him, I use the calculus of probabilities; but I shall not guarantee success. That is what I shall call *subjective probability*. In this case we might be content with the explanation of which I have just given a sketch. But assume that an observer is present at the play, that he knows of the *coup*, and that play goes on for a long time, and that he makes a summary of his notes. He will find that events have taken place in conformity with the laws of the calculus of probabilities. That is what I shall call *objective probability*, and it is this phenomenon which has to be explained. There are numerous Insurance Societies which apply the rules of the calculus of probabilities, and they distribute to their shareholders dividends, the objective reality of which cannot be contested. In order to explain them, we must do more than invoke our ignorance and the necessity of action. Thus, absolute skepticism is not admissible. We may distrust, but we cannot condemn *en bloc*. Discussion is necessary.

I. Classification of the Problems of Probability. In order to classify the problems which are presented to us with reference to probabilities, we must look at them from different points of view, and first of all, from that of *generality.* I said above that probability is the ratio of the number of favourable to the number of possible cases. What for want of a better term I call generality will increase with the number of possible cases. This number may be finite, as, for instance, if we take a throw of the dice in which the number of possible cases is 36. That is the first degree of generality. But if we ask, for instance, what is the probability that a point within a circle is within the inscribed square, there are as many possible cases as there are points in the circle—that is to say, an infinite number. This is the second degree of generality. Generality can be pushed further still. We may ask the probability that a function will satisfy a given condition. There are then as many possible cases as one can imagine different functions. This is the third degree of generality, which we reach, for instance, when we try to find the most probable law after a finite number of observations. Yet we may place ourselves at a quite different point of view. If we were not ignorant there would be no probability, there could only be certainty. But our ignorance cannot be absolute, for then there would be no longer any probability at all. Thus the problems of probability may be classed according to the greater or lesser depth of this ignorance. In mathematics we may set ourselves problems in probability. What is the probability that the fifth decimal of a logarithm taken at random from a table is a 9? There is no hesitation in answering that this probability is 1-10th. Here we possess all the data of the problem. We can calculate our logarithm without having recourse to the table, but we need not give ourselves the trouble. This is the first degree of ignorance. In the physical sciences our ignorance is already greater. The state of a system at a given moment depends on two things—its initial state, and the law according to which that state varies. If we know both this law and this initial state, we have a simple mathematical problem to solve, and we fall back upon our first degree of ignorance. Then it often happens that we know the law and do not know the initial state. It may be asked, for instance, what is the present distribution of the minor planets? We know that from all time they have obeyed the laws of Kepler, but we do not know what was their initial distribution. In the kinetic theory of gases we assume that the gaseous molecules follow rectilinear paths and obey the laws of impact and elastic bodies; yet as we know nothing of their initial velocities, we know nothing of their present velocities. The calculus of probabilities alone enables us to predict the mean phenomena which will result from a combination of these velocities. This is the second degree of ignorance. Finally it is possible that not only the initial conditions but the laws themselves are unknown. We then reach the third degree of ignorance, and in general we can no longer affirm anything at all as to the probability of a phenomenon. It often happens that instead of trying to discover an event by means of a more or less imperfect knowledge of the law, the events may be known, and we want to find the law; or that, instead of deducing effects from causes, we wish to deduce the causes from the effects. Now, these problems are classified as *probability of causes*, and are the most interesting of all from their scientific applications. I play at *écarté* with a gentleman whom I know to be perfectly honest. What is the chance that he turns up the king? It is $\frac{1}{8}$. This is a problem of the probability of effects. I play with a gentleman whom I do not know. He has dealt ten times, and he has turned the king up six times. What is the chance that he is a sharper? This is a problem in the probability of causes. It may be said that it is the essential problem of the experimental method. I have observed *n* values of *x* and the corresponding values of *y*. I have found that the ratio of the latter to the former is practically constant. There is the event; what is the cause? Is it probable that there is a general law according to which *y* would be proportional to *x*, and that small divergencies are due to errors of observation? This is the type of question that we are ever asking, and which we unconsciously solve whenever we are engaged in scientific work. I am now going to pass in review these different categories of problems by discussing in succession what I have called subjective and objective probability.

II. Probability in Mathematics. The impossibility of squaring the circle was shown in 1885, but before that date all geometers considered this impossibility as so "probable" that the Académie des

Sciences rejected without examination the, alas! too numerous memoirs on this subject that a few unhappy madmen sent in every year. Was the Académie wrong? Evidently not, and it knew perfectly well that by acting in this manner it did not run the least risk of stifling a discovery of moment. The Académie could not have proved that it was right, but it knew quite well that its instinct did not deceive it. If you had asked the academicians, they would have answered: "We have compared the probability that an unknown scientist should have found out what has been vainly sought for so long, with the probability that there is one madman the more on the earth, and the latter has appeared to us the greater." These are very good reasons, but there is nothing mathematical about them; they are purely psychological. If you had pressed them further, they would have added: "Why do you expect a particular value of a transcendental function to be an algebraical number; if π be the root of an algebraical equation, why do you expect this root to be a period of the function *sin 2x*, and why is it not the same with the other roots of the same equation?" To sum up, they would have invoked the principle of sufficient reason in its vaguest form. Yet what information could they draw from it? At most a rule of conduct for the employment of their time, which would be more usefully spent at their ordinary work than in reading a lucubration that inspired in them a legitimate distrust. But what I called above objective probability has nothing in common with this first problem. It is otherwise with the second. Let us consider the first 10,000 logarithms that we find in a table. Among these 10,000 logarithms I take one at random. What is the probability that its third decimal is an even number? You will say without any hesitation that the probability is $\frac{1}{2}$, and in fact if you pick out in a table the third decimals in these 10,000 numbers you will find nearly as many even digits as odd. Or, if you prefer it, let us write 10,000 numbers corresponding to our 10,000 logarithms, writing down for each of these numbers +1 if the third decimal of the corresponding logarithm is even, and −1 if odd; and then let us take the mean of these 10,000 numbers. I do not hesitate to say that the mean of these 10,000 units is probably zero, and if I were to calculate it practically, I would verify that it is extremely small. But this verification is needless. I might have rigorously proved that this mean is smaller than 0.003. To prove this result I should have had to make a rather long calculation for which there is no room here, and for which I may refer the reader to an article that I published in the *Revue générale des sciences,* April 15th, 1899. The only point to which I wish to draw attention is the following. In this calculation I had occasion to rest my case on only two facts—namely, that the first and second derivatives of the logarithm remain, in the interval considered, between certain limits. Hence our first conclusion is that the property is not only true of the logarithm but of any continuous function whatever, since the derivatives of every continuous function are limited. If I was certain beforehand of the result, it is because I have often observed analogous facts for other continuous functions; and next, it is because I went through in my mind in a more or less unconscious and imperfect manner the reasoning which led me to the preceding inequalities, just as a skilled calculator before finishing his multiplication takes into account what it ought to come to approximately. And besides, since what I call my intuition was only an incomplete summary of a piece of true reasoning, it is clear that observation has confirmed my predictions, and that the objective and subjective probabilities are in agreement. As a third example I shall choose the following: the number u is taken at random and n is a given very large integer. What is the mean value of $\sin nu$? This problem has no meaning by itself. To give it one, a convention is required—namely, we agree that the probability for the number u to lie between a and $a + da$ is $\phi(a)da$; that it is therefore proportional to the infinitely small interval da, and is equal to this multiplied by a function $\phi(a)$, only depending on a. As for this function I choose it arbitrarily, but I must assume it to be continuous. The value of $\sin nu$ remaining the same when u increases by 2π, I may without loss of generality assume that u lies between 0 and 2π, and I shall thus be led to suppose that $\phi(a)$ is a periodic function whose period is 2π. The mean value that we seek is readily expressed by a simple integral, and it is easy to show that this integral is smaller than $\frac{2\pi M_K}{n^k}$, M_K being the maximum value of the kth derivative of $\phi(u)$. We see then that if the kth derivative is finite, our mean value will tend towards zero when n increases indefinitely, and that more rapidly than $\frac{1}{n^{k-1}}$. The mean value of sin nu when n is very large is therefore zero. To define this value I required a convention, but the result remains the same *whatever that convention may be.* I have imposed upon myself but slight restrictions when I assumed that the function $\phi(a)$ is continuous and periodic, and these hypotheses are so natural that we may ask ourselves how they can be escaped. Examination of the three preceding examples, so different in all respects, has already given us a glimpse on the one hand of the role of what philosophers call the principle of sufficient reason, and on the other hand of the importance of the fact that certain properties are common to all continuous functions. The study of probability in the physical sciences will lead us to the same result.

III. Probability in the Physical Sciences. We now come to the problems which are connected with what I have called the second degree of ignorance—namely, those in which we know the law but do not know the initial state of the system. I could multiply examples, but I shall take only one. What is the probable present distribution of the minor planets on the zodiac? We know they obey the laws of Kepler. We may even, without changing the nature of the problem, suppose that their orbits are circular and situated in the same plane, a plane which we are given. On the other hand, we know absolutely nothing about their initial distribution. However, we do not hesitate to affirm that this distribution is now nearly uniform. Why? Let b be the longitude of a minor planet in the initial epoch—that is to say, the epoch zero. Let a be its mean motion. Its longitude at the present time—*i.e.,* at the time t will be $at + b$. To say that the present distribution is uniform is to say that the mean value of the sines and cosines of multiples of $at + b$ is zero. Why do we assert this? Let us represent our minor planet by a point in a plane—namely, the point whose co-ordinates are a and b. All these representative points will be contained in a certain region of the plane, but as they are very numerous this region will appear dotted with points. We know nothing else about the distribution of the points. Now what do we do when we apply the calculus of probabilities to such a question as this? What is the probability that one or more representative points may be found in a certain portion of the plane? In our ignorance we are compelled to make an arbitrary hypothesis. To explain the nature of this hypothesis I may be allowed to use, instead of a mathematical formula, a crude but concrete image. Let us suppose that over the surface of our plane has been spread imaginary matter, the density of which is variable, but varies continuously. We shall then agree to say that the probable number of representative points to be found on a certain portion of the plane is proportional to the quantity of this imaginary matter which is found there. If there are, then, two regions of the plane of the same extent, the probabilities that a representative point of one of our minor planets is in one or other of these regions will be as the mean densities of the imaginary matter in one or other of the regions. Here then are two distributions, one real, in which the representative points are very numerous, very close together, but discrete like the molecules of matter in the atomic hypothesis; the other remote from reality, in which our representative points are replaced by imaginary continuous matter. We know that the latter cannot be real, but we are forced to adopt it through our ignorance. If, again, we had some idea of the real distribution of the representative points, we could arrange it so that in a region of some extent the density of this imaginary continuous matter may be nearly proportional to the number of representative points, or, if it is preferred, to the number of atoms which are contained in that region. Even that is impossible, and our ignorance is so great that we are forced to choose arbitrarily the function which defines the density of our imaginary matter. We shall be compelled to adopt a hypothesis from which we can hardly get away; we shall suppose that this function is continuous. That is sufficient, as we shall see, to enable us to reach our conclusion.

What is at the instant t the probable distribution of the minor planets—or rather, what is the mean value of the sine of the longitude at the moment *t*—*i.e.,* of sin $(at+b)$? We made at the outset an arbitrary convention, but if we adopt it, this probable value is entirely defined. Let us decompose the plane into elements of surface. Consider the value of sin $(at+b)$ at the centre of each of these elements. Multiply this value by the surface of the element and by the corresponding density of the imaginary matter. Let us then take the sum for all the elements of the plane. This sum, by definition, will be the probable mean value we seek, which will thus be expressed by a double integral. It may be thought at first that this mean value depends on the choice of the function ϕ which defines the density of the imaginary matter, and as this function ϕ is arbitrary, we can, according to the arbitrary choice which we make, obtain a certain mean value. But this is not the case. A simple calculation shows us that our double integral decreases very rapidly as *t* increases. Thus, I cannot tell what hypothesis to make as to the probability of this or that initial distribution, but when once the hypothesis is made the result will be the same, and this gets me out of my difficulty. Whatever the function ϕ may be, the mean value tends towards zero as *t* increases, and as the minor planets have certainly accomplished a very large number of revolutions, I may assert that this mean value is very small. I may give to ϕ any value I choose, with one restriction: this function must be continuous; and, in fact, from the point of view of subjective probability, the choice of a discontinuous function would have been unreasonable. What reason could I have, for instance, for supposing that the initial longitude might be exactly 0°, but that it could not lie between 0° and 1°?

The difficulty reappears if we look at it from the point of view of objective probability; if we pass from our imaginary distribution in which the supposititious matter was assumed to be continuous, to the real distribution in which our representative points are formed as discrete atoms. The mean value of sin $(at+b)$ will be represented quite simply by

$$\frac{1}{n} \Sigma \sin (at+b),$$

n being the number of minor planets. Instead of a double integral referring to a continuous function, we shall have a sum of discrete terms. However, no one will seriously doubt that this mean value is practically very small. Our representative points being very close together, our discrete sum will in general differ very little from an integral. An integral is the limit towards which a sum of terms tends when the number of these terms is indefinitely increased. If the terms are very numerous, the sum will differ very little from its limit—that is to say, from the integral, and what I said of the latter will still be true of the sum itself. But there are exceptions. If, for instance, for all the minor planets $b = \frac{\pi}{2} - at$, the longitude of all the planets at the time t would be $\frac{\pi}{2}$, and the mean value in question would be evidently unity. For this to be the case at the time o, the minor planets must have all been lying on a kind of spiral of peculiar form, with its spires very close together. All will admit that such an initial distribution is extremely improbable (and even if it were realised, the distribution would not be uniform at the present time—for example, on the 1st January 1900; but it would become so a few years later). Why, then, do we think this initial distribution improbable? This must be explained, for if we are wrong in rejecting as improbable this absurd hypothesis, our inquiry breaks down, and we can no longer affirm anything on the subject of the probability of this or that present distribution. Once more we shall invoke the principle of sufficient reason, to which we must always recur. We might admit that at the beginning the planets were distributed almost in a straight line. We might admit that they were irregularly distributed. But it seems to us that there is no sufficient reason for the unknown cause that gave them birth to have acted along a curve so regular and yet so complicated, which would appear to have been expressly chosen so that the distribution at the present day would not be uniform.

IV. Rouge et Noir. The questions raised by games of chance, such as roulette, are, fundamentally, quite analogous to those we have just treated. For example, a wheel is divided into thirty-seven equal compartments, alternately red and black. A ball is spun round the wheel, and after having moved round a number of times, it stops in front of one of these sub-divisions. The probability that the division is red is obviously $\frac{1}{2}$. The needle describes an angle θ, including several complete revolutions. I do not know what is the probability that the ball is spun with such a force that this angle should lie between θ and $\theta + d\theta$, but I can make a convention. I can suppose that this probability is $\phi(\theta)d\theta$. As for the function $\phi(\theta)$, I can choose it in an entirely arbitrary manner. I have nothing to guide me in my choice, but I am naturally induced to suppose the function to be continuous. Let ϵ be a length (measured on the circumference of the circle of radius unity) of each red and black compartment. We have to calculate the integral of $\phi(\theta)d\theta$, extending it on the one hand to all the red, and on the other hand to all the black compartments, and to compare the results. Consider an interval 2ϵ comprising two consecutive red and black compartments. Let M and m be the maximum and minimum values of the function $\phi(\theta)$ in this interval. The integral extended to the red compartments will be smaller than $\Sigma M\epsilon$; extended to the black it will be greater than $\Sigma m\epsilon$. The difference will therefore be smaller than $\Sigma(M-m)\epsilon$. But if the function ϕ is supposed continuous, and if on the other hand the interval ϵ is very small with respect to the total angle described by the needle, the difference $M-m$ will be very small. The difference of the two integrals will be therefore very small, and the probability will be very nearly $\frac{1}{2}$. We see that without knowing anything of the function ϕ we must act as if the probability were $\frac{1}{2}$. And on the other hand it explains why, from the objective point of view, if I watch a certain number of *coups*, observation will give me almost as many black *coups* as red. All the players know this objective law; but it leads them into a remarkable error, which has often been exposed, but into which they are always falling. When the red has won, for example, six times running, they bet on black, thinking that they are playing an absolutely safe game, because they say it is a very rare thing for the red to win seven times running. In reality their probability of winning is still $\frac{1}{2}$. Observation shows, it is true, that the series of seven consecutive reds is very rare, but series of six reds followed by a black are also very rare. They have noticed the rarity of the series of seven reds; if they have not remarked the rarity of six reds and a black, it is only because such series strike the attention less.

V. The Probability of Causes. We now come to the problems of the probability of causes, the most important from the point of view of scientific applications. Two stars, for instance, are very close together on the celestial sphere. Is this apparent contiguity a mere effect of chance? Are these stars, although almost on the same visual ray, situated at very different distances from the earth, and therefore very far indeed from one another? or does the apparent correspond to a real contiguity? This is a problem on the probability of causes.

First of all, I recall that at the outset of all problems of probability of effects that have occupied our attention up to now, we have had to use a convention which was more or less justified; and if in most cases the result was to a certain extent independent of this convention, it was only the condition of certain hypotheses which enabled us *à priori* to reject discontinuous functions, for example, or certain absurd conventions. We shall again find something analogous to this when we deal with the probability of causes. An effect may be produced by the cause *a* or by the cause *b.* The effect has just been observed. We ask the probability that it is due to the cause *a.* This is an *à posteriori* probability of cause. But I could not calculate it, if a convention more or less justified did not tell me in advance what is the *à priori* probability for the cause *a* to come into play—I mean the probability of this event to someone who had not observed the effect. To make my meaning clearer, I go back to the game of *écarté* mentioned before. My adversary deals for the first time and turns up a king. What is the probability that he is a sharper? The formulæ ordinarily taught give $\frac{8}{9}$, a result which is obviously rather surprising. If we look at it closer, we see that the conclusion is arrived at as if, before sitting down at the table, I had considered that there was one chance in two that my adversary was not honest. An absurd hypothesis, because in that case I should certainly not have played with him; and this explains the absurdity of the conclusion. The function on the *à priori* probability was unjustified, and that is why the conclusion of the *à posteriori* probability led me into an inadmissible result. The importance of this preliminary convention is obvious. I shall even add that if none were made, the problem of the *à posteriori* probability would have no meaning. It must be always made either explicitly or tacitly.

Let us pass on to an example of a more scientific character. I require to determine an experimental law; this law, when discovered, can be represented by a curve. I make a certain number of isolated observations, each of which may be represented by a point. When I have obtained these different points, I draw a curve between them as carefully as possible, giving my curve a regular form, avoiding sharp angles, accentuated inflexions, and any sudden variation of the radius of curvature. This curve will represent to me the probable law, and not only will it give me the values of the functions intermediary to those which have been observed, but it also gives me the observed values more accurately than direct observation does; that is why I make the curve pass near the points and not through the points themselves.

Here, then, is a problem in the probability of causes. The effects are the measurements I have recorded; they depend on the combination of two causes—the true law of the phenomenon and errors of observation. Knowing the effects, we have to find the probability that the phenomenon shall obey this law or that, and that the observations have been accompanied by this or that error. The most probable law, therefore, corresponds to the curve we have traced, and the most probable error is represented by the distance of the corresponding point from that curve. But the problem has no meaning if before the observations I had an *à priori* idea of the probability of this law or that, or of the chances of error to which I am exposed. If my instruments are good (and I knew whether this was so or not before beginning the observations), I shall not draw the curve far from the points which represent the rough measurements. If they are inferior, I may draw it a little farther from the points, so that I may get a less sinuous curve; much will be sacrificed to regularity.

Why, then, do I draw a curve without sinuosities? Because I consider *à priori* a law represented by a continuous function (or function the derivatives of which to a high order are small) as more probable than a law not satisfying those conditions. But for this conviction the problem would have no meaning; interpolation would be impossible; no law could be deduced from a finite number of observations; science would cease to exist.

Fifty years ago physicists considered, other things being equal, a simple law as more probable than a complicated law. This principle was even invoked in favour of Mariotte's law as against that of Regnault. But this belief is now repudiated; and yet, how many times are we compelled to act as though we still held it! However that may be, what remains of this tendency is the belief in continuity, and as we have just seen, if the belief in continuity were to disappear, experimental science would become impossible.

VI. The Theory of Errors. We are thus brought to consider the theory of errors which is directly connected with the problem of the probability of causes. Here again we find *effects*—to wit, a certain number of irreconcilable observations, and we try to find the *causes* which are, on the one hand, the true value of the quantity to be measured, and, on the other, the error made in each isolated observation. We must calculate the probable *à posteriori* value of each error, and therefore the probable value of the quantity to be measured. But, as I have just explained, we cannot undertake this calculation unless we admit *à priori*—i.e., before any observations are made—that there is a law of the probability of errors. Is there a law of errors? The law to which all calculators assent is Gauss's law, that is represented by a certain transcendental curve known as the "bell."

But it is first of all necessary to recall the classic distinction between systematic and accidental errors. If the metre with which we measure a length is too long, the number we get will be too small, and it will be no use to measure several times—that is a systematic error. If we measure with an accurate metre, we may make a mistake, and find the length sometimes too large and sometimes too small, and when we take the mean of a large number of measurements, the error will tend to grow small. These are accidental errors.

It is clear that systematic errors do not satisfy Gauss's law, but do accidental errors satisfy it? Numerous proofs have been attempted, almost all of them crude paralogisms. But starting from the following hypotheses we may prove Gauss's law: the error is the result of a very large number of partial and independent errors; each partial error is very small and obeys any law of probability whatever, provided the probability of a positive error is the same as that of an equal negative error. It is clear that these conditions will be often, but not always, fulfilled, and we may reserve the name of accidental for errors which satisfy them.

We see that the method of least squares is not legitimate in every case; in general, physicists are more distrustful of it than astronomers. This is no doubt because the latter, apart from the systematic errors to which they and the physicists are subject alike, have to contend with an extremely important source of error which is entirely accidental—I mean atmospheric undulations. So it is very curious to hear a discussion between a physicist and an astronomer about a method of observation. The physicist, persuaded that one good measurement is worth more than many bad ones, is preeminently concerned with the elimination by means of every precaution of the final systematic errors; the astronomer retorts: "But you can only observe a small number of stars, and accidental errors will not disappear."

What conclusion must we draw? Must we continue to use the method of least squares? We must distinguish. We have eliminated all the systematic errors of which we have any suspicion; we are quite certain that there are others still, but we cannot detect them; and yet we must make up our minds and adopt a definitive value which will be regarded as the probable value; and for that purpose it is clear that the best thing we can do is to apply Gauss's law. We have only applied a practical rule referring to subjective probability. And there is no more to be said.

Yet we want to go further and say that not only the probable value is so much, but that the probable error in the result is so much. *This is absolutely invalid:* it would be true only if we were sure that all the systematic errors were eliminated, and of that we know absolutely nothing. We have two series of observations; by applying the law of least squares we find that the probable error in the first series is twice as small as in the second. The second series may, however, be more accurate than the first, because the first is perhaps affected by a large systematic error. All that we can say is that the first series is *probably* better than the second because its accidental error is smaller, and that we have no reason for affirming that the systematic error is greater for one of the series than for the other, our ignorance on this point being absolute.

VII. Conclusions. In the preceding lines I have set several problems, and have given no solutions. I do not regret this, for perhaps they will invite the reader to reflect on these delicate questions.

However that may be, there are certain points which seem to be well established. To undertake the calculation of any probability, and even for that calculation to have any meaning at all, we must admit, as a point of departure, an hypothesis or convention which has always something arbitrary about it. In the choice of this convention we can be guided only by the principle of sufficient reason. Unfortunately, this principle is very vague and very elastic, and in the cursory examination we have just made we have seen it assume different forms. The form under which we meet it most often is the belief in continuity, a belief which it would be difficult to justify by apodeictic reasoning, but without which all science would be impossible. Finally, the problems to which the calculus of probabilities may be applied with profit are those in which the result is independent of the hypothesis made at the outset, provided only that this hypothesis satisfies the condition of continuity.

CHAPTER XII*

OPTICS AND ELECTRICITY

Fresnel's Theory. The best example that can be chosen is the theory of light and its relations to the theory of electricity. It is owing to Fresnel that the science of optics is more advanced than any other branch of physics. The theory called the theory of undulations forms a complete whole, which is satisfying to the mind; but we must not ask from it what it cannot give us. The object of mathematical theories is not to reveal to us the real nature of things; that would be an unreasonable claim. Their only object is to co-ordinate the physical laws with which physical experiment makes us acquainted, the enunciation of which, without the aid of mathematics, we should be unable to effect. Whether the ether exists or not matters little—let us leave that to the metaphysicians; what is essential for us is, that everything happens as if it existed, and that this hypothesis is found to be suitable for the explanation of phenomena. After all, have we any other reason for believing in the existence of material objects? That, too, is only a convenient hy-

---

*This chapter is mainly taken from the prefaces of two of my books—*Théorie mathématique de la lumière* (Paris: Naud, 1889) and *Électricité et optique* (Paris: Naud, 1901).

pothesis; only, it will never cease to be so, while some day, no doubt, the ether will be thrown aside as useless.

But at the present moment the laws of optics, and the equations which translate them into the language of analysis, hold good—at least as a first approximation. It will therefore be always useful to study a theory which brings these equations into connection.

The undulatory theory is based on a molecular hypothesis; this is an advantage to those who think they can discover the cause under the law. But others find in it a reason for distrust; and this distrust seems to me as unfounded as the illusions of the former. These hypotheses play but a secondary role. They may be sacrificed, and the sole reason why this is not generally done is that it would involve a certain loss of lucidity in the explanation. In fact, if we look at it a little closer we shall see that we borrow from molecular hypotheses but two things—the principle of the conservation of energy, and the linear form of the equations, which is the general law of small movements as of all small variations. This explains why most of the conclusions of Fresnel remain unchanged when we adopt the electro-magnetic theory of light.

Maxwell's Theory. We all know that it was Maxwell who connected by a slender tie two branches of physics—optics and electricity— until then unsuspected of having anything in common. Thus blended in a larger aggregate, in a higher harmony, Fresnel's theory of optics did not perish. Parts of it are yet alive, and their mutual relations are still the same. Only, the language which we use to express them has changed; and, on the other hand, Maxwell has revealed to us other relations, hitherto unsuspected, between the different branches of optics and the domain of electricity.

The first time a French reader opens Maxwell's book, his admiration is tempered with a feeling of uneasiness, and often of distrust.

It is only after prolonged study, and at the cost of much effort, that this feeling disappears. Some minds of high calibre never lose this feeling. Why is it so difficult for the ideas of this English scientist to become acclimatised among us? No doubt the education received by most enlightened Frenchmen predisposes them to appreciate precision and logic more than any other qualities. In this respect the old theories of mathematical physics gave us complete satisfaction. All our masters, from Laplace to Cauchy, proceeded along the same lines. Starting with clearly enunciated hypotheses, they deduced from them all their consequences with mathematical rigour, and then compared them with experiment. It seemed to be their aim to give to each of the branches of physics the same precision as to celestial mechanics.

A mind accustomed to admire such models is not easily satisfied with a theory. Not only will it not tolerate the least appearance of contradiction, but it will expect the different parts to be logically connected with one another, and will require the number of hypotheses to be reduced to a minimum.

This is not all; there will be other demands which appear to me to be less reasonable. Behind the matter of which our senses are aware, and which is made known to us by experiment, such a thinker will expect to see another kind of matter—the only true matter in its opinion—which will no longer have anything but purely geometrical qualities, and the atoms of which will be mathematical points subject to the laws of dynamics alone. And yet he will try to represent to himself, by an unconscious contradiction, these invisible and colourless atoms, and therefore to bring them as close as possible to ordinary matter.

Then only will he be thoroughly satisfied, and he will then imagine that he has penetrated the secret of the universe. Even if the satisfaction is fallacious, it is nonetheless difficult to give it up. Thus, on opening the pages of Maxwell, a Frenchman expects to find a theoretical whole, as logical and as precise as the physical optics that is founded on the hypothesis of the ether. He is thus preparing for himself a disappointment which I should like the reader to avoid; so I will warn him at once of what he will find and what he will not find in Maxwell.

Maxwell does not give a mechanical explanation of electricity and magnetism; he confines himself to showing that such an explanation is possible. He shows that the phenomena of optics are only a particular case of electro-magnetic phenomena. From the whole theory of electricity a theory of light can be immediately deduced. Unfortunately the converse is not true; it is not always easy to find a complete explanation of electrical phenomena. In particular it is not easy if we take as our starting-point Fresnel's theory; to do so, no doubt, would be impossible; but nonetheless we must ask ourselves if we are compelled to surrender admirable results which we thought we had definitively acquired. That seems a step backwards, and many sound intellects will not willingly allow of this.

Should the reader consent to set some bounds to his hopes, he will still come across other difficulties. The English scientist does not try to erect a unique, definitive, and well-arranged building; he seems to raise rather a large number of provisional and independent constructions, between which communication is difficult and sometimes impossible. Take, for instance, the chapter in which electrostatic attractions are explained by the pressures and tensions of the dielectric medium. This chapter might be suppressed without the rest of the book's being thereby less clear or less complete, and yet it contains a theory which is self-sufficient, and which can be understood without reading a word of what precedes or follows. But it is not only independent of the rest of the book; it is difficult to reconcile it with the fundamental ideas of the volume. Maxwell does not even attempt to reconcile it; he merely says: "I have not been able to make the next step—namely, to account by mechanical considerations for these stresses in the dielectric."

This example will be sufficient to show what I mean; I could quote many others. Thus, who would suspect on reading the pages devoted to magnetic rotatory polarisation that there is an identity between optical and magnetic phenomena?

We must not flatter ourselves that we have avoided every contradiction, but we ought to make up our minds. Two contradictory theories, provided that they are kept from overlapping, and that we do not look to find in them the explanation of things, may, in fact, be very useful instruments of research; and perhaps the reading of Maxwell would be less suggestive if he had not opened up to us so many new and divergent ways. But the fundamental idea is masked, as it were. So far this is the case, that in most works that are popularised, this idea is the only point which is left completely untouched. To show the importance of this, I think I ought to explain in what this fundamental idea consists; but for that purpose a short digression is necessary.

The Mechanical Explanation of Physical Phenomena. In every physical phenomenon there is a certain number of parameters which are reached directly by experiment, and which can be measured. I shall call them the parameters q. Observation next teaches us the laws of the variations of these parameters, and these laws can be generally stated in the form of differential equations which connect together the parameters q and time. What can be done to give a mechanical interpretation to such a phenomenon? We may endeavour to explain it, either by the movements of ordinary matter, or by those of one or more hypothetical fluids. These fluids will be considered as formed of a very large number of isolated molecules m. When may we say that we have a complete mechanical explanation of the phenomenon? It will be, on the one hand, when we know the differential equations which are satisfied by the co-ordinates of these hypothetical molecules m, equations which must, in addition, conform to the laws of dynamics; and, on the other hand, when we know the relations which define the co-ordinates of the molecules m as functions of the parameters q, attainable by experiment. These equations, as I have said, should conform to the principles of dynamics, and, in particular, to the principle of the conservation of energy, and to that of least action.

The first of these two principles teaches us that the total energy is constant, and may be divided into two parts:

(1) Kinetic energy, or *vis viva*, which depends on the masses of the hypothetical molecules m, and on their velocities. This I shall call T. (2) The potential energy which depends only on the co-ordinates of these molecules, and this I shall call U. It is the sum of the energies T and U that is constant.

Now what are we taught by the principle of least action? It teaches us that to pass from the initial position occupied at the instant t_0 to the final position occupied at the instant t_1, the system must describe such a path that in the interval of time between the instant t_0 and t_1, the mean value of the action—*i.e.,* the *difference* between the two energies T and U, must be as small as possible. The first of these two principles is, moreover, a consequence of the second. If we know the functions T and U, this second principle is sufficient to determine the equations of motion.

Among the paths which enable us to pass from one position to another, there is clearly one for which the mean value of the action is smaller than for all the others. In addition, there is only one such path; and it follows from this, that the principle of least action is sufficient to determine the path followed, and therefore the equations of motion. We thus obtain what are called the equations of Lagrange. In these equations the independent variables are the coordinates of the hypothetical molecules m; but I now assume that we take for the variables the parameters q, which are directly accessible to experiment.

The two parts of the energy should then be expressed as a function of the parameters q and their derivatives; it is clear that it is under this form that they will appear to the experimenter. The latter will naturally endeavour to define kinetic and potential energy by the aid of quantities he can directly observe.* If this be granted, the system will always proceed from one position to another by such a path that the mean value of the action is a minimum. It matters little that T and U are now expressed by the aid of the parameters q and their derivatives; it matters little that it is also by the aid of these parameters that we define the initial and final positions; the principle of least action will always remain true.

Now here again, of the whole of the paths which lead from one position to another, there is one and only one for which the mean action is a minimum. The principle of least action is therefore suf-

---

* We may add that U will depend only on the q parameters, that T will depend on them and their derivatives with respect to time, and will be a homogeneous polynomial of the second degree with respect to these derivatives.

ficient for the determination of the differential equations which define the variations of the parameters q. The equations thus obtained are another form of Lagrange's equations.

To form these equations we need not know the relations which connect the parameters q with the co-ordinates of the hypothetical molecules, nor the masses of the molecules, nor the expression of U as a function of the co-ordinates of these molecules. All we need know is the expression of U as a function of the parameters q, and that of T as a function of the parameters q and their derivatives— *i.e.*, the expressions of the kinetic and potential energy in terms of experimental data.

One of two things must now happen. Either for a convenient choice of T and U the Lagrangian equations, constructed as we have indicated, will be identical with the differential equations deduced from experiment, or there will be no functions T and U for which this identity takes place. In the latter case it is clear that no mechanical explanation is possible. The *necessary* condition for a mechanical explanation to be possible is therefore this: that we may choose the functions T and U so as to satisfy the principle of least action, and of the conservation of energy. Besides, this condition is *sufficient*. Suppose, in fact, that we have found a function U of the parameters q, which represents one of the parts of energy, and that the part of the energy which we represent by T is a function of the parameters q and their derivatives; that it is a polynomial of the second degree with respect to its derivatives, and finally that the Lagrangian equations formed by the aid of these two functions T and U are in conformity with the data of the experiment. How can we deduce from this a mechanical explanation? U must be regarded as the potential energy of a system of which T is the kinetic energy. There is no difficulty as far as U is concerned, but can T be regarded as the *vis viva* of a material system?

It is easily shown that this is always possible, and in an unlimited number of ways. I will be content with referring the reader to the pages of the preface of my *Électricité et optique* for further details. Thus, if the principle of least action cannot be satisfied, no mechanical explanation is possible; if it can be satisfied, there is not only one explanation, but an unlimited number, whence it follows that since there is one there must be an unlimited number.

One more remark. Among the quantities that may be reached by experiment directly we shall consider some as the co-ordinates of our hypothetical molecules, some will be our parameters q, and the rest will be regarded as dependent not only on the co-ordinates but on the velocities—or what comes to the same thing, we look on them as derivatives of the parameters q, or as combinations of these parameters and their derivatives.

Here then a question occurs: among all these quantities measured experimentally which shall we choose to represent the parameters q? and which shall we prefer to regard as the derivatives of these parameters? This choice remains arbitrary to a large extent, but a mechanical explanation will be possible if it is done so as to satisfy the principle of least action.

Next, Maxwell asks: Can this choice and that of the two energies T and U be made so that electric phenomena will satisfy this principle? Experiment shows us that the energy of an electro-magnetic field decomposes into electro-static and electro-dynamic energy. Maxwell recognised that if we regard the former as the potential energy U, and the latter as the kinetic energy T, and that if on the other hand we take the electro-static charges of the conductors as the parameters q, and the intensity of the currents as derivatives of other parameters q—under these conditions, Maxwell has recognised that electric phenomena satisfy the principle of least action. He was then certain of a mechanical explanation. If he had expounded this theory at the beginning of his first volume, instead of relegating it to a corner of the second, it would not have escaped the attention of most readers. If therefore a phenomenon allows of a complete mechanical explanation, it allows of an unlimited number of others, which will equally take into account all the particulars revealed by experiment. And this is confirmed by the history of every branch of physics. In optics, for instance, Fresnel believed vibration to be perpendicular to the plane of polarisation; Neumann holds that it is parallel to that plane. For a long time an *experimen-*

tum crucis was sought for, which would enable us to decide between these two theories, but in vain. In the same way, without going out of the domain of electricity, we find that the theory of two fluids and the single fluid theory equally account in a satisfactory manner for all the laws of electro-statics. All these facts are easily explained, thanks to the properties of the Lagrange equations.

It is easy now to understand Maxwell's fundamental idea. To demonstrate the possibility of a mechanical explanation of electricity we need not trouble to find the explanation itself; we need only know the expression of the two functions T and U, which are the two parts of energy, and to form with these two functions Lagrange's equations, and then to compare these equations with the experimental laws.

How shall we choose from all the possible explanations one in which the help of experiment will be wanting? The day will perhaps come when physicists will no longer concern themselves with questions which are inaccessible to positive methods, and will leave them to the metaphysicians. That day has not yet come; man does not so easily resign himself to remaining forever ignorant of the causes of things. Our choice cannot be therefore any longer guided by considerations in which personal appreciation plays too large a part. There are, however, solutions which all will reject because of their fantastic nature, and others which all will prefer because of their simplicity. As far as magnetism and electricity are concerned, Maxwell abstained from making any choice. It is not that he has a systematic contempt for all that positive methods cannot reach, as may be seen from the time he has devoted to the kinetic theory of gases. I may add that if in his *magnum opus* he develops no complete explanation, he has attempted one in an article in the *Philosophical Magazine.* The strangeness and the complexity of the hypotheses he found himself compelled to make, led him afterwards to withdraw it.

The same spirit is found throughout his whole work. He throws into relief the essential—*i.e.,* what is common to all theories; everything that suits only a particular theory is passed over almost in silence. The reader therefore finds himself in the presence of form nearly devoid of matter, which at first he is tempted to take as a fugitive and unassailable phantom. But the efforts he is thus compelled to make force him to think, and eventually he sees that there is often something rather artificial in the theoretical "aggregates" which he once admired.

CHAPTER XIII

Electro-Dynamics

The history of electro-dynamics is very instructive from our point of view. The title of Ampère's immortal work is *Théorie des phénomènes electrodynamiques, uniquement fondée sur expérience.* He therefore imagined that he had made no hypotheses; but as we shall not be long in recognising, he was mistaken; only, of these hypotheses he was quite unaware. On the other hand, his successors see them clearly enough, because their attention is attracted by the weak points in Ampère's solution. They made fresh hypotheses, but this time deliberately. How many times they had to change them before they reached the classic system, which is perhaps even now not quite definitive, we shall see.

I. Ampère's Theory. In Ampère's experimental study of the mutual action of currents, he has operated, and he could operate only, with closed currents. This was not because he denied the existence or possibility of open currents. If two conductors are positively and negatively charged and brought into communication by a wire, a current is set up which passes from one to the other until the two potentials are equal. According to the ideas of Ampère's time, this was considered to be an open current; the current was known to pass from the first conductor to the second, but they did not know it returned from the second to the first. All currents of this kind were therefore considered by Ampère to be open currents—for instance, the currents of discharge of a condenser; he was unable to experiment on them, their duration being too short. Another kind of open current may be imagined. Suppose we have two conductors A and B connected by a wire AMB. Small conducting masses in motion are first of all placed in contact with the conductor B, receive an electric charge, and leaving B are set in motion along a path BNA, carrying their charge with them. On coming into contact with A they lose their charge, which then returns to B along the wire AMB. Now here we have, in a sense, a closed circuit, since the electricity describes the closed circuit BNAMB; but the two parts of the current are quite different. In the wire AMB the electricity is displaced *through* a fixed conductor like a voltaic current, overcoming an ohmic resistance and developing heat; we say that it is displaced by *conduction*. In the part BNA the electricity is *carried* by a moving conductor, and is said to be displaced by *convection*. If therefore the convection current is considered to be perfectly analogous to the conduction current, the circuit BNAMB is closed; if on the contrary the convection current is not a "true current," and, for instance, does not act on the magnet, there is only the conduction current AMB, which is *open*. For example, if we connect by a wire the poles of a Holtz machine, the charged rotating disc transfers the electricity by convection from one pole to the other, and it returns to the first pole by conduction through the wire. But currents of this kind are very difficult to produce with appreciable intensity; in fact, with the means at Ampère's disposal we may almost say it was impossible.

To sum up, Ampère could conceive of the existence of two kinds of open currents, but he could experiment on neither, because they were not strong enough, or because their duration was too short. Experiment therefore could only show him the action of a closed current on a closed current—or more accurately, the action of a closed current on a portion of current, because a current can be made to describe a *closed* circuit, of which part may be in motion and the other part fixed. The displacements of the moving part may be studied under the action of another closed current. On the other hand, Ampère had no means of studying the action of an open current either on a closed or on another open current.

1. THE CASE OF CLOSED CURRENTS. In the case of the mutual action of two closed currents, experiment revealed to Ampère remarkably simple laws. The following will be useful to us in the sequel:

1. *If the intensity of the currents is kept constant*, and if the two circuits, after having undergone any displacements and deformations whatever, return finally to their initial positions, the total work done by the electro-dynamical actions is zero. In other words, there is an *electro-dynamical potential* of the two circuits proportional to the product of their intensities, and depending on the form and relative positions of the circuits; the work done by the electro-dynamical actions is equal to the change of this potential.

2. The action of a closed solenoid is zero.

3. The action of a circuit C on another voltaic circuit C′ depends only on the "magnetic field" developed by the circuit C. At each point in space we can, in fact, define in magnitude and direction a certain force called "magnetic force," which enjoys the following properties:

(*a*) The force exercised by C on a magnetic pole is applied to that pole, and is equal to the magnetic force multiplied by the magnetic mass of the pole.

(*b*) A very short magnetic needle tends to take the direction of the magnetic force, and the couple to which it tends to reduce is proportional to the product of the magnetic force, the magnetic moment of the needle, and the sine of the dip of the needle.

(*c*) If the circuit C′ is displaced, the amount of the work done by the electro-dynamic action of C on C′ will be equal to the increment of "flow of magnetic force" which passes through the circuit.

2. ACTION OF A CLOSED CURRENT ON A PORTION OF CURRENT. Ampère being unable to produce the open current properly

so called, had only one way of studying the action of a closed current on a portion of current. This was by operating on a circuit C composed of two parts, one movable and the other fixed. The movable part was, for instance, a movable wire $\alpha\beta$, the ends α and β of which could slide along a fixed wire. In one of the positions of the movable wire the end α rested on the point A, and the end β on the point B of the fixed wire. The current ran from α to β—*i.e.*, from A to B along the movable wire, and then from B to A along the fixed wire. *This current was therefore closed.*

In the second position, the movable wire having slipped, the points α and β were respectively at A′ and B′ on the fixed wire. The current ran from α to β—*i.e.*, from A′ to B′ on the movable wire, and returned from B′ to B, and then from B to A, and then from A to A′—all on the fixed wire. This current was also closed. If a similar circuit be exposed to the action of a closed current C, the movable part will be displaced just as if it were acted on by a force. Ampère *admits* that the force, apparently acting on the movable part A B, representing the action of C on the portion $\alpha\beta$ of the current, remains the same whether an open current runs through $\alpha\beta$, stopping at α and β, or whether a closed current runs first to β, and then returns to α through the fixed portion of the circuit. This hypothesis seemed natural enough, and Ampère innocently assumed it; nevertheless the hypothesis *is not a necessity,* for we shall presently see that Helmholtz rejected it. However that may be, it enabled Ampère, although he had never produced an open current, to lay down the laws of the action of a closed current on an open current, or even on an element of current. They are simple:

1. The force acting on an element of current is applied to that element; it is normal to the element and to the magnetic force, and proportional to that component of the magnetic force which is normal to the element.

2. The action of a closed solenoid on an element of current is zero. But the electro-dynamic potential has disappeared—*i.e.*, when a closed and an open current of constant intensities return to their initial positions, the total work done is not zero.

3. CONTINUOUS ROTATIONS. The most remarkable electro-dynamical experiments are those in which continuous rotations are produced, and which are called *unipolar induction* experiments. A magnet may turn about its axis; a current passes first through a fixed wire and then enters the magnet by the pole N, for instance, passes through half the magnet, and emerges by a sliding contact and re-enters the fixed wire. The magnet then begins to rotate continuously. This is Faraday's experiment. How is it possible? If it were a question of two circuits of invariable form, C fixed and C' movable about an axis, the latter would never take up a position of continuous rotation; in fact, there is an electro-dynamical potential; there must therefore be a position of equilibrium when the potential is a maximum. Continuous rotations are therefore possible only when the circuit C' is composed of two parts—one fixed, and the other movable about an axis, as in the case of Faraday's experiment. Here again it is convenient to draw a distinction. The passage from the fixed to the movable part, or *vice versa*, may take place either by simple contact, the same point of the movable part remaining constantly in contact with the same point of the fixed part, or by sliding contact, the same point of the movable part coming successively into contact with the different points of the fixed part.

It is only in the second case that there can be continuous rotation. This is what then happens: the system tends to take up a position of equilibrium; but, when at the point of reaching that position, the sliding contact puts the moving part in contact with a fresh point in the fixed part; it changes the connexions and therefore the conditions of equilibrium, so that as the position of equilibrium is ever eluding, so to speak, the system which is trying to reach it, rotation may take place indefinitely.

Ampère admits that the action of the circuit on the movable part of C' is the same as if the fixed part of C' did not exist, and therefore as if the current passing through the movable part were an open current. He concluded that the action of a closed on an open current, or *vice versa*, that of an open current on a fixed current, may give rise to continuous rotation. But this conclusion depends on the hypothesis which I have enunciated, and to which, as I said above, Helmholtz declined to subscribe.

4. MUTUAL ACTION OF TWO OPEN CURRENTS. As far as the mutual action of two open currents, and in particular that of two elements of current, is concerned, all experiment breaks down. Ampère falls back on hypothesis. He assumes: (1) that the mutual action of two elements reduces to a force acting along their *join;* (2) that the action of two closed currents is the resultant of the mutual actions of their different elements, which are the same as if these elements were isolated.

The remarkable thing is that here again Ampère makes two hypotheses without being aware of it. However that may be, these two hypotheses, together with the experiments on closed currents, suffice to determine completely the law of mutual action of two elements. But then, most of the simple laws we have met in the case of closed currents are no longer true. In the first place, there is no electro-dynamical potential; nor was there any, as we have seen, in the case of a closed current acting on an open current. Next, there is, properly speaking, no magnetic force; and we have above defined this force in three different ways: (1) By the action on a magnetic pole; (2) by the director couple which orientates the magnetic needle; (3) by the action on an element of current.

In the case with which we are immediately concerned, not only are these three definitions not in harmony, but each has lost its meaning:

1. A magnetic pole is no longer acted on by a unique force applied to that pole. We have seen, in fact, the action of an element of current on a pole is not applied to the pole but to the element; it may, moreover, be replaced by a force applied to the pole and by a couple.

2. The couple which acts on the magnetic needle is no longer a simple director couple, for its moment with respect to the axis of the needle is not zero. It decomposes into a director couple, properly so called, and a supplementary couple which tends to produce the continuous rotation of which we have spoken above.

3. Finally, the force acting on an element of a current is not normal to that element. In other words, *the unity of the magnetic force has disappeared.*

Let us see in what this unity consists. Two systems which exercise the same action on a magnetic pole will also exercise the same action on an indefinitely small magnetic needle, or on an element of current placed at the point in space at which the pole is. Well, this is true if the two systems only contain closed currents, and according to Ampère it would not be true if the systems contained open currents. It is sufficient to remark, for instance, that if a magnetic pole is placed at A and an element at B, the direction of the element being in AB produced, this element, which will exercise no action on the pole, will exercise an action either on a magnetic needle placed at A, or on an element of current at A.

5. INDUCTION. We know that the discovery of electro-dynamical induction followed not long after the immortal work of Ampère. As long as it is only a question of closed currents there is no difficulty, and Helmholtz has even remarked that the principle of the conservation of energy is sufficient for us to deduce the laws of induction from the electro-dynamical laws of Ampère. But on the condition, as Bertrand has shown—that we make a certain number of hypotheses.

The same principle again enables this deduction to be made in the case of open currents, although the result cannot be tested by experiment, since such currents cannot be produced.

If we wish to compare this method of analysis with Ampère's theorem on open currents, we get results which are calculated to surprise us. In the first place, induction cannot be deduced from the variation of the magnetic field by the well-known formula of scientists and practical men; in fact, as I have said, properly speaking, there is no magnetic field. But further, if a circuit C is subjected to the induction of a variable voltaic system S, and if this system S be displaced and deformed in any way whatever, so that the intensity of the currents of this system varies according to any law whatever, then so long as after these variations the system eventually returns to its initial position, it seems natural to suppose that the *mean* electro-motive force induced in the current C is zero. This is true if the circuit C is closed, and if the system S only contains closed currents. It is no longer true if we accept the theory of Ampère, since there would be open currents. So that not only will induction no longer be the variation of the flow of magnetic force in any of the usual senses of the word, but it cannot be represented by the variation of that force whatever it may be.

II. Helmholtz's Theory. I have dwelt upon the consequences of Ampère's theory and on his method of explaining the action of open currents. It is difficult to disregard the paradoxical and artificial character of the propositions to which we are thus led. We feel bound to think "it cannot be so." We may imagine then that Helmholtz has been led to look for something else. He rejects the fundamental hypothesis of Ampère—namely, that the mutual action of two elements of current reduces to a force along their join. He admits that an element of current is not acted upon by a single force but by a force and a couple, and this is what gave rise to the celebrated polemic between Bertrand and Helmholtz. Helmholtz replaces Ampère's hypothesis by the following: two elements of current always admit of an electro-dynamic potential, depending solely upon their position and orientation; and the work of the forces that they exercise one on the other is equal to the variation of this potential. Thus Helmholtz can no more do without hypothesis than Ampère, but at least he does not do so without explicitly announcing it. In the case of closed currents, which alone are accessible to experiment, the two theories agree; in all other cases they differ. In the first place, contrary to what Ampère supposed, the force which seems to act on the movable portion of a closed current is not the same as that acting on the movable portion if it were isolated and if it constituted an open current. Let us return to the circuit C', of which we spoke above, and which was formed of a movable wire sliding on a fixed wire. In the only experiment that can be made the movable portion $\alpha\beta$ is not isolated, but is part of a closed circuit. When it passes from AB to A′B′, the total electro-dynamic potential varies for two reasons. First, it has a slight increment because the potential of A′B′ with respect to the circuit C is not the same as that of AB; secondly, it has a second increment because it must be increased by the potentials of the elements AA′ and B′B with respect to C. It is this *double* increment which represents the work of the force acting upon the portion AB. If, on the contrary, αβ be isolated, the potential would only have the first increment, and this first increment alone would measure the work of the force acting on AB. In the second place, there could be no continuous rotation without sliding contact, and in fact, that, as we have seen in the case of closed currents, is an immediate consequence of the existence of an electro-dynamic potential. In Faraday's experiment, if the magnet is fixed, and if the part of the current external to the magnet runs along a movable wire, that movable wire may undergo continuous rotation. But it does not mean that, if the contacts of the weir with the magnet were suppressed, and an open current were to run along the wire, the wire would still have a movement of continuous rotation. I have just said, in fact, that an isolated element is not acted on in the same way as a movable element making part of a closed circuit. But there is another difference. The action of a solenoid on a closed current is zero according to experiment and according to the two theories. Its action on an open current would be zero according to Ampère, and it would not be zero according to Helmholtz. From this follows an important consequence. We have given above three definitions of the magnetic force. The third has no meaning here, since an element of current is no longer acted upon by a single force. Nor has the first any meaning. What, in fact, is a magnetic pole? It is the extremity of an indefinite linear magnet. This magnet may be replaced by an indefinite solenoid. For the definition of magnetic force to have any meaning, the action exercised by an open current on an indefinite solenoid would only depend on the position of the extremity of that solenoid—*i.e.,* that the action of a closed solenoid is zero. Now we have just seen that this is not the case. On the other hand, there is nothing to prevent us from adopting the second definition which is founded on the measurement of the director couple which tends to orientate the magnetic needle; but, if it is adopted, neither the effects of induction nor electro-dynamic effects will depend solely on the distribution of the lines of force in this magnetic field.

III. Difficulties Raised by These Theories. Helmholtz's theory is an advance on that of Ampère; it is necessary, however, that every difficulty should be removed. In both, the name "magnetic field" has no meaning, or, if we give it one by a more or less artificial convention, the ordinary laws so familiar to electricians no longer apply; and it is thus that the electro-motive force induced in a wire is no longer measured by the number of lines of force met by that wire. And our objections do not proceed only from the fact that it is difficult to give up deeply rooted habits of language and thought. There is something more. If we do not believe in actions at a distance, electro-dynamic phenomena must be explained by a modification of the medium. And this medium is precisely what we call "magnetic field," and then the electro-magnetic effects should only depend on that field. All these difficulties arise from the hypothesis of open currents.

IV. Maxwell's Theory. Such were the difficulties raised by the current theories, when Maxwell with a stroke of the pen caused them to vanish. To his mind, in fact, all currents are closed currents. Maxwell admits that if in a dielectric, the electric field happens to vary, this dielectric becomes the seat of a particular phenomenon acting on the galvanometer like a current and called the *current of displacement*. If, then, two conductors bearing positive and negative charges are placed in connection by means of a wire, during the discharge there is an open current of conduction in that wire; but there are produced at the same time in the surrounding dielectric currents of displacement which close this current of conduction. We know that Maxwell's theory leads to the explanation of optical phenomena which would be due to extremely rapid electrical oscillations. At that period such a conception was only a daring hypothesis which could be supported by no experiment; but after twenty years Maxwell's ideas received the confirmation of experiment. Hertz succeeded in producing systems of electric oscillations which reproduce all the properties of light, and only differ by the length of their wave—that is to say, as violet differs from red. In some measure he made a synthesis of light. It might be said that Hertz has not directly proved Maxwell's fundamental idea of the action of the current of displacement on the galvanometer. That is true in a sense. What he has shown directly is that electro-magnetic induction is not instantaneously propagated, as was supposed, but its speed is the speed of light. Yet, to suppose there is no current of displacement, and that induction is with the speed of light; or, rather, to suppose that the currents of displacement produce inductive effects, and that the induction takes place instantaneously—*comes to the same thing.* This cannot be seen at the first glance, but it is proved by an analysis of which I must not even think of giving even a summary here.

V. Rowland's Experiment. But, as I have said above, there are two kinds of open conduction currents. There are first the currents of discharge of a condenser, or of any conductor whatever. There are also cases in which the electric charges describe a closed contour, being displaced by conduction in one part of the circuit and by convection in the other part. The question might be regarded as solved for open currents of the first kind; they were closed by currents of displacement. For open currents of the second kind the solution appeared still more simple.

It seemed that if the current were closed it could only be by the current of convection itself. For that purpose it was sufficient to admit that a "convection current"—*i.e.,* a charged conductor in motion—could act on the galvanometer. But experimental confirmation was lacking. It appeared difficult, in fact, to obtain a sufficient intensity even by increasing as much as possible the charge and the velocity of the conductors. Rowland, an extremely skilful experimentalist, was the first to triumph, or to seem to triumph, over these difficulties. A disc received a strong electrostatic charge and a very high speed of rotation. An astatic magnetic system placed beside the disc underwent deviations. The experiment was made twice by Rowland, once in Berlin and once in Baltimore. It was afterwards repeated by Himstedt. These physicists even believed that they could announce that they had succeeded in making quantitative measurements. For twenty years Rowland's law was admitted without objection by all physicists, and, indeed, everything seemed to confirm it. The spark certainly does produce a magnetic effect, and does it not seem extremely likely that the spark discharged is due to particles taken from one of the electrodes and transferred to the other electrode with their charge? Is not the very spectrum of the spark, in which we recognise the lines of the metal of the electrode, a proof of it? The spark would then be a real current of induction.

On the other hand, it is also admitted that in an electrolyte the electricity is carried by the ions in motion. The current in an electrolyte would therefore also be a current of convection; but it acts on the magnetic needle. And in the same way for cathodic rays; Crookes attributed these rays to very subtle matter charged with negative electricity and moving with very high velocity. He looked upon them, in other words, as currents of convection. Now, these cathodic rays are deviated by the magnet. In virtue of the principle of action and re-action, they should in their turn deviate the magnetic needle. It is true that Hertz believed he had proved that the cathodic rays do not carry negative electricity, and that they do not act on the magnetic needle; but Hertz was wrong. First of all, Perrin succeeded in collecting the electricity carried by these rays—electricity of which Hertz denied the existence; the German scientist appears to have been deceived by the effects due to the action of the X-rays, which were not yet discovered. Afterwards, and quite recently, the action of the cathodic rays on the magnetic needle has been brought to light. Thus all these phenomena looked upon as currents of convection, electric sparks, electrolytic currents, cathodic rays, act in the same manner on the galvanometer and in conformity to Rowland's law.

VI. Lorentz's Theory. We need not go much further. According to Lorentz's theory, currents of conduction would themselves be true convection currents. Electricity would remain indissolubly connected with certain material particles called *electrons*. The circulation of these electrons through bodies would produce voltaic currents, and what would distinguish conductors from insulators would be that the one could be traversed by these electrons, while the others would check the movement of the electrons. Lorentz's theory is very attractive. It gives a very simple explanation of certain phenomena, which the earlier theories—even Maxwell's in its primitive form—could only deal with in an unsatisfactory manner; for example, the aberration of light, the partial impulse of luminous waves, magnetic polarisation, and Zeeman's experiment.

A few objections still remained. The phenomena of an electric system seemed to depend on the absolute velocity of translation of the centre of gravity of this system, which is contrary to the idea that we have of the relativity of space. Supported by M. Crémieu, M. Lippman has presented this objection in a very striking form. Imagine two charged conductors with the same velocity of translation. They are relatively at rest. However, each of them being equivalent to a current of convection, they ought to attract one another, and by measuring this attraction we could measure their absolute velocity. "No!" replied the partisans of Lorentz. "What we could measure in that way is not their absolute velocity, but their relative velocity *with respect to the ether,* so that the principle of relativity is safe." Whatever there may be in these objections, the edifice of electro-dynamics seemed, at any rate in its broad lines, definitively constructed. Everything was presented under the most satisfactory aspect. The theories of Ampère and Helmholtz, which were made for the open currents that no longer existed, seem to have no more than purely historic interest, and the inextricable complications to which these theories led have been almost forgotten. This quiescence has been recently disturbed by the experiments of M. Crémieu, which have contradicted, or at least have seemed to contradict, the results formerly obtained by Rowland. Numerous investigators have endeavoured to solve the question, and fresh experiments have been undertaken. What result will they give? I shall take care not to risk a prophecy which might be falsified between the day this book is ready for the press and the day on which it is placed before the public.

THE VALUE

OF

SCIENCE

Author's Essay Prefatory to the Translation

THE CHOICE OF FACTS

Tolstoi somewhere explains why "science for its own sake" is in his eyes an absurd conception. We cannot know *all* facts, since their number is practically infinite. It is necessary to choose; then we may let this choice depend on the pure caprice of our curiosity; would it not be better to let ourselves be guided by utility, by our practical and above all by our moral needs; have we nothing better to do than count the number of lady-bugs on our planet?

It is clear the word *utility* has not for him the sense men of affairs give it, and following them most of our contemporaries. Little cares he for industrial applications, for the marvels of electricity or of automobilism, which he regards rather as obstacles to moral progress; utility for him is solely what can make man better.

For my part, it need scarce be said, I could never be content with either the one or the other ideal; I want neither that plutocracy grasping and mean, nor that democracy goody and mediocre, occupied solely in turning the other cheek, where would dwell sages without curiosity, who, shunning excess, would not die of disease, but would surely die of ennui. But that is a matter of taste and is not what I wish to discuss.

The question nevertheless remains and should fix our attention; if our choice can only be determined by caprice or by immediate utility, there can be no science for its own sake, and consequently no science. But is that true? That a choice must be made is incontestable; whatever be our activity, facts go quicker than we, and we cannot catch them; while the scientist discovers one fact, there happen milliards of milliards in a cubic millimetre of his body. To wish to comprise nature in science would be to want to put the whole into the part.

But scientists believe there is a hierarchy of facts and that among them may be made a judicious choice. They are right, since otherwise there would be no science, yet science exists. One need only open the eyes to see that the conquests of industry which have enriched so many practical men would never have seen the light, if these practical men alone had existed and if they had not been preceded by unselfish devotees who died poor, who never thought of utility, and yet had a guide far other than caprice.

As Mach says, these devotees have spared their successors the trouble of thinking. Those who might have worked solely in view of an immediate application would have left nothing behind them, and, in face of a new need, all must have been begun over again. Now most men do not love to think, and this is perhaps fortunate when instinct guides them, for most often, when they pursue an aim which is immediate and ever the same, instinct guides them better than reason would guide a pure intelligence. But instinct is routine, and if thought did not fecundate it, it would no more progress in man than in the bee or ant. It is needful then to think for those who love not thinking and, as they are numerous, it is needful that each of our thoughts be as often useful as possible, and this is why a law will be the more precious the more general it is.

This shows us how we should choose: the most interesting facts are those which may serve many times; these are the facts which have a chance of coming up again. We have been so fortunate as to be born in a world where there are such. Suppose that instead of 60 chemical elements there were 60 milliards of them, that they were not, some common, the others rare, but that they were uniformly distributed. Then, every time we picked up a new pebble there would be great probability of its being formed of some unknown substance; all that we knew of other pebbles would be worthless for it; before each new object we should be as the new-born babe; like it we could only obey our caprices or our needs. Biologists would be just as much at a loss if there were only individuals and no species and if heredity did not make sons like their fathers.

In such a world there would be no science; perhaps thought and even life would be impossible, since evolution could not there develop the preservational instincts. Happily it is not so; like all good fortune to which we are accustomed, this is not appreciated at its true worth.

Which then are the facts likely to reappear? They are first the simple facts. It is clear that in a complex fact a thousand circumstances are united by chance, and that only a chance still much less probable could reunite them anew. But are there any simple facts? And if there are, how recognize them? What assurance is there that a thing we think simple does not hide a dreadful complexity? All we can say is that we ought to prefer the facts which *seem* simple to those where our crude eye discerns unlike elements. And then one of two things: either this simplicity is real, or else the elements are so intimately mingled as not to be distinguishable. In the first case there is chance of our meeting anew this same simple fact, either in all its purity or entering itself as element in a complex manifold. In the second case this intimate mixture has likewise more chances of recurring than a heterogeneous assemblage; chance knows how to mix, it knows not how to disentangle, and to make with multiple elements a well-ordered edifice in which something is distinguishable, it must be made expressly. The facts which appear simple, even if they are not so, will therefore be more easily revived by chance. This it is which justifies the method instinctively adopted by the scientist, and what justifies it still better, perhaps, is that oft-recurring facts appear to us simple, precisely because we are used to them.

But where is the simple fact? Scientists have been seeking it in the two extremes, in the infinitely great and in the infinitely small. The astronomer has found it because the distances of the stars are immense, so great that each of them appears but as a point, so great that the qualitative differences are effaced, and because a point is simpler than a body which has form and qualities. The physicist, on the other hand, has sought the elementary phenomenon in fictively cutting up bodies into infinitesimal cubes, because the conditions of the problem, which undergo slow and continuous variation in passing from one point of the body to another, may be regarded as constant in the interior of each of these little cubes. In the same way the biologist has been instinctively led to regard the cell as more interesting than the whole animal, and the outcome has shown his wisdom, since cells belonging to organisms the most different are more alike, for the one who can recognize their resemblances, than are these organisms themselves. The sociologist is more embarrassed; the elements, which for him are men, are too unlike, too variable, too capricious in a word, too complex; besides, history never begins over again. How then choose the interesting fact, which is that which begins again? Method is precisely the choice of facts; it is needful then to be occupied first with creating a method, and many have been imagined, since none imposes itself, so that sociology is the science which has the most methods and the fewest results.

Therefore it is by the regular facts that it is proper to begin; but after the rule is well established, after it is beyond all doubt, the facts in full conformity with it are ere long without interest since they no longer teach us anything new. It is then the exception which becomes important. We cease to seek resemblances; we devote ourselves above all to the differences, and among the differences are chosen first the most accentuated, not only because they are the most striking, but because they will be the most instructive. A simple example will make my thought plainer: suppose one wishes to determine a curve by observing some of its points. The practician who concerns himself only with immediate utility would observe only the points he might need for some special object.

These points would be badly distributed on the curve; they would be crowded in certain regions, rare in others, so that it would be impossible to join them by a continuous line, and they would be unavailable for other applications. The scientist will proceed differently; as he wishes to study the curve for itself, he will distribute regularly the points to be observed, and when enough are known he will join them by a regular line and then he will have the entire curve. But for that how does he proceed? If he has determined an extreme point of the curve, he does not stay near this extremity, but goes first to the other end; after the two extremities the most instructive point will be the mid-point and so on.

So when a rule is established we should first seek the cause where this rule has the greatest chance of failing. Thence, among other reasons, come the interest of astronomic facts and the interest of the geologic past; by going very far away in space or very far away in time, we may find our usual rules entirely overturned, and these grand overturnings aid us the better to see or the better to understand the little changes which may happen nearer to us, in the little corner of the world where we are called to live and act. We shall better know this corner for having traveled in distant countries with which we have nothing to do.

But what we ought to aim at is less the ascertainment of resemblances and differences than the recognition of likenesses hidden under apparent divergences. Particular rules seem at first discordant, but looking more closely we see in general that they resemble each other; different as to matter, they are alike as to form, as to the order of their parts. When we look at them with this bias, we shall see them enlarge and tend to embrace everything. And this it is which makes the value of certain facts which come to complete an assemblage and to show that it is the faithful image of other known assemblages.

I will not further insist, but these few words suffice to show that the scientist does not choose at random the facts he observes. He does not, as Tolstoi says, count the lady-bugs, because, however interesting lady-bugs may be, their number is subject to capricious variations. He seeks to condense much experience and much thought into a slender volume; and that is why a little book on physics contains so many past experiences and a thousand times as many possible experiences whose result is known beforehand.

But we have as yet looked at only one side of the question. The scientist does not study nature because it is useful; he studies it because he delights in it, and he delights in it because it is beautiful. If nature were not beautiful, it would not be worth knowing, and if nature were not worth knowing, life would not be worth living. Of course I do not here speak of that beauty which strikes the senses, the beauty of qualities and of appearances; not that I undervalue such beauty, far from it, but it has nothing to do with science; I mean that profounder beauty which comes from the harmonious order of the parts and which a pure intelligence can grasp. This it is which gives body, a structure so to speak, to the iridescent appearances which flatter our senses, and without this support the beauty of these fugitive dreams would be only imperfect, because it would be vague and always fleeting. On the contrary, intellectual beauty is sufficient unto itself, and it is for its sake, more perhaps than for the future good of humanity, that the scientist devotes himself to long and difficult labors.

It is, therefore, the quest of this especial beauty, the sense of the harmony of the cosmos, which makes us choose the facts most fitting to contribute to this harmony, just as the artist chooses from among the features of his model those which perfect the picture and give it character and life. And we need not fear that this instinctive and unavowed prepossession will turn the scientist aside from the search for the true. One may dream a harmonious world, but how far the real world will leave it behind! The greatest artists that ever lived, the Greeks, made their heavens; how shabby it is beside the true heavens, ours!

And it is because simplicity, because grandeur, is beautiful, that we preferably seek simple facts, sublime facts, that we delight now to follow the majestic course of the stars, now to examine with the microscope that prodigious littleness which is also a grandeur, now to seek in geologic time the traces of a past which attracts because it is far away.

We see too that the longing for the beautiful leads us to the same choice as the longing for the useful. And so it is that this economy of thought, this economy of effort, which is, according to Mach, the constant tendency of science, is at the same time a source of beauty and a practical advantage. The edifices that we admire are those where the architect has known how to proportion the means to the end, where the columns seem to carry gaily, without effort, the weight placed upon them, like the gracious cariatids of the Erechtheion.

Whence comes this concordance? Is it simply that the things which seem to us beautiful are those which best adapt themselves to our intelligence, and that consequently they are at the same time the implement this intelligence knows best how to use? Or is there here a play of evolution and natural selection? Have the peoples whose ideal most conformed to their highest interest exterminated the others and taken their place? All pursued their ideals without reference to consequences, but while this quest led some to destruction, to others it gave empire. One is tempted to believe it. If the Greeks triumphed over the barbarians and if Europe, heir of Greek thought, dominates the world, it is because the savages loved loud colors and the clamorous tones of the drum which occupied only their senses, while the Greeks loved the intellectual beauty which hides beneath sensuous beauty, and that this intellectual beauty it is which makes intelligence sure and strong.

Doubtless such a triumph would horrify Tolstoi, and he would not like to acknowledge that it might be truly useful. But this disinterested quest of the true for its own beauty is sane also and able to make man better. I well know that there are mistakes, that the thinker does not always draw thence the serenity he should find therein, and even that there are scientists of bad character. Must we, therefore, abandon science and study only morals? What! Do you think the moralists themselves are irreproachable when they come down from their pedestal?

INTRODUCTION

The search for truth should be the goal of our activities; it is the sole end worthy of them. Doubtless we should first bend our efforts to assuage human suffering, but why? Not to suffer is a negative ideal more surely attained by the annihilation of the world. If we wish more and more to free man from material cares, it is that he may be able to employ the liberty obtained in the study and contemplation of truth.

But sometimes truth frightens us. And in fact we know that it is sometimes deceptive, that it is a phantom never showing itself for a moment except to ceaselessly flee, that it must be pursued further and ever further without ever being attained. Yet to work one must stop, as some Greek, Aristotle or another, has said. We also know how cruel the truth often is, and we wonder whether illusion is not more consoling, yea, even more bracing, for illusion it is which gives confidence. When it shall have vanished, will hope remain and shall we have the courage to achieve? Thus would not the horse harnessed to his treadmill refuse to go, were his eyes not bandaged? And then to seek truth it is necessary to be independent, wholly independent. If on the contrary we wish to act, to be strong, we should be united. This is why many of us fear truth; we consider it a cause of weakness. Yet truth should not be feared, for it alone is beautiful.

When I speak here of truth, assuredly I refer first to scientific truth; but I also mean moral truth, of which what we call justice is only one aspect. It may seem that I am misusing words, that I combine thus under the same name two things having nothing in common; that scientific truth, which is demonstrated, can in no way be likened to moral truth, which is felt. And yet I cannot separate them, and whosoever loves the one cannot help loving the other. To find the one, as well as to find the other, it is necessary to free the soul completely from prejudice and from passion; it is necessary to attain absolute sincerity. These two sorts of truth when discovered give the same joy; each when perceived beams with the same splendour, so that we must see it or close our eyes. Lastly, both attract us and flee from us; they are never fixed: when we think to have reached them, we find that we have still to advance, and he who pursues them is condemned never to know repose. It must be added that those who fear the one will also fear the other; for they are the ones who in everything are concerned above all with consequences. In a word, I liken the two truths, because the same reasons make us love them and because the same reasons make us fear them.

If we ought not to fear moral truth, still less should we dread scientific truth. In the first place it cannot conflict with ethics. Ethics and science have their own domains, which touch but do not interpenetrate. The one shows us to what goal we should aspire, the other, given the goal, teaches us how to attain it. So they can never conflict since they can never meet. There can no more be immoral science than there can be scientific morals.

But if science is feared, it is above all because it cannot give us happiness. Of course it cannot. We may even ask whether the beast does not suffer less than man. But can we regret that earthly paradise where man, brute-like, was really immortal in knowing not that he must die? When we have tasted the apple, no suffering can make us forget its savour. We always come back to it. Could it be otherwise? As well ask if one who has seen and is blind will not long for the light. Man, then, cannot be happy through science, but today he can much less be happy without it.

But if truth be the sole aim worth pursuing, may we hope to attain it? It may well be doubted. Readers of my little book *Science and Hypothesis* already know what I think about the question. The truth we are permitted to glimpse is not altogether what most men call by that name. Does this mean that our most legitimate, most imperative aspiration is at the same time the most vain? Or can we, despite all, approach truth on some side? This it is which must be investigated.

In the first place, what instrument have we at our disposal for this conquest? Is not human intelligence, more specifically the intelligence of the scientist, susceptible of infinite variation? Volumes could be written without exhausting this subject; I, in a few brief pages, have only touched it lightly. That the geometer's mind is not like the physicist's or the naturalist's, all the world would agree; but mathematicians themselves do not resemble each other; some recognize only implacable logic, others appeal to intuition and see in it the only source of discovery. And this would be a reason for distrust. To minds so unlike can the mathematical theorems themselves appear in the same light? Truth which is not the same for all, is it truth? But looking at things more closely, we see how these very different workers collaborate in a common task which could not be achieved without their cooperation. And that already reassures us.

Next must be examined the frames in which nature seems enclosed and which are called time and space. In *Science and Hypothesis* I have already shown how relative their value is; it is not nature which imposes them upon us, it is we who impose them upon nature because we find them convenient. But I have spoken of scarcely more than space, and particularly quantitative space, so to say, that is of the mathematical relations whose aggregate constitutes geometry. I should have shown that it is the same with time as with space and still the same with "qualitative space"; in particular,

I should have investigated why we attribute three dimensions to space. I may be pardoned then for taking up again these important questions.

Is mathematical analysis then, whose principal object is the study of these empty frames, only a vain play of the mind? It can give to the physicist only a convenient language; is this not a mediocre service, which, strictly speaking, could be done without; and even is it not to be feared that this artificial language may be a veil interposed between reality and the eye of the physicist? Far from it; without this language most of the intimate analogies of things would have remained forever unknown to us; and we should forever have been ignorant of the internal harmony of the world, which is, we shall see, the only true objective reality.

The best expression of this harmony is law. Law is one of the most recent conquests of the human mind; there still are people who live in the presence of a perpetual miracle and are not astonished at it. On the contrary, we it is who should be astonished at nature's regularity. Men demand of their gods to prove their existence by miracles; but the eternal marvel is that there are not miracles without cease. The world is divine because it is a harmony. If it were ruled by caprice, what could prove to us it was not ruled by chance?

This conquest of law we owe to astronomy, and just this makes the grandeur of the science rather than the material grandeur of the objects it considers. It was altogether natural then that celestial mechanics should be the first model of mathematical physics; but since then this science has developed; it is still developing, even rapidly developing. And it is already necessary to modify in certain points the scheme I outlined in 1900 and from which I drew two chapters of *Science and Hypothesis*. In an address at the St. Louis exposition in 1904, I sought to survey the road travelled; the result of this investigation the reader shall see farther on.

The progress of science has seemed to imperil the best established principles, those even which were regarded as fundamental. Yet nothing shows they will not be saved; and if this comes about only imperfectly, they will still subsist even though they are modified. The advance of science is not comparable to the changes of a city, where old edifices are pitilessly torn down to give place to new, but to the continuous evolution of zoologic types which develop ceaselessly and end by becoming unrecognizable to the common sight, but where an expert eye finds always traces of the prior work of the centuries past. One must not think then that the old-fashioned theories have been sterile and vain.

Were we to stop there, we should find in these pages some reasons for confidence in the value of science, but many more for distrusting it; an impression of doubt would remain; it is needful now to set things to rights.

Some people have exaggerated the role of convention in science; they have even gone so far as to say that law, that scientific fact itself, was created by the scientist. This is going much too far in the direction of nominalism. No, scientific laws are not artificial creations; we have no reason to regard them as accidental, though it be impossible to prove they are not.

Does the harmony the human intelligence thinks it discovers in nature exist outside of this intelligence? No, beyond doubt, a reality completely independent of the mind which conceives it, sees or feels it, is an impossibility. A world as exterior as that, even if it existed, would for us be forever inaccessible. But what we call objective reality is, in the last analysis, what is common to many thinking beings, and could be common to all; this common part, we shall see, can only be the harmony expressed by mathematical laws. It is this harmony then which is the sole objective reality, the only truth we can attain; and when I add that the universal harmony of the world is the source of all beauty, it will be understood what price we should attach to the slow and difficult progress which little by little enables us to know it better.

PART I

THE
MATHEMATICAL
SCIENCES

CHAPTER I

INTUITION AND LOGIC IN
MATHEMATICS

I

It is impossible to study the works of the great mathematicians, or even those of the lesser, without noticing and distinguishing two opposite tendencies, or rather two entirely different kinds of minds. The one sort are above all preoccupied with logic; to read their works, one is tempted to believe they have advanced only step by step, after the manner of a Vauban who pushes on his trenches against the place besieged, leaving nothing to chance. The other sort are guided by intuition and at the first stroke make quick but sometimes precarious conquests, like bold cavalrymen of the advance guard.

The method is not imposed by the matter treated. Though one often says of the first that they are *analysts* and calls the others *geometers,* that does not prevent the one sort from remaining analysts even when they work at geometry, while the others are still geometers even when they occupy themselves with pure analysis. It is the very nature of their mind which makes them logicians or intuitionalists, and they cannot lay it aside when they approach a new subject.

Nor is it education which has developed in them one of the two tendencies and stifled the other. The mathematician is born, not made, and it seems he is born a geometer or an analyst. I should like to cite examples and there are surely plenty; but to accentuate the contrast I shall begin with an extreme example, taking the liberty of seeking it in two living mathematicians.

M. Méray wants to prove that a binomial equation always has a root, or, in ordinary words, that an angle may always be subdivided. If there is any truth that we think we know by direct intuition, it is this. Who could doubt that an angle may always be divided into any number of equal parts? M. Méray does not look at it that way; in his eyes this proposition is not at all evident and to prove it he needs several pages.

On the other hand, look at Professor Klein: he is studying one of the most abstract questions of the theory of functions to determine whether on a given Riemann surface there always exists a function admitting of given singularities. What does the celebrated German geometer do? He replaces his Riemann surface by a metallic surface whose electric conductivity varies according to certain laws. He connects two of its points with the two poles of a battery. The current, says he, must pass, and the distribution of this current on the surface will define a function whose singularities will be precisely those called for by the enunciation.

Doubtless Professor Klein well knows he has given here only a sketch: nevertheless he has not hesitated to publish it; and he would probably believe he finds in it, if not a rigorous demonstration, at least a kind of moral certainty. A logician would have rejected with horror such a conception, or rather he would not have had to reject it, because in his mind it would never have originated.

Again, permit me to compare two men, the honor of French science, who have recently been taken from us, but who both entered long ago into immortality. I speak of M. Bertrand and M. Hermite. They were scholars of the same school at the same time; they had the same education, were under the same influences; and yet what a difference! Not only does it blaze forth in their writings; it is in their teaching, in their way of speaking, in their very look. In the memory of all their pupils these two faces are stamped in deathless lines; for all who have had the pleasure of following their teaching, this remembrance is still fresh; it is easy for us to evoke it.

While speaking, M. Bertrand is always in motion; now he seems in combat with some outside enemy, now he outlines with a gesture of the hand the figures he studies. Plainly he sees and he is eager to paint, this is why he calls gesture to his aid. With M. Hermite, it is just the opposite; his eyes seem to shun contact with the world; it is not without, it is within he seeks the vision of truth.

Among the German geometers of this century, two names above all are illustrious, those of the two scientists who have founded the general theory of functions, Weierstrass and Riemann. Weierstrass leads everything back to the consideration of series and their ana-lytic transformations; to express it better, he reduces analysis to a sort of prolongation of arithmetic; you may turn through all his books without finding a figure. Riemann, on the contrary, at once calls geometry to his aid; each of his conceptions is an image that no one can forget, once he has caught its meaning.

More recently, Lie was an intuitionalist; this might have been doubted in reading his books, no one could doubt it after talking with him; you saw at once that he thought in pictures. Madame Ko-valevski was a logician.

Among our students we notice the same differences; some pre-fer to treat their problems "by analysis," others "by geometry." The first are incapable of "seeing in space," the others are quickly tired of long calculations and become perplexed.

The two sorts of minds are equally necessary for the progress of science; both the logicians and the intuitionalists have achieved great things that others could not have done. Who would venture to say whether he preferred that Weierstrass had never written or that there had never been a Riemann? Analysis and synthesis have then both their legitimate roles. But it is interesting to study more closely in the history of science the part which belongs to each.

II

Strange! If we read over the works of the ancients we are tempted to class them all among the intuitionalists. And yet nature is always the same; it is hardly probable that it has begun in this century to create minds devoted to logic. If we could put ourselves into the flow of ideas which reigned in their time, we should recognize that many of the old geometers were in tendency analysts. Euclid, for example, erected a scientific structure wherein his contemporaries could find no fault. In this vast construction, of which each piece however is due to intuition, we may still today, without much effort, recognize the work of a logician.

It is not minds that have changed, it is ideas; the intuitional minds have remained the same; but their readers have required of them greater concessions.

What is the cause of this evolution? It is not hard to find. Intuition cannot give us rigour, nor even certainty; this has been recognized more and more. Let us cite some examples. We know there exist continuous functions lacking derivatives. Nothing is more shocking to intuition than this proposition which is imposed upon us by logic. Our fathers would not have failed to say: "It is evident that every continuous function has a derivative, since every curve has a tangent."

How can intuition deceive us on this point? It is because when we seek to imagine a curve, we cannot represent it to ourselves without width; just so, when we represent to ourselves a straight line, we see it under the form of a rectilinear band of a certain breadth. We well know these lines have no width; we try to imagine them narrower and narrower and thus to approach the limit; so we do in a certain measure, but we shall never attain this limit. And then it is clear we can always picture these two narrow bands, one straight, one curved, in a position such that they encroach slightly one upon the other without crossing. We shall thus be led, unless warned by a rigorous analysis, to conclude that a curve always has a tangent.

I shall take as a second example Dirichlet's principle on which rest so many theorems of mathematical physics; today we establish it by reasonings very rigorous but very long; heretofore, on the contrary, we were content with a very summary proof. A certain integral depending on an arbitrary function can never vanish. Hence it is concluded that it must have a minimum. The flaw in this reasoning strikes us immediately, since we use the abstract term *function* and are familiar with all the singularities functions can present when the word is understood in the most general sense.

But it would not be the same had we used concrete images, had we, for example, considered this function as an electric potential; it would have been thought legitimate to affirm that electrostatic equilibrium can be attained. Yet perhaps a physical comparison would have awakened some vague distrust. But if care had been taken to translate the reasoning into the language of geometry, intermediate between that of analysis and that of physics, doubtless this distrust would not have been produced, and perhaps one might thus, even today, still deceive many readers not forewarned.

Intuition, therefore, does not give us certainty. This is why the evolution had to happen; let us now see how it happened.

It was not slow in being noticed that rigour could not be introduced in the reasoning unless first made to enter into the definitions. For the most part the objects treated of by mathematicians were long ill defined; they were supposed to be known because represented by means of the senses or the imagination; but one had only a crude image of them and not a precise idea on which reasoning could take hold. It was there first that the logicians had to direct their efforts.

So, in the case of incommensurable numbers. The vague idea of continuity, which we owe to intuition, resolved itself into a complicated system of inequalities referring to whole numbers.

By that means the difficulties arising from passing to the limit, or from the consideration of infinitesimals, are finally removed. Today in analysis only whole numbers are left or systems, finite or infinite, of whole numbers bound together by a net of equality or inequality relations. Mathematics, as they say, is arithmetized.

III

A first question presents itself. Is this evolution ended? Have we finally attained absolute rigour? At each stage of the evolution our fathers also thought they had reached it. If they deceived themselves, do we not likewise cheat ourselves?

We believe that in our reasonings we no longer appeal to intuition; the philosophers will tell us this is an illusion. Pure logic could never lead us to anything but tautologies; it could create nothing new; not from it alone can any science issue. In one sense these philosophers are right; to make arithmetic, as to make geometry, or to make any science, something else than pure logic is necessary. To designate this something else we have no word other than *intuition*. But how many different ideas are hidden under this same word?

Compare these four axioms: (1) Two quantities equal to a third are equal to one another; (2) if a theorem is true of the number 1 and if we prove that it is true of $n + 1$ if true for n, then will it be true of all whole numbers; (3) if on a straight line the point C is between A and B and the point D between A and C, then the point D will be between A and B; (4) through a given point there is not more than one parallel to a given straight line.

All four are attributed to intuition, and yet the first is the enunciation of one of the rules of formal logic; the second is a real synthetic *à priori* judgment, it is the foundation of rigorous mathematical induction; the third is an appeal to the imagination; the fourth is a disguised definition.

Intuition is not necessarily founded on the evidence of the senses; the senses would soon become powerless; for example, we cannot represent to ourselves a chiliagon, and yet we reason by intuition on polygons in general, which include the chiliagon as a particular case.

You know what Poncelet understood by the *principle of continuity*. What is true of a real quantity, said Poncelet, should be true of an imaginary quantity; what is true of the hyperbola whose asymptotes are real, should then be true of the ellipse whose asymptotes are imaginary. Poncelet was one of the most intuitive minds of this century; he was passionately, almost ostentatiously, so; he regarded the principle of continuity as one of his boldest conceptions, and yet this principle did not rest on the evidence of the senses. To assimilate the hyperbola to the ellipse was rather to contradict this evidence. It was only a sort of precocious and instinctive generalization which, moreover, I have no desire to defend.

We have then many kinds of intuition; first, the appeal to the senses and the imagination; next, generalization by induction, copied, so to speak, from the procedures of the experimental sciences; finally, we have the intuition of pure number, whence arose the second of the axioms just enunciated, which is able to create the real mathematical reasoning. I have shown above by examples that the first two cannot give us certainty; but who will seriously doubt the third, who will doubt arithmetic?

Now in the analysis of today, when one cares to take the trouble to be rigorous, there can be nothing but syllogisms or appeals to this intuition of pure number, the only intuition which cannot deceive us. It may be said that today absolute rigour is attained.

IV

The philosophers make still another objection: "What you gain in rigour," they say, "you lose in objectivity. You can rise toward your logical ideal only by cutting the bonds which attach you to reality. Your science is infallible, but it can only remain so by imprisoning itself in an ivory tower and renouncing all relation with the external world. From this seclusion it must go out when it would attempt the slightest application."

For example, I seek to show that some property pertains to some object whose concept seems to me at first indefinable, because it is intuitive. At first I fail or must content myself with approximate proofs; finally I decide to give to my object a precise definition, and this enables me to establish this property in an irreproachable manner.

"And then," say the philosophers, "it still remains to show that the object which corresponds to this definition is indeed the same made known to you by intuition; or else that some real and concrete object whose conformity with your intuitive idea you believe you immediately recognize corresponds to your new definition. Only then could you affirm that it has the property in question. You have only displaced the difficulty."

That is not exactly so; the difficulty has not been displaced, it has been divided. The proposition to be established was in reality composed of two different truths, at first not distinguished. The first was a mathematical truth, and it is now rigorously established. The second was an experimental verity. Experience alone can teach us that some real and concrete object corresponds or does not correspond to some abstract definition. This second verity is not mathematically demonstrated, but neither can it be, no more than can the empirical laws of the physical and natural sciences. It would be unreasonable to ask more.

Well, is it not a great advance to have distinguished what long was wrongly confused? Does this mean that nothing is left of this objection of the philosophers? That I do not intend to say; in becoming rigorous, mathematical science takes a character so artificial as to strike everyone; it forgets its historical origins; we see how the questions can be answered, we no longer see how and why they are put.

This shows us that logic is not enough; that the science of demonstration is not all science and that intuition must retain its role as complement, I was about to say, as counter-poise or as antidote of logic.

I have already had occasion to insist on the place intuition should hold in the teaching of the mathematical sciences. Without it young minds could not make a beginning in the understanding of mathematics; they could not learn to love it and would see in it only a vain logomachy; above all, without intuition they would never become capable of applying mathematics. But now I wish before all to speak of the role of intuition in science itself. If it is useful to the student, it is still more so to the creative scientist.

V

We seek reality, but what is reality? The physiologists tell us that organisms are formed of cells; the chemists add that cells themselves are formed of atoms. Does this mean that these atoms or these cells constitute reality, or rather the sole reality? The way in which these cells are arranged and from which results the unity of the individual, is not it also a reality much more interesting than that of the isolated elements, and should a naturalist who had never studied the elephant except by means of the microscope think himself sufficiently acquainted with that animal?

Well, there is something analogous to this in mathematics. The logician cuts up, so to speak, each demonstration into a very great number of elementary operations; when we have examined these operations one after the other and ascertained that each is correct, are we to think we have grasped the real meaning of the demonstration? Shall we have understood it even when, by an effort of memory, we have become able to repeat this proof by reproducing all these elementary operations in just the order in which the inventor had arranged them? Evidently not; we shall not yet possess the entire reality; that I know not what which makes the unity of the demonstration will completely elude us.

Pure analysis puts at our disposal a multitude of procedures whose infallibility it guarantees; it opens to us a thousand different ways on which we can embark in all confidence; we are assured of meeting there no obstacles; but of all these ways, which will lead us most promptly to our goal? Who shall tell us which to choose? We need a faculty which makes us see the end from afar, and intuition is this faculty. It is necessary to the explorer for choosing his route; it is not less so to the one following his trail who wants to know why he chose it.

If you are present at a game of chess, it will not suffice, for the understanding of the game, to know the rules for moving the pieces. That will only enable you to recognize that each move has been made conformably to these rules, and this knowledge will truly have very little value. Yet this is what the reader of a book on mathematics would do if he were a logician only. To understand the game is wholly another matter; it is to know why the player moves this piece rather than that other which he could have moved without breaking the rules of the game. It is to perceive the inward reason which makes of this series of successive moves a sort of organized whole. This faculty is still more necessary for the player himself, that is, for the inventor.

Let us drop this comparison and return to mathematics. For example, see what has happened to the idea of continuous function. At the outset this was only a sensible image, for example, that of a continuous mark traced by the chalk on a blackboard. Then it became little by little more refined; ere long it was used to construct a complicated system of inequalities, which reproduced, so to speak, all the lines of the original image; this construction finished, the centering of the arch, so to say, was removed, that crude representation which had temporarily served as support and which was afterwards useless was rejected; there remained only the construction itself, irreproachable in the eyes of the logician. And yet if the primitive image had totally disappeared from our recollection, how could we divine by what caprice all these inequalities were erected in this fashion one upon another?

Perhaps you think I use too many comparisons; yet pardon still another. You have doubtless seen those delicate assemblages of silicious needles which form the skeleton of certain sponges. When the organic matter has disappeared, there remains only a frail and elegant lace-work. True, nothing is there except silica, but what is interesting is the form this silica has taken, and we could not understand it if we did not know the living sponge which has given it precisely this form. Thus it is that the old intuitive notions of our fathers, even when we have abandoned them, still imprint their form upon the logical constructions we have put in their place.

This view of the aggregate is necessary for the inventor; it is equally necessary for whoever wishes really to comprehend the inventor. Can logic give it to us? No; the name mathematicians give it would suffice to prove this. In mathematics logic is called *analysis* and analysis means *division, dissection*. It can have, therefore, no tool other than the scalpel and the microscope.

Thus logic and intuition have each their necessary role. Each is indispensable. Logic, which alone can give certainty, is the instrument of demonstration; intuition is the instrument of invention.

VI

But at the moment of formulating this conclusion I am seized with scruples. At the outset I distinguished two kinds of mathematical minds, the one sort logicians and analysts, the others intuitionalists and geometers. Well, the analysts also have been inventors. The names I have just cited make my insistence on this unnecessary.

Here is a contradiction, at least apparently, which needs explanation. And first, do you think these logicians have always proceeded from the general to the particular, as the rules of formal logic would seem to require of them? Not thus could they have extended the boundaries of science; scientific conquest is to be made only by generalization.

In one of the chapters of *Science and Hypothesis*, I have had occasion to study the nature of mathematical reasoning, and I have shown how this reasoning, without ceasing to be absolutely rigorous, could lift us from the particular to the general by a procedure I have called *mathematical induction*. It is by this procedure that the analysts have made science progress, and if we examine the detail itself of their demonstrations, we shall find it there at each instant beside the classic syllogism of Aristotle. We, therefore, see already that the analysts are not simply makers of syllogisms after the fashion of the scholastics.

Besides, do you think they have always marched step by step with no vision of the goal they wished to attain? They must have divined the way leading thither, and for that they needed a guide. This guide is, first, analogy. For example, one of the methods of demonstration dear to analysts is that founded on the employment of dominant functions. We know it has already served to solve a multitude of problems; in what consists then the role of the inven-

.or who wishes to apply it to a new problem? At the outset he must recognize the analogy of this question with those which have already been solved by this method; then he must perceive in what way this new question differs from the others, and thence deduce the modifications necessary to apply to the method.

But how does one perceive these analogies and these differences? In the example just cited they are almost always evident, but I could have found others where they would have been much more deeply hidden; often a very uncommon penetration is necessary for their discovery. The analysts, not to let these hidden analogies escape them, that is, in order to be inventors, must, without the aid of the senses and imagination, have a direct sense of what constitutes the unity of a piece of reasoning, of what makes, so to speak, its soul and inmost life.

When one talked with M. Hermite, he never evoked a sensuous image, and yet you soon perceived that the most abstract entities were for him like living beings. He did not see them, but he perceived that they are not an artificial assemblage, and that they have some principle of internal unity.

But, one will say, that still is intuition. Shall we conclude that the distinction made at the outset was only apparent, that there is only one sort of mind and that all the mathematicians are intuitionalists, at least those who are capable of inventing?

No, our distinction corresponds to something real. I have said above that there are many kinds of intuition. I have said how much the intuition of pure number, whence comes rigorous mathematical induction, differs from sensible intuition to which the imagination, properly so called, is the principal contributor.

Is the abyss which separates them less profound than it at first appeared? Could we recognize with a little attention that this pure intuition itself could not do without the aid of the senses? This is the affair of the psychologist and the metaphysician and I shall not discuss the question. But the thing's being doubtful is enough to justify me in recognizing and affirming an essential difference between the two kinds of intuition; they have not the same object and seem to call into play two different faculties of our soul; one would think of two search-lights directed upon two worlds strangers to one another.

It is the intuition of pure number, that of pure logical forms, which illumines and directs those we have called *analysts*. This it is which enables them not alone to demonstrate, but also to invent. By it they perceive at a glance the general plan of a logical edifice, and that too without the senses appearing to intervene. In rejecting the aid of the imagination, which, as we have seen, is not always infallible, they can advance without fear of deceiving themselves. Happy, therefore, are those who can do without this aid! We must admire them; but how rare they are!

Among the analysts there will then be inventors, but they will be few. The majority of us, if we wished to see afar by pure intuition alone, would soon feel ourselves seized with vertigo. Our weakness has need of a staff more solid, and, despite the exceptions of which we have just spoken, it is nonetheless true that sensible intuition is in mathematics the most usual instrument of invention.

Àpropos of these reflections, a question comes up that I have not the time either to solve or even to enunciate with the developments it would admit of. Is there room for a new distinction, for distinguishing among the analysts those who above all use this pure intuition and those who are first of all preoccupied with formal logic?

M. Hermite, for example, whom I have just cited, cannot be classed among the geometers who make use of the sensible intuition; but neither is he a logician, properly so called. He does not conceal his aversion to purely deductive procedures which start from the general and end in the particular.

CHAPTER II

THE MEASURE OF TIME

I

So long as we do not go outside the domain of consciousness, the notion of time is relatively clear. Not only do we distinguish without difficulty present sensation from the remembrance of past sensations or the anticipation of future sensations, but we know perfectly well what we mean when we say that, of two conscious phenomena which we remember, one was anterior to the other; or that, of two foreseen conscious phenomena, one will be anterior to the other.

When we say that two conscious facts are simultaneous, we mean that they profoundly interpenetrate, so that analysis cannot separate them without mutilating them.

The order in which we arrange conscious phenomena does not admit of any arbitrariness. It is imposed upon us and of it we can change nothing.

I have only a single observation to add. For an aggregate of sensations to have become a remembrance capable of classification in time, it must have ceased to be actual, we must have lost the sense of its infinite complexity, otherwise it would have remained present. It must, so to speak, have crystallized around a center of associations of ideas which will be a sort of label. It is only when they thus have lost all life that we can classify our memories in time as a botanist arranges dried flowers in his herbarium.

But these labels can only be finite in number. On that score, psychologic time should be discontinuous. Whence comes the feeling that between any two instants there are others? We arrange our recollections in time, but we know that there remain empty compartments. How could that be, if time were not a form preexistent in our mind? How could we know there were empty compartments, if these compartments were revealed to us only by their content?

II

But that is not all; into this form we wish to put not only the phenomena of our own consciousness, but those of which other consciousnesses are the theater. But more, we wish to put there physical facts, these I know not what with which we people space and which no consciousness sees directly. This is necessary because without it science could not exist. In a word, psychologic time is given to us and must needs create scientific and physical time. There the difficulty begins, or rather the difficulties, for there are two.

Think of two consciousnesses, which are like two worlds impenetrable one to the other. By what do we strive to put them into the same mould, to measure them by the same standard? Is it not as if one strove to measure length with a gramme or weight with a metre? And besides, why do we speak of measuring? We know perhaps that some fact is anterior to some other, but not *by how much* it is anterior.

Therefore two difficulties: (1) Can we transform psychologic time, which is qualitative, into a quantitative time? (2) Can we reduce to one and the same measure facts which transpire in different worlds?

III

The first difficulty has long been noticed; it has been the subject of long discussions and one may say the question is settled. *We have not a direct intuition of the equality of two intervals of time.* The persons who believe they possess this intuition are dupes of an illusion. When I say, from noon to one the same time passes as from two to three, what meaning has this affirmation?

The least reflection shows that by itself it has none at all. It will only have that which I choose to give it, by a definition which will certainly possess a certain degree of arbitrariness. Psychologists could have done without this definition; physicists and astronomers could not; let us see how they have managed.

To measure time they use the pendulum and they suppose by definition that all the beats of this pendulum are of equal duration. But this is only a first approximation; the temperature, the resistance of the air, the barometric pressure, make the pace of the pendulum vary. If we could escape these sources of error, we should obtain a much closer approximation, but it would still be only an approximation. New causes, hitherto neglected, electric, magnetic or others, would introduce minute perturbations.

In fact, the best chronometers must be corrected from time to time, and the corrections are made by the aid of astronomic observations; arrangements are made so that the sidereal clock marks the same hour when the same star passes the meridian. In other words, it is the sidereal day, that is, the duration of the rotation of the earth, which is the constant unit of time. It is supposed, by a new definition substituted for that based on the beats of the pendulum, that two complete rotations of the earth about its axis have the same duration.

However, the astronomers are still not content with this definition. Many of them think that the tides act as a check on our globe, and that the rotation of the earth is becoming slower and slower. Thus would be explained the apparent acceleration of the motion of the moon, which would seem to be going more rapidly than theory permits because our watch, which is the earth, is going slow.

IV

All this is unimportant, one will say; doubtless our instruments of measurement are imperfect, but it suffices that we can conceive a perfect instrument. This ideal cannot be reached, but it is enough to have conceived it and so to have put rigour into the definition of the unit of time.

The trouble is that there is no rigour in the definition. When we use the pendulum to measure time, what postulate do we implicitly admit? *It is that the duration of two identical phenomena is the same;* or, if you prefer, that the same causes take the same time to produce the same effects.

And at first blush, this is a good definition of the equality of two durations. But take care. Is it impossible that experiment may some day contradict our postulate?

Let me explain myself. I suppose that at a certain place in the world the phenomenon α happens, causing as consequence at the end of a certain time the effect α'. At another place in the world very far away from the first, happens the phenomenon β, which causes as consequence the effect β'. The phenomena α and β are simultaneous, as are also the effects α' and β'.

Later, the phenomenon α is reproduced under approximately the same conditions as before, and *simultaneously* the phenomenon β is also reproduced at a very distant place in the world and almost under the same circumstances. The effects α' and β' also take place. Let us suppose that the effect α' happens perceptibly before the effect β'.

If experience made us witness such a sight, our postulate would be contradicted. For experience would tell us that the first duration $\alpha\alpha'$ is equal to the first duration $\beta\beta'$ and that the second duration $\alpha\alpha'$ is less than the second duration $\beta\beta'$. On the other hand, our postulate would require that the two durations $\alpha\alpha'$ should be equal to each other, as likewise the two durations $\beta\beta'$. The equality and the inequality deduced from experience would be incompatible with the two equalities deduced from the postulate.

Now can we affirm that the hypotheses I have just made are

absurd? They are in no ways contrary to the principle of contradiction. Doubtless they could not happen without the principle of sufficient reason seeming violated. But to justify a definition so fundamental I should prefer some other guarantee.

V

But that is not all. In physical reality one cause does not produce a given effect, but a multitude of distinct causes contribute to produce it, without our having any means of discriminating the part of each of them.

Physicists seek to make this distinction; but they make it only approximately, and, however they progress, they never will make it except approximately. It is approximately true that the motion of the pendulum is due solely to the earth's attraction; but in all rigour every attraction, even of Sirius, acts on the pendulum.

Under these conditions, it is clear that the causes which have produced a certain effect will never be reproduced except approximately. Then we should modify our postulate and our definition. Instead of saying: "The same causes take the same time to produce the same effects," we should say: "Causes almost identical take almost the same time to produce almost the same effects."

Our definition therefore is no longer anything but approximate. Besides, as M. Calinon very justly remarks in a recent memoir:[*]

> One of the circumstances of any phenomenon is the velocity of the earth's rotation; if this velocity of rotation varies, it constitutes in the reproduction of this phenomenon a circumstance which no longer remains the same. But to suppose this velocity of rotation constant is to suppose that we know how to measure time.

Our definition is therefore not yet satisfactory; it is certainly not that which the astronomers of whom I spoke above implicitly

---

[*] *Etude sur les diverses grandeurs,* Paris, Gauthier-Villars, 1897.

adopt, when they affirm that the terrestrial rotation is slowing down.

What meaning according to them has this affirmation? We can only understand it by analysing the proofs they give of their proposition. They say first that the friction of the tides producing heat must destroy *vis viva.* They invoke therefore the principle of *vis viva,* or of the conservation of energy.

They say next that the secular acceleration of the moon, calculated according to Newton's law, would be less than that deduced from observations unless the correction relative to the slowing down of the terrestrial rotation were made. They invoke therefore Newton's law. In other words, they define duration in the following way: time should be so defined that Newton's law and that of *vis viva* may be verified. Newton's law is an experimental truth; as such it is only approximate, which shows that we still have only a definition by approximation.

If now it be supposed that another way of measuring time is adopted, the experiments on which Newton's law is founded would nonetheless have the same meaning. Only the enunciation of the law would be different, because it would be translated into another language; it would evidently be much less simple. So that the definition implicitly adopted by the astronomers may be summed up thus: time should be so defined that the equations of mechanics may be as simple as possible. In other words, there is not one way of measuring time more true than another; that which is generally adopted is only more *convenient.* Of two watches, we have no right to say that the one goes true, the other wrong; we can only say that it is advantageous to conform to the indications of the first.

The difficulty which has just occupied us has been, as I have said, often pointed out; among the most recent works in which it is considered, I may mention, besides M. Calinon's little book, the treatise on mechanics of M. Andrade.

VI

The second difficulty has up to the present attracted much less attention; yet it is altogether analogous to the preceding; and even, logically, I should have spoken of it first.

Two psychological phenomena happen in two different consciousnesses; when I say they are simultaneous, what do I mean? When I say that a physical phenomenon, which happens outside of every consciousness, is before or after a psychological phenomenon, what do I mean?

In 1572, Tycho Brahe noticed in the heavens a new star. An immense conflagration had happened in some far distant heavenly body; but it had happened long before; at least two hundred years were necessary for the light from that star to reach our earth. This conflagration therefore happened before the discovery of America. Well, when considering this gigantic phenomenon, which perhaps had no witness, since the satellites of that star were perhaps uninhabited, I say this phenomenon is anterior to the formation of the visual image of the isle of Española in the consciousness of Christopher Columbus, what do I mean?

A little reflection is sufficient to understand that all these affirmations have by themselves no meaning. They can have one only as the outcome of a convention.

VII

We should first ask ourselves how one could have had the idea of putting into the same frame so many worlds impenetrable to each other. We should like to represent to ourselves the external universe, and only by so doing could we feel that we understood it. We know we never can attain this representation: our weakness is too great. But at least we desire the ability to conceive an infinite intelligence for which this representation would be possible, a sort of great consciousness which should see all, and which should classify all *in its time*, as we classify, *in our time*, the little we see.

This hypothesis is indeed crude and incomplete, because this supreme intelligence would be only a demigod; infinite in one sense, it would be limited in another, since it would have only an imperfect recollection of the past; and it could have no other, since otherwise all recollections would be equally present to it and for it there would be no time. And yet when we speak of time, for all which happens outside of us, do we not unconsciously adopt this hypothesis; do we not put ourselves in the place of this imperfect god; and do not even the atheists put themselves in the place where god would be if he existed?

What I have just said shows us, perhaps, why we have tried to put all physical phenomena into the same frame. But that cannot pass for a definition of simultaneity, since this hypothetical intelligence, even if it existed, would be for us impenetrable. It is therefore necessary to seek something else.

VIII

The ordinary definitions which are proper for psychologic time would suffice us no better. Two simultaneous psychologic facts are so closely bound together that analysis cannot separate without mutilating them. Is it the same with two physical facts? Is not my present nearer my past of yesterday than the present of Sirius?

It has also been said that two facts should be regarded as simultaneous when the order of their succession may be inverted at will. It is evident that this definition would not suit two physical facts which happen far from one another, and that, in what concerns them, we no longer even understand what this reversibility would be; besides, succession itself must first be defined.

IX

Let us then seek to give an account of what is understood by simultaneity or antecedence, and for this let us analyse some examples.

I write a letter; it is afterward read by the friend to whom I have addressed it. There are two facts which have had for their theater two different consciousnesses. In writing this letter I have had the visual image of it, and my friend has had in his turn this same visual image in reading the letter. Though these two facts happen in impenetrable worlds, I do not hesitate to regard the first as anterior to the second, because I believe it is its cause.

I hear thunder, and I conclude there has been an electric discharge; I do not hesitate to consider the physical phenomenon as anterior to the auditory image perceived in my consciousness, because I believe it is its cause.

Behold then the rule we follow, and the only one we can follow: when a phenomenon appears to us as the cause of another, we regard it as anterior. It is therefore by cause that we define time; but most often, when two facts appear to us bound by a constant relation, how do we recognize which is the cause and which the effect? We assume that the anterior fact, the antecedent, is the cause of the other, of the consequent. It is then by time that we define cause. How save ourselves from this *petitio principii?*

We say now *post hoc, ergo propter hoc;* now *propter hoc, ergo post hoc;* shall we escape from this vicious circle?

X

Let us see, not how we succeed in escaping, for we do not completely succeed, but how we try to escape.

I execute a voluntary act A and I feel afterward a sensation D, which I regard as a consequence of the act A; on the other hand, for whatever reason, I infer that this consequence is not immediate, but that outside my consciousness two facts B and C, which I have not witnessed, have happened, and in such a way that B is the effect of A, that C is the effect of B, and D of C.

But why? If I think I have reason to regard the four facts A, B, C, D, as bound to one another by a causal connection, why range them in the causal order $A\,B\,C\,D$, and at the same time in the chronologic order $A\,B\,C\,D$, rather than in any other order?

I clearly see that in the act A I have the feeling of having been active, while in undergoing the sensation D, I have that of having been passive. This is why I regard *A* as the initial cause and *D* as th ultimate effect; this is why I put *A* at the beginning of the chain and *D* at the end; but why put *B* before *C* rather than *C* before *B*?

If this question is put, the reply ordinarily is: we know that it is *B* which is the cause of *C* because we *always* see *B* happen before *C*. These two phenomena, when witnessed, happen in a certain order; when analogous phenomena happen without witness, there is no reason to invert this order.

Doubtless, but take care; we never know directly the physical phenomena *B* and *C*. What we know are sensations *B'* and *C'* produced respectively by *B* and *C*. Our consciousness tells us immediately that *B'* precedes *C'* and we *suppose* that *B* and *C* succeed one another in the same order.

This rule appears in fact very natural, and yet we are often led to depart from it. We hear the sound of the thunder only some seconds after the electric discharge of the cloud. Of two flashes of lightning, the one distant, the other near, cannot the first be anterior to the second, even though the sound of the second comes to us before that of the first?

XI

Another difficulty; have we really the right to speak of the cause of a phenomenon? If all the parts of the universe are interchained in a certain measure, any one phenomenon will not be the effect of a single cause, but the resultant of causes infinitely numerous; it is, one often says, the consequence of the state of the universe a moment before. How enunciate rules applicable to circumstances so complex? And yet it is only thus that these rules can be general and rigorous.

Not to lose ourselves in this infinite complexity let us make a simpler hypothesis. Consider three stars, for example, the sun, Jupiter and Saturn; but, for greater simplicity, regard them as reduced to material points and isolated from the rest of the world. The positions and the velocities of three bodies at a given instant suffice to determine their positions and velocities at the following

.nstant, and consequently at any instant. Their positions at the instant t determine their positions at the instant $t + h$ as well as their positions at the instant $t - h$.

Even more; the position of Jupiter at the instant t, together with that of Saturn at the instant $t + a$, determines the position of Jupiter at any instant and that of Saturn at any instant.

The aggregate of positions occupied by Jupiter at the instant $t + e$ and Saturn at the instant $t + a + e$ is bound to the aggregate of positions occupied by Jupiter at the instant t and Saturn at the instant $t + a$, by laws as precise as that of Newton, though more complicated. Then why not regard one of these aggregates as the cause of the other, which would lead to considering as simultaneous the instant t of Jupiter and the instant $t + a$ of Saturn?

In answer there can only be reasons, very strong, it is true, of convenience and simplicity.

XII

But let us pass to examples less artificial; to understand the definition implicitly supposed by the savants, let us watch them at work and look for the rules by which they investigate simultaneity.

I will take two simple examples, the measurement of the velocity of light and the determination of longitude.

When an astronomer tells me that some stellar phenomenon, which his telescope reveals to him at this moment, happened nevertheless fifty years ago, I seek his meaning, and to that end I shall ask him first how he knows it, that is, how he has measured the velocity of light.

He has begun by *supposing* that light has a constant velocity, and in particular that its velocity is the same in all directions. That is a postulate without which no measurement of this velocity could be attempted. This postulate could never be verified directly by experiment; it might be contradicted by it if the results of different measurements were not concordant. We should think ourselves fortunate that this contradiction has not happened and that the slight discordances which may happen can be readily explained.

The postulate, at all events, resembling the principle of sufficient reason, has been accepted by everybody; what I wish to emphasize is that it furnishes us with a new rule for the investigation of simultaneity, entirely different from that which we have enunciated above.

This postulate assumed, let us see how the velocity of light has been measured. You know that Roemer used eclipses of the satellites of Jupiter, and sought how much the event fell behind its prediction. But how is this prediction made? It is by the aid of astronomic laws, for instance Newton's law.

Could not the observed facts be just as well explained if we attributed to the velocity of light a little different value from that adopted, and supposed Newton's law only approximate? Only this would lead to replacing Newton's law by another more complicated. So for the velocity of light a value is adopted, such that the astronomic laws compatible with this value may be as simple as possible. When navigators or geographers determine a longitude, they have to solve just the problem we are discussing; they must, without being at Paris, calculate Paris time. How do they accomplish it? They carry a chronometer set for Paris. The qualitative problem of simultaneity is made to depend upon the quantitative problem of the measurement of time. I need not take up the difficulties relative to this latter problem, since above I have emphasized them at length.

Or else they observe an astronomic phenomenon, such as an eclipse of the moon, and they suppose that this phenomenon is perceived simultaneously from all points of the earth. That is not altogether true, since the propagation of light is not instantaneous; if absolute exactitude were desired, there would be a correction to make according to a complicated rule.

Or else finally they use the telegraph. It is clear first that the reception of the signal at Berlin, for instance, is after the sending of this same signal from Paris. This is the rule of cause and effect analysed above. But how much after? In general, the duration of the transmission is neglected and the two events are regarded as simultaneous. But, to be rigorous, a little correction would still have to be made by a complicated calculation; in practise it is not made, because it would be well within the errors of observation; its theoretic necessity is nonetheless from our point of view, which is that of a rigorous definition. From this discussion, I wish to emphasize two things: (1) The rules applied are exceedingly various. (2) It is difficult to separate the qualitative problem of simultaneity from the quantitative problem of the measurement of time; no matter whether a chronometer is used, or whether account must be taken of a velocity of transmission, as that of light, because such a velocity could not be measured without *measuring* a time.

XIII

To conclude: We have not a direct intuition of simultaneity, nor of the equality of two durations. If we think we have this intuition, this is an illusion. We replace it by the aid of certain rules which we apply almost always without taking count of them.

But what is the nature of these rules? No general rule, no rigorous rule; a multitude of little rules applicable to each particular case.

These rules are not imposed upon us and we might amuse ourselves in inventing others; but they could not be cast aside without greatly complicating the enunciation of the laws of physics, mechanics and astronomy.

We therefore choose these rules, not because they are true, but because they are the most convenient, and we may recapitulate them as follows: "The simultaneity of two events, or the order of their succession, the equality of two durations, are to be so defined that the enunciation of the natural laws may be as simple as possible. In other words, all these rules, all these definitions are only the fruit of an unconscious opportunism."

CHAPTER III

THE NOTION OF SPACE

1. INTRODUCTION

In the articles I have heretofore devoted to space I have above all emphasized the problems raised by non-Euclidean geometry, while leaving almost completely aside other questions more difficult of approach, such as those which pertain to the number of dimensions. All the geometries I considered had thus a common basis, that tridimensional continuum which was the same for all and which differentiated itself only by the figures one drew in it or when one aspired to measure it.

In this continuum, primitively amorphous, we may imagine a network of lines and surfaces, we may then convene to regard the meshes of this net as equal to one another, and it is only after this convention that this continuum, become measurable, becomes Euclidean or non-Euclidean space. From this amorphous continuum can therefore arise indifferently one or the other of the two spaces, just as on a blank sheet of paper may be traced indifferently a straight or a circle.

In space we know rectilinear triangles the sum of whose angles is equal to two right angles; but equally we know curvilinear triangles the sum of whose angles is less than two right angles. The existence of the one sort is not more doubtful than that of the other. To give the name of straights to the sides of the first is to adopt Euclidean geometry; to give the name of straights to the sides of the latter is to adopt the non-Euclidean geometry. So that to ask what geometry it is proper to adopt is to ask, to what line is it proper to give the name straight?

It is evident that experiment cannot settle such a question; one would not ask, for instance, experiment to decide whether I should call *AB* or *CD* a straight. On the other hand, neither can I say that I have not the right to give the name of straights to the sides of non-Euclidean triangles because they are not in conformity with the eternal idea of straight which I have by intuition. I grant, indeed, that I have the intuitive idea of the side of the Euclidean triangle, but I have equally the intuitive idea of the side of the non-Euclidean triangle. Why should I have the right to apply the name of straight to the first of these ideas and not to the second? Wherein does this syllable form an integrant part of this intuitive idea? Evidently when we say that the Euclidean straight is a *true* straight and that the non-Euclidean straight is not a true straight, we simply mean that the first intuitive idea corresponds to a *more noteworthy* object than the second. But how do we decide that this object is more noteworthy? This question I have investigated in *Science and Hypothesis.*

It is here that we saw experience come in. If the Euclidean straight is more noteworthy than the non-Euclidean straight, it is so chiefly because it differs little from certain noteworthy natural objects from which the non-Euclidean straight differs greatly. But, it will be said, the definition of the non-Euclidean straight is artificial; if we for a moment adopt it, we shall see that two circles of different radius both receive the name of non-Euclidean straights, while of two circles of the same radius one can satisfy the definition without the other being able to satisfy it, and then if we transport one of these so-called straights without deforming it, it will cease to be a straight. But by what right do we consider as equal these two figures which the Euclidean geometers call two circles with the same radius? It is because by transporting one of them without deforming it we can make it coincide with the other. And why do we say this transportation is effected without deformation? It is impossible to give a good reason for it. Among all the motions conceivable, there are some of which the Euclidean geometers say that they are not accompanied by deformation; but there are others of which the non-Euclidean geometers would say that they are not accompanied by deformation. In the first, called Euclidean motions, the Euclidean straights remain Euclidean straights, and the non-Euclidean straights do not remain non-Euclidean straights; in the motions of the second sort, or non-Euclidean motions, the non-Euclidean straights remain non-Euclidean straights and the Euclidean straights do not remain Euclidean straights. It has, therefore, not been demonstrated that it was unreasonable to call straights the sides of non-Euclidean triangles; it has only been shown that that would be unreasonable if one continued to call the Euclidean motions motions without deformation; but it has at the same time been shown that it would be just as unreasonable to call straights the sides of Euclidean triangles if the non-Euclidean motions were called motions without deformation.

Now when we say that the Euclidean motions are the *true* motions without deformation, what do we mean? We simply mean that they are *more noteworthy* than the others. And why are they more noteworthy? It is because certain noteworthy natural bodies, the solid bodies, undergo motions almost similar.

And then when we ask: Can one imagine non-Euclidean space? that means: Can we imagine a world where there would be noteworthy natural objects affecting almost the form of non-Euclidean straights, and noteworthy natural bodies frequently undergoing motions almost similar to the non-Euclidean motions? I have shown in *Science and Hypothesis* that to this question we must answer yes.

It has often been observed that if all the bodies in the universe were dilated simultaneously and in the same proportion, we should have no means of perceiving it, since all our measuring instruments would grow at the same time as the objects themselves which they serve to measure. The world, after this dilatation, would continue on its course without anything apprising us of so considerable an event. In other words, two worlds similar to one another (understanding the word *similitude* in the sense of Euclid, Book VI) would be absolutely indistinguishable. But more; worlds will be indistinguishable not only if they are equal or similar, that is, if we can pass from one to the other by changing the axes of co-ordinates, or by changing the scale to which lengths are referred; but they will still be indistinguishable if we can pass from one to the other by any "point-transformation" whatever. I will explain my meaning. I suppose that to each point of one corresponds one point of the other and only one, and inversely; and besides that the co-ordinates of a point are continuous functions, *otherwise altogether arbitrary*, of the corresponding point. I suppose besides that to each object of the first world corresponds in the second an object of the same nature placed precisely at the corresponding point. I suppose finally that this correspondence fulfilled at the initial instant is maintained indefinitely. We should have no means of distinguishing these two worlds one from the other. The relativity of space is not ordinarily understood in so broad a sense; it is thus, however, that it would be proper to understand it.

If one of these universes is our Euclidean world, what its inhabitants will call straight will be our Euclidean straight; but what the inhabitants of the second world will call straight will be a curve which will have the same properties in relation to the world they inhabit and in relation to the motions that they will call motions without deformation. Their geometry will, therefore, be Euclidean geometry, but their straight will not be our Euclidean straight. It will be its transform by the point-transformation which carries over from our world to theirs. The straights of these men will not be our straights, but they will have among themselves the same relations as our straights to one another. It is in this sense I say their geometry will be ours. If then we wish after all to proclaim that they deceive themselves, that their straight is not the true straight, if we still are unwilling to admit that such an affirmation has no meaning, at least we must confess that these people have no means whatever of recognizing their error.

2. QUALITATIVE GEOMETRY

All that is relatively easy to understand, and I have already so often repeated it that I think it needless to expatiate further on the matter. Euclidean space is not a form imposed upon our sensibility, since we can imagine non-Euclidean space; but the two spaces, Euclidean and non-Euclidean, have a common basis, that amorphous continuum of which I spoke in the beginning. From this continuum we can get either Euclidean space or Lobachevskian space, just as we can, by tracing upon it a proper graduation, transform an ungraduated thermometer into a Fahrenheit or a Réaumur thermometer.

And then comes a question: Is not this amorphous continuum that our analysis has allowed to survive a form imposed upon our sensibility? If so, we should have enlarged the prison in which this sensibility is confined, but it would always be a prison.

This continuum has a certain number of properties, exempt from all idea of measurement. The study of these properties is the object of a science which has been cultivated by many great geometers and in particular by Riemann and Betti and which has received the name of analysis situs. In this science abstraction is made of every quantitative idea and, for example, if we ascertain that on a line the point B is between the points A and C, we shall be content with this ascertainment and shall not trouble to know whether the line ABC is straight or curved, nor whether the length AB is equal to the length BC, or whether it is twice as great.

The theorems of analysis situs have, therefore, this peculiarity that they would remain true if the figures were copied by an inexpert draftsman who should grossly change all the proportions and replace the straights by lines more or less sinuous. In mathematical

228 · *The Value of Science*

terms, they are not altered by any "point-transformation" whatsoever. It has often been said that metric geometry was quantitative, while projective geometry was purely qualitative. That is not altogether true. The straight is still distinguished from other lines by properties which remain quantitative in some respects. The real qualitative geometry is, therefore, analysis situs.

The same questions which came up *àpropos* of the truths of Euclidean geometry, come up anew *àpropos* of the theorems of analysis situs. Are they obtainable by deductive reasoning? Are they disguised conventions? Are they experimental verities? Are they the characteristics of a form imposed either upon our sensibility or upon our understanding?

I wish simply to observe that the last two solutions exclude each other. We cannot admit at the same time that it is impossible to imagine space of four dimensions and that experience proves to us that space has three dimensions. The experimenter puts to nature a question: Is it this or that? and he cannot put it without imagining the two terms of the alternative. If it were impossible to imagine one of these terms, it would be futile and besides impossible to consult experience. There is no need of observation to know that the hand of a watch is not marking the hour 15 on the dial, because we know beforehand that there are only 12, and we could not look at the mark 15 to see if the hand is there, because this mark does not exist.

Note likewise that in analysis situs the empiricists are disembarrassed of one of the gravest objections that can be levelled against them, of that which renders absolutely vain in advance all their efforts to apply their thesis to the verities of Euclidean geometry. These verities are rigorous and all experimentation can only be approximate. In analysis situs approximate experiments may suffice to give a rigorous theorem and, for instance, if it is seen that space cannot have either two or fewer than two dimensions, nor four or more than four, we are certain that it has exactly three, since it could not have two and a half or three and a half.

Of all the theorems of analysis situs, the most important is that which is expressed in saying that space has three dimensions. This it is that we are about to consider, and we shall put the question in these terms: When we say that space has three dimensions, what do we mean?

3. The Physical Continuum of Several Dimensions

I have explained in *Science and Hypothesis* whence we derive the notion of physical continuity and how that of mathematical continuity has arisen from it. It happens that we are capable of distinguishing two impressions one from the other, while each is indistinguishable from a third. Thus we can readily distinguish a weight of 12 grammes from a weight of 10 grammes, while a weight of 11 grammes could neither be distinguished from the one nor the other. Such a statement, translated into symbols, may be written:

$$A=B, B=C, A<C.$$

This would be the formula of the physical continuum, as crude experience gives it to us, whence arises an intolerable contradiction that has been obviated by the introduction of the mathematical continuum. This is a scale of which the steps (commensurable or incommensurable numbers) are infinite in number, but are exterior to one another instead of encroaching on one another as do the elements of the physical continuum, in conformity with the preceding formula.

The physical continuum is, so to speak, a nebula not resolved; the most perfect instruments could not attain to its resolution. Doubtless if we measured the weights with a good balance instead of judging them by the hand, we could distinguish the weight of 11 grammes from those of 10 and 12 grammes, and our formula would become:

$$A<B, B<C, A<C.$$

But we should always find between A and B and between B and C new elements D and E, such that

$$A=D, D=B, A<B; B=E, E=C, B<C,$$

and the difficulty would only have receded and the nebula would always remain unresolved; the mind alone can resolve it and the mathematical continuum it is which is the nebula resolved into stars.

Yet up to this point we have not introduced the notion of the number of dimensions. What is meant when we say that a mathematical continuum or that a physical continuum has two or three dimensions?

First we must introduce the notion of cut, studying first physical continua. We have seen what characterizes the physical continuum. Each of the elements of this continuum consists of a manifold of impressions; and it may happen either that an element cannot be discriminated from another element of the same continuum, if this new element corresponds to a manifold of impressions not sufficiently different, or, on the contrary, that the discrimination is possible; finally it may happen that two elements indistinguishable from a third, may, nevertheless, be distinguished one from the other.

That postulated, if A and B are two distinguishable elements of a continuum C, a series of elements may be found, $E_1, E_2, \cdots, E_n$, all belonging to this same continuum C and such that each of them is indistinguishable from the preceding, that E_1 is indistinguishable from A and E_n indistinguishable from B. Therefore we can go from A to B by a continuous route and without quitting C. If this condition is fulfilled for any two elements A and B of the continuum C, we may say that this continuum C is all in one piece. Now let us distinguish certain of the elements of C which may either be all distinguishable from one another, or themselves form one or several continua. The assemblage of the elements thus chosen arbitrarily among all those of C will form what I shall call the *cut* or the *cuts*.

Take on C any two elements A and B. Either we can also find a series of elements $E_1, E_2, \cdots, E_m$ such: (1) that they all belong to C; (2) that each of them is indistinguishable from the following, E_1 indistinguishable from A and E_n from B; (3) *and besides that none of the elements E is indistinguishable from any element of the cut.* Or else, on the contrary, in each of the series $E_1, E_2, \cdots, E_n$ satisfying the first two conditions, there will be an element E indistinguishable from one of the elements of the cut. In the first case we can go from *A* to *B* by a continuous route without quitting *C* and *without meeting the cuts;* in the second case that is impossible.

If then for any two elements *A* and *B* of the continuum *C,* it is always the first case which presents itself, we shall say that *C* remains all in one piece despite the cuts.

Thus, if we choose the cuts in a certain way, otherwise arbitrary, it may happen either that the continuum remains all in one piece or that it does not remain all in one piece; in this latter hypothesis we shall then say that it is *divided* by the cuts.

It will be noticed that all these definitions are constructed in setting out solely from this very simple fact, that two manifolds of impressions sometimes can be discriminated, sometimes cannot be. That postulated, if, to *divide* a continuum, it suffices to consider as cuts a certain number of elements all distinguishable from one another, we say that this continuum *is of one dimension;* if, on the contrary, to divide a continuum, it is necessary to consider as cuts a system of elements themselves forming one or several continua, we shall say that this continuum is *of several dimensions.*

If to divide a continuum *C,* cuts forming one or several continua of one dimension suffice, we shall say that *C* is a continuum *of two dimensions;* if cuts suffice which form one or several continua of two dimensions at most, we shall say that *C* is a continuum *of three dimensions;* and so on.

To justify this definition it is proper to see whether it is in this way that geometers introduce the notion of three dimensions at the beginning of their works. Now, what do we see? Usually they begin by defining surfaces as the boundaries of solids or pieces of space, lines as the boundaries of surfaces, points as the boundaries of lines, and they affirm that the same procedure cannot be pushed further.

This is just the idea given above: to divide space, cuts that are called surfaces are necessary; to divide surfaces, cuts that are called lines are necessary; to divide lines, cuts that are called points are necessary; we can go no further, the point cannot be divided, so the point is not a continuum. Then lines which can be divided by cuts which are not continua will be continua of one dimension; surfaces which can be divided by continuous cuts of one dimension will be continua of two dimensions; finally space which can be divided by continuous cuts of two dimensions will be a continuum of three dimensions.

Thus the definition I have just given does not differ essentially from the usual definitions; I have only endeavoured to give it a form applicable not to the mathematical continuum, but to the physical continuum, which alone is susceptible of representation, and yet to retain all its precision. Moreover, we see that this definition applies not alone to space; that in all which falls under our senses we find the characteristics of the physical continuum, which would allow of the same classification; that it would be easy to find there examples of continua of four, of five, dimensions, in the sense of the preceding definition; such examples occur of themselves to the mind.

I should explain finally, if I had the time, that this science, of which I spoke above and to which Riemann gave the name of analysis situs, teaches us to make distinctions among continua of the same number of dimensions and that the classification of these continua rests also on the consideration of cuts.

From this notion has arisen that of the mathematical continuum of several dimensions in the same way that the physical continuum of one dimension engendered the mathematical continuum of one dimension. The formula

$$A > C, A = B, B = C,$$

which summed up the data of crude experience, implied an intolerable contradiction. To get free from it it was necessary to introduce a new notion while still respecting the essential characteristics of the physical continuum of several dimensions. The mathematical continuum of one dimension admitted of a scale whose divisions, infinite in number, corresponded to the different values, commensurable or not, of one same magnitude. To have the mathematical continuum of n dimensions, it will suffice to take n like scales whose divisions correspond to different values of n independent magnitudes called co-ordinates. We thus shall have an image of the physical continuum of n dimensions, and this image will be as faithful as it can be after the determination not to allow the contradiction of which I spoke above.

4. THE NOTION OF POINT

It seems now that the question we put to ourselves at the start is answered. When we say that space has three dimensions, it will be said, we mean that the manifold of points of space satisfies the definition we have just given of the physical continuum of three dimensions. To be content with that would be to suppose that we know what is the manifold of points of space, or even one point of space.

Now that is not as simple as one might think. Everyone believes he knows what a point is, and it is just because we know it too well that we think there is no need of defining it. Surely we cannot be required to know how to define it, because in going back from definition to definition a time must come when we must stop. But at what moment should we stop?

We shall stop first when we reach an object which falls under our senses or that we can represent to ourselves; definition then will become useless; we do not define the sheep to a child; we say to him: *See* the sheep.

So, then, we should ask ourselves if it is possible to represent to ourselves a point of space. Those who answer yes do not reflect that they represent to themselves in reality a white spot made with the chalk on a blackboard or a black spot made with a pen on white paper, and that they can represent to themselves only an object or rather the impressions that this object made on their senses.

When they try to represent to themselves a point, they represent the impressions that very little objects made them feel. It is needless to add that two different objects, though both very little, may produce extremely different impressions, but I shall not dwell on this difficulty, which would still require some discussion.

But it is not a question of that; it does not suffice to represent *one* point, it is necessary to represent *a certain* point and to have the means of distinguishing it from an *other* point. And in fact, that we may be able to apply to a continuum the rule I have above expounded and by which one may recognize the number of its dimensions, we must rely upon the fact that two elements of this continuum sometimes can and sometimes cannot be distinguished. It is necessary therefore that we should in certain cases know how to represent to ourselves *a specific* element and to distinguish it from an *other* element.

The question is to know whether the point that I represented to myself an hour ago is the same as this that I now represent to myself, or whether it is a different point. In other words, how do we know whether the point occupied by the object A at the instant α is the same as the point occupied by the object B at the instant β, or still better, what this means?

I am seated in my room; an object is placed on my table; during a second I do not move, no one touches the object. I am tempted to say that the point A which this object occupied at the beginning of this second is identical with the point B which it occupies at its end. Not at all; from the point A to the point B is 30 kilometres, because the object has been carried along in the motion of the earth. We cannot know whether an object, be it large or small, has not changed its absolute position in space, and not only can we not affirm it, but this affirmation has no meaning and in any case cannot correspond to any representation.

But then we may ask ourselves if the relative position of an object with regard to other objects has changed or not, and first whether the relative position of this object with regard to our body has changed. If the impressions this object makes upon us have not changed, we shall be inclined to judge that neither has this relative position changed; if they have changed, we shall judge that this object has changed either in state or in relative position. It remains to decide which of the *two*. I have explained in *Science and Hypothesis* how we have been led to distinguish the changes of position. Moreover, I shall return to that further on. We come to know, therefore, whether the relative position of an object with regard to our body has or has not remained the same.

If now we see that two objects have retained their relative position with regard to our body, we conclude that the relative position of these two objects with regard to one another has not changed; but we reach this conclusion only by indirect reasoning. The only thing that we know directly is the relative position of the objects with regard to our body. *A fortiori* it is only by indirect reasoning that we think we know (and, moreover, this belief is delusive) whether the absolute position of the object has changed.

In a word, the system of co-ordinate axes to which we naturally refer all exterior objects is a system of axes invariably bound to our body, and carried around with us.

It is impossible to represent to oneself absolute space; when I try to represent to myself simultaneously objects and myself in motion in absolute space, in reality I represent to myself my own self motionless and seeing move around me different objects and a man that is exterior to me, but that I convene to call me.

Will the difficulty be solved if we agree to refer everything to these axes bound to our body? Shall we know then what is a point thus defined by its relative position with regard to ourselves? Many persons will answer yes and will say that they "localize" exterior objects.

What does this mean? To localize an object simply means to represent to oneself the movements that would be necessary to reach it. I will explain myself. It is not a question of representing the movements themselves in space, but solely of representing to oneself the muscular sensations which accompany these movements and which do not presuppose the preexistence of the notion of space.

If we suppose two different objects which successively occupy the same relative position with regard to ourselves, the impressions that these two objects make upon us will be very different; if we localize them at the same point, this is simply because it is necessary to make the same movements to reach them; apart from that, one cannot just see what they could have in common.

But, given an object, we can conceive many different series of movements which equally enable us to reach it. If then we represent to ourselves a point by representing to ourselves the series of muscular sensations which accompany the movements which enable us to reach this point, there will be many ways entirely different of representing to oneself the same point. If one is not satisfied with this solution, but wishes, for instance, to bring in the visual sensations along with the muscular sensations, there will be one or two more ways of representing to oneself this same point and the difficulty will only be increased. In any case the following question comes up: Why do we think that all these representations so different from one another still represent the same point?

Another remark: I have just said that it is to our own body that we naturally refer exterior objects; that we carry about everywhere with us a system of axes to which we refer all the points of space, and that this system of axes seems to be invariably bound to our body. It should be noticed that rigorously we could not speak of axes invariably bound to the body unless the different parts of this body were themselves invariably bound to one another. As this is not the case, we ought, before referring exterior objects to these fictitious axes, to suppose our body brought back to the initial attitude.

5. The Notion of Displacement

I have shown in *Science and Hypothesis* the preponderant role played by the movements of our body in the genesis of the notion of space. For a being completely immovable there would be neither space nor geometry; in vain would exterior objects be displaced about him, the variations which these displacements would make in his impressions would not be attributed by this being to changes of position, but to simple changes of state; this being would have no means of distinguishing these two sorts of changes, and this distinction, fundamental for us, would have no meaning for him.

The movements that we impress upon our members have as effect the varying of the impressions produced on our senses by external objects; other causes may likewise make them vary; but we are led to distinguish the changes produced by our own motions and we easily discriminate them for two reasons: (1) because they are voluntary; (2) because they are accompanied by muscular sensations.

So we naturally divide the changes that our impressions may undergo into two categories to which perhaps I have given an inappropriate designation: (1) the internal changes, which are voluntary and accompanied by muscular sensations; (2) the external changes, having the opposite characteristics.

We then observe that among the external changes are some which can be corrected, thanks to an internal change which brings everything back to the primitive state; others cannot be corrected in this way (it is thus that when an exterior object is displaced, we may then by changing our own position replace ourselves as regards this object in the same relative position as before, so as to reestablish the original aggregate of impressions; if this object was not displaced, but changed its state, that is impossible). Thence comes a new distinction among external changes: those which may be so corrected we call changes of position; and the others, changes of state.

Think, for example, of a sphere with one hemisphere blue and the other red; it first presents to us the blue hemisphere, then it so revolves as to present the red hemisphere. Now think of a spherical vase containing a blue liquid which becomes red in consequence of a chemical reaction. In both cases the sensation of red has replaced that of blue; our senses have experienced the same impressions which have succeeded each other in the same order, and yet these two changes are regarded by us as very different; the first is a displacement, the second a change of state. Why? Because in the first case it is sufficient for me to go around the sphere to place myself opposite the blue hemisphere and reestablish the original blue sensation.

Still more; if the two hemispheres, in place of being red and blue, had been yellow and green, how should I have interpreted the revolution of the sphere? Before, the red succeeded the blue, now the green succeeds the yellow; and yet I say that the two spheres have undergone the same revolution, that each has turned about its axis; yet I cannot say that the green is to yellow as the red is to blue;

how then am I led to decide that the two spheres have undergone the *same* displacement? Evidently because, in one case as in the other, I am able to re-establish the original sensation by going around the sphere, by making the same movements, and I know that I have made the same movements because I have felt the same muscular sensations; to know it, I do not need, therefore, to know geometry in advance and to represent to myself the movements of my body in geometric space.

Another example: an object is displaced before my eye; its image was first formed at the center of the retina; then it is formed at the border; the old sensation was carried to me by a nerve fiber ending at the center of the retina; the new sensation is carried to me by *another* nerve fiber starting from the border of the retina; these two sensations are qualitatively different; otherwise, how could I distinguish them?

Why then am I led to decide that these two sensations, qualitatively different, represent the same image, which has been displaced? It is because I *can follow the object with the eye* and by a displacement of the eye, voluntary and accompanied by muscular sensations, bring back the image to the center of the retina and re-establish the primitive sensation.

I suppose that the image of a red object has gone from the center *A* to the border *B* of the retina, then that the image of a blue object goes in its turn from the center *A* to the border *B* of the retina; I shall decide that these two objects have undergone the *same* displacement. Why? Because in both cases I shall have been able to re-establish the primitive sensation, and that to do it I shall have had to execute the *same* movement of the eye, and I shall know that my eye has executed the same movement because I shall have felt the *same* muscular sensations.

If I could not move my eye, should I have any reason to suppose that the sensation of red at the center of the retina is to the sensation of red at the border of the retina as that of blue at the center is to that of blue at the border? I should only have four sensations qualitatively different, and if I were asked if they are connected by the proportion I have just stated, the question would seem to me ridiculous, just as if I were asked if there is an analogous proportion between an auditory sensation, a tactile sensation and an olfactory sensation.

Let us now consider the internal changes, that is, those which are produced by the voluntary movements of our body and which are accompanied by muscular changes. They give rise to the two following observations, analogous to those we have just made on the subject of external changes.

1. I may suppose that my body has moved from one point to another but that the same *attitude* is retained; all the parts of the body have therefore retained or resumed the same *relative* situation, although their absolute situation in space may have varied. I may suppose that not only has the position of my body changed, but that its attitude is no longer the same, that, for instance, my arms which before were folded are now stretched out.

I should therefore distinguish the simple changes of position without change of attitude, and the changes of attitude. Both would appear to me under form of muscular sensations. How then am I led to distinguish them? It is that the first may serve to correct an external change, and that the others cannot, or at least can only give an imperfect correction.

This fact I proceed to explain as I would explain it to someone who already knew geometry, but it need not thence be concluded that it is necessary already to know geometry to make this distinction; before knowing geometry I ascertain the fact (experimentally, so to speak), without being able to explain it. But merely to make the distinction between the two kinds of change, I do not need to *explain* the fact, it suffices me *to ascertain* it.

However that may be, the explanation is easy. Suppose that an exterior object is displaced; if we wish the different parts of our body to resume with regard to this object their initial relative position, it is necessary that these different parts should have resumed likewise their initial relative position with regard to one another. Only the internal changes which satisfy this latter condition will be capable of correcting the external change produced by the displacement of that object. If, therefore, the relative position of my eye with regard to my finger has changed, I shall still be able to replace the eye in its initial relative situation with regard to the object and re-establish thus the primitive visual sensations, but then the relative position of the finger with regard to the object will have changed and the tactile sensations will not be re-established.

2. We ascertain likewise that the same external change may be corrected by two internal changes corresponding to different muscular sensations. Here again I can ascertain this without knowing geometry: and I have no need of anything else; but I proceed to give the explanation of the fact employing geometrical language. To go from the position A to the position B I may take several routes. To the first of these routes will correspond a series S of muscular sensations; to a second route will correspond another series S'' of muscular sensations which generally will be completely different, since other muscles will be used.

How am I led to regard these two series S and S'' as corresponding to the same displacement AB? It is because these two series are capable of correcting the same external change. Apart from that, they have nothing in common.

Let us now consider two external changes: α and β, which shall be, for instance, the rotation of a sphere half blue, half red, and that of a sphere half yellow, half green; these two changes have nothing in common, since the one is for us the passing of blue into red and the other the passing of yellow into green. Consider, on the other hand, two series of internal changes S and S''; like the others, they will have nothing in common. And yet I say that α and β correspond to the same displacement, and that S and S'' correspond also to the same displacement. Why? Simply because S can correct β as well as α and because α can be corrected by S'' as well as by S. And then a question suggests itself: If I have ascertained that S corrects α and β and that S'' corrects α, am I certain that S'' likewise corrects β? Experiment alone can teach us whether this law is verified. If it were not verified, at least approximately, there would be no geometry, there would be no space, because we should have no more interest in classifying the internal and external changes as I have just done, and, for instance, in distinguishing changes of state from changes of position.

It is interesting to see what has been the role of experience in all this. It has shown me that a certain law is approximately verified. It has not told me *wherefore* space is, and that it satisfies the condition in question. I knew in fact, before all experience, that space satisfied this condition or that it would not be; nor have I any right to say that experience told me that geometry is possible; I very well see that geometry is possible, since it does not imply contradiction; experience only tells me that geometry is useful.

6. VISUAL SPACE

Although motor impressions have had, as I have just explained, an altogether preponderant influence in the genesis of the notion of space, which never would have taken birth without them, it will not be without interest to examine also the role of visual impressions and to investigate how many dimensions "visual space" has, and for that purpose to apply to these impressions the definition of § 3.

A first difficulty presents itself: consider a red color sensation affecting a certain point of the retina; and on the other hand a blue color sensation affecting the same point of the retina. It is necessary that we have some means of recognizing that these two sensations, qualitatively different, have something in common. Now, according to the considerations expounded in the preceding paragraph, we have been able to recognize this only by the movements of the eye and the observations to which they have given rise. If the eye were immovable, or if we were unconscious of its movements, we should not have been able to recognize that these two sensations, of different quality, had something in common; we should not have been able to disengage from them what gives them a geometric character. The visual sensations, without the muscular sensations, would have nothing geometric, so that it may be said there is no pure visual space.

To do away with this difficulty, consider only sensations of the same nature, red sensations for instance, differing one from another only as regards the point of the retina that they affect. It is clear that I have no reason for making such an arbitrary choice among all the possible visual sensations, for the purpose of uniting in the same class all the sensations of the same color, whatever may be the point of the retina affected. I should never have dreamt of it, had I not before learned, by the means we have just seen, to distinguish changes of state from changes of position, that is, if my eye were immovable. Two sensations of the same color affecting two different parts of the retina would have appeared to me as qualitatively distinct, just as two sensations of different color.

In restricting myself to red sensations, I therefore impose upon myself an artificial limitation and I neglect systematically one whole side of the question; but it is only by this artifice that I am able to analyse visual space without mingling any motor sensation.

Imagine a line traced on the retina and dividing in two its surface; and set apart the red sensations affecting a point of this line, or those differing from them too little to be distinguished from them. The aggregate of these sensations will form a sort of cut that I shall call C, and it is clear that this cut suffices to divide the manifold of possible red sensations, and that if I take two red sensations affecting two points situated on one side and the other of the line, I cannot pass from one of these sensations to the other in a continuous way without passing at a certain moment through a sensation belonging to the cut.

If, therefore, the cut has n dimensions, the total manifold of my red sensations, or, if you wish, the whole visual space, will have $n+1$.

Now, I distinguish the red sensations affecting a point of the cut C. The assemblage of these sensations will form a new cut C'. It is clear that this *will divide* the cut C, always giving to the word divide the same meaning.

If, therefore, the cut C' has n dimensions, the cut C will have $n+1$ and the whole of visual space $n+2$.

If all the red sensations affecting the same point of the retina were regarded as identical, the cut C' reducing to a single element would have 0 dimension, and visual space would have 2.

And yet most often it is said that the eye gives us the sense of a third dimension, and enables us in a certain measure to recognize the distance of objects. When we seek to analyse this feeling, we ascertain that it reduces either to the consciousness of the convergence of the eyes, or to that of the effort of accommodation which the ciliary muscle makes to focus the image.

Two red sensations affecting the same point of the retina will therefore be regarded as identical only if they are accompanied by the same sensation of convergence and also by the same sensation of effort of accommodation or at least by sensations of convergence and accommodation so slightly different as to be indistinguishable.

On this account the cut C' is itself a continuum and the cut C has more than one dimension.

But it happens precisely that experience teaches us that when two visual sensations are accompanied by the same sensation of convergence, they are likewise accompanied by the same sensation of accommodation. If then we form a new cut C'' with all those of the sensations of the cut C', which are accompanied by a certain sensation of convergence, in accordance with the preceding law they will all be indistinguishable and may be regarded as identical. Therefore C'' will not be a continuum and will have zero dimension; and as C'' divides C' it will thence result that C' has one, C two and *the whole visual space three dimensions.*

But would it be the same if experience had taught us the contrary and if a certain sensation of convergence were not always accompanied by the same sensation of accommodation? In this case two sensations affecting the same point of the retina and accompanied by the same sense of convergence, two sensations which consequently would both appertain to the cut C'' could nevertheless be distinguished since they would be accompanied by two different sensations of accommodation. Therefore C'' would be in its turn a continuum and would have one dimension (at least); then C' would have two, *C* three and *the whole visual space would have four dimensions.*

Will it then be said that it is experience which teaches us that space has three dimensions, since it is in setting out from an experimental law that we have come to attribute three to it? But we have therein performed, so to speak, only an experiment in physiology; and as also it would suffice to fit over the eyes glasses of suitable construction to put an end to the accord between the feelings of convergence and of accommodation, are we to say that putting on spectacles is enough to make space have four dimensions and that the optician who constructed them has given one more dimension to space? Evidently not; all we can say is that experience has taught us that it is convenient to attribute three dimensions to space.

But visual space is only one part of space, and in even the notion of this space there is something artificial, as I have explained at the beginning. The real space is motor space and this it is that we shall examine in the following chapter.

CHAPTER IV

SPACE AND ITS THREE
DIMENSIONS

I. THE GROUP OF DISPLACEMENTS

Let us sum up briefly the results obtained. We proposed to investigate what was meant in saying that space has three dimensions and we have asked first what is a physical continuum and when it may be said to have *n* dimensions. If we consider different systems of impressions and compare them with one another, we often recognize that two of these systems of impressions are indistinguishable (which is ordinarily expressed in saying that they are too close to one another, and that our senses are too crude, for us to distinguish them) and we ascertain besides that two of these systems can sometimes be discriminated from one another though indistinguishable from a third system. In that case we say the manifold of these systems of impressions forms a physical continuum *C*. And each of these systems is called an *element* of the continuum *C*.

How many dimensions has this continuum? Take first two elements *A* and *B* of *C*, and suppose there exists a series Σ of elements, all belonging to the continuum *C*, of such a sort that *A* and *B* are the two extreme terms of this series and that each term of the series is indistinguishable from the preceding. If such a series Σ can be

246 · *The Value of Science*

found, we say that A and B are joined to one another; and if any two elements of C are joined to one another, we say that C is all of one piece.

Now take on the continuum C a certain number of elements in a way altogether arbitrary. The aggregate of these elements will be called a *cut*. Among the various series Σ which join A to B, we shall distinguish those of which an element is indistinguishable from one of the elements of the cut (we shall say that these are they which *cut* the cut) and those of which *all* the elements are distinguishable from all those of the cut. If *all* the series Σ which join A to B cut the cut, we shall say that A and B are *separated* by the cut, and that the cut *divides* C. If we cannot find on C two elements which are separated by the cut, we shall say that the cut *does not divide C*.

These definitions laid down, if the continuum C can be divided by cuts which do not themselves form a continuum, this continuum C has only one dimension; in the contrary case it has several. If a cut forming a continuum of 1 dimension suffices to divide C, C will have 2 dimensions; if a cut forming a continuum of 2 dimensions suffices, C will have 3 dimensions, etc. Thanks to these definitions, we can always recognize how many dimensions any physical continuum has. It only remains to find a physical continuum which is, so to speak, equivalent to space, of such a sort that to every point of space corresponds an element of this continuum, and that to points of space very near one another correspond indistinguishable elements. Space will have then as many dimensions as this continuum.

The intermediation of this physical continuum, capable of representation, is indispensable; because we cannot represent space to ourselves, and that for a multitude of reasons. Space is a mathematical continuum, it is infinite, and we can represent to ourselves only physical continua and finite objects. The different elements of space, which we call points, are all alike, and, to apply our definition, it is necessary that we know how to distinguish the elements from one another, at least if they are not too close. Finally absolute space is nonsense, and it is necessary for us to begin by referring space to a system of axes invariably bound to our body (which we must always suppose put back in the initial attitude).

Then I have sought to form with our visual sensations a physical continuum equivalent to space; that certainly is easy and this example is particularly appropriate for the discussion of the number of dimensions; this discussion has enabled us to see in what measure it is allowable to say that "visual space" has three dimensions. Only this solution is incomplete and artificial. I have explained why, and it is not on visual space, but on motor space that it is necessary to bring our efforts to bear. I have then recalled what is the origin of the distinction we make between changes of position and changes of state. Among the changes which occur in our impressions, we distinguish, first the *internal* changes, voluntary and accompanied by muscular sensations, and the *external* changes, having opposite characteristics. We ascertain that it may happen that an external change may be *corrected* by an internal change which reestablishes the primitive sensations. The external changes capable of being corrected by an internal change are called *changes of position*, those not capable of it are called *changes of state*. The internal changes capable of correcting an external change are called *displacements of the whole body*; the others are called *changes of attitude*.

Now let α and β be two external changes, α' and β' two internal changes. Suppose that α may be corrected either by α' or by β', and that α' can correct either α or β; experience tells us then that β' can likewise correct β. In this case we say that α and β correspond to the *same* displacement and also that α' and β' correspond to the *same* displacement. That postulated, we can imagine a physical continuum which we shall call *the continuum or group of displacements* and which we shall define in the following manner. The elements of this continuum shall be the internal changes capable of correcting an external change. Two of these internal changes α' and β' shall be regarded as indistinguishable: (1) if they are so naturally, that is, if they are too close to one another; (2) if α' is capable of correcting the same external change as a third internal change naturally indistinguishable from β'. In this second case, they will be, so to speak, indistinguishable by convention, I mean by agreeing to disregard circumstances which might distinguish them.

Our continuum is now entirely defined, since we know its elements and have fixed under what conditions they may be regarded as indistinguishable. We thus have all that is necessary to apply our definition and determine how many dimensions this continuum has. We shall recognize that it has *six*. The continuum of displacements is, therefore, not equivalent to space, since the number of dimensions is not the same; it is only related to space. Now how do we know that this continuum of displacements has six dimensions? We know it *by experience*.

It would be easy to describe the experiments by which we could arrive at this result. It would be seen that in this continuum cuts can be made which divide it and which are continua; that these cuts themselves can be divided by other cuts of the second order which yet are continua, and that this would stop only after cuts of the sixth order which would no longer be continua. From our definitions that would mean that the group of displacements has six dimensions.

That would be easy, I have said, but that would be rather long; and would it not be a little superficial? This group of displacements, we have seen, is related to space, and space could be deduced from it, but it is not equivalent to space, since it has not the same number of dimensions; and when we shall have shown how the notion of this continuum can be formed and how that of space may be deduced from it, it might always be asked why space of three dimensions is much more familiar to us than this continuum of six dimensions, and consequently doubted whether it was by this detour that the notion of space was formed in the human mind.

II. IDENTITY OF TWO POINTS

What is a point? How do we know whether two points of space are identical or different? Or, in other words, when I say: the object *A* occupied at the instant α the point which the object *B* occupies at the instant β, what does that mean?

Such is the problem we set ourselves in the preceding chapter, §4. As I have explained it, it is not a question of comparing the positions of the objects *A* and *B* in absolute space; the question then would manifestly have no meaning. It is a question of comparing the positions of these two objects with regard to axes invariably bound to my body, supposing always this body replaced in the same attitude.

I suppose that between the instants α and β I have moved neither my body nor my eye, as I know from my muscular sense. Nor have I moved either my head, my arm or my hand. I ascertain that at the instant α impressions that I attributed to the object *A* were transmitted to me, some by one of the fibers of my optic nerve, the others by one of the sensitive tactile nerves of my finger; I ascertain that at the instant β other impressions which I attribute to the object *B* are transmitted to me, some by this same fiber of the optic nerve, the others by this same tactile nerve.

Here I must pause for an explanation; how am I told that this impression which I attribute to *A*, and that which I attribute to *B*, impressions which are qualitatively different, are transmitted to me by the same nerve? Must we suppose, to take for example the visual sensations, that *A* produces two simultaneous sensations, a sensation purely luminous α and a colored sensation α', that *B* produces in the same way simultaneously a luminous sensation *b* and a colored sensation *b'*, that if these different sensations are transmitted to me by the same retinal fiber, *a* is identical with *b*, but that in general the colored sensations *a'* and *b'* produced by different bodies are different? In that case it would be the identity of the sensation *a* which accompanies *a'* with the sensation *b* which accompanies *b'*, which would tell that all these sensations are transmitted to me by the same fiber.

However it may be with this hypothesis and although I am led to prefer it to others considerably more complicated, it is certain that we are told in some way that there is something in common between these sensations $a + a'$ and $b + b'$, without which we should have no means of recognizing that the object *B* has taken the place of the object *A*.

Therefore I do not further insist and I recall the hypothesis I have just made: I suppose that I have ascertained that the impressions which I attribute to B are transmitted to me at the instant β by the same fibers, optic as well as tactile, which, at the instant α, had transmitted to me the impressions that I attributed to A. If it is so, we shall not hesitate to declare that the point occupied by B at the instant β is identical with the point occupied by A at the instant α.

I have just enunciated two conditions for these points being identical; one is relative to sight, the other to touch. Let us consider them separately. The first is necessary, but is not sufficient. The second is at once necessary and sufficient. A person knowing geometry could easily explain this in the following manner: Let O be the point of the retina where is formed at the instant α the image of the body A; let M be the point of space occupied at the instant α by this body A; let M' be the point of space occupied at the instant β by the body B. For this body B to form its image in O, it is not necessary that the points M and M' coincide; since vision acts at a distance, it suffices for the thee points $O\,M\,M'$ to be in a straight line. This condition that the two objects form their image on O is therefore necessary, but not sufficient for the points M and M' to coincide. Let now P be the point occupied by my finger and where it remains, since it does not budge. As touch does not act at a distance, if the body A touches my finger at the instant α, it is because M and P coincide; if B touches my finger at the instant β, it is because M' and P coincide. Therefore M and M' coincide. Thus this condition that if A touches my finger at the instant α, B touches it at the instant β, is at once necesarry and sufficient for M and M' to coincide.

But we who, as yet, do not know geometry cannot reason thus; all that we can do is to ascertain experimentally that the first condition relative to sight may be fulfilled without the second, which is relative to touch, but that the second cannot be fulfilled without the first.

Suppose experience had taught us the contrary, as might well be; this hypothesis contains nothing absurd. Suppose, therefore, that we had ascertained experimentally that the condition relative to touch may be fulfilled without that of sight being fulfilled, and that, on the contrary, that of sight cannot be fulfilled without that of touch being also. It is clear that if this were so we should conclude that it is touch which may be exercised at a distance, and that sight does not operate at a distance.

But this is not all; up to this time I have supposed that to determine the place of an object, I have made use only of my eye and a single finger; but I could just as well have employed other means, for example, all my other fingers.

I suppose that my first finger receives at the instant α a tactile impression which I attribute to the object A. I make a series of movements, corresponding to a series S of muscular sensations. After these movements, at the instant α, my *second* finger receives a tactile impression that I attribute likewise to A. Afterwards, at the instant β, without my having budged, as my muscular sense tells me, this same second finger transmits to me anew a tactile impression which I attribute this time to the object B; I then make a series of movements, corresponding to a series S' of muscular sensations. I know that this series S' is the inverse of the series S and corresponds to contrary movements. I know this because many previous experiences have shown me that if I made successively the two series of movements corresponding to S and to S', the primitive impressions would be re-established, in other words, that the two series mutually compensate. That settled, should I expect that at the instant β', when the second series of movements is ended, my *first finger* would feel a tactile impression attributable to the object B?

To answer this question, those already knowing geometry would reason as follows: there are chances that the object A has not budged, between the instants α and α', nor the object B between the instants β and β'; assume this. At the instant α, the object A occupied a certain point M of space. Now at this instant it touched my first finger, and *as touch does not operate at a distance,* my first finger was likewise at the point M. I afterwards made the series S of movements and at the end of this series, at the instant α', I ascertained that the object A touched my second finger. I thence conclude that this second finger was then at M, that is, that the movements S had the result of bringing the second finger to the place of the first. At the instant β the object B has come in contact with my second finger: as I have not budged, this second finger has remained at M;

therefore the object B has come to M; by hypothesis it does not budge up to the instant β'. But between the instants β and β' I have made the movements S'; as these movements are the inverse of the movements S, they must have for effect bringing the first finger in the place of the second. At the instant β' this first finger will, therefore, be at M; and as the object B, is likewise at M, this object B will touch my first finger. To the question put, the answer should, therefore, be yes.

We who do not yet know geometry cannot reason thus; but we ascertain that this anticipation is ordinarily realized; and we can always explain the exceptions by saying that the object A has moved between the instants α and α', or the object B between the instants β and β'.

But could not experience have given a contrary result? Would this contrary result have been absurd in itself? Evidently not. What should we have done then if experience had given this contrary result? Would all geometry thus have become impossible? Not the least in the world. We should have contented ourselves with concluding that *touch can operate at a distance.*

When I say, touch does not operate at a distance, but sight operates at a distance, this assertion has only one meaning, which is as follows: to recognize whether B occupies at the instant β the point occupied by A at the instant α, I can use a multitude of different criteria. In one my eye intervenes, in another my first finger, in another my second finger, etc. Well, it is sufficient for the criterion relative to one of my fingers to be satisfied in order that all the others should be satisfied, but it is not sufficient that the criterion relative to the eye should be. This is the sense of my assertion, I content myself with affirming an experimental fact which is ordinarily verified.

At the end of the preceding chapter we analysed visual space; we saw that to engender this space it is necessary to bring in the retinal sensations, the sensation of convergence and the sensation of accommodation; that if these last two were not always in accord, visual space would have four dimensions in place of three; we also saw that if we brought in only the retinal sensations, we should obtain "simple visual space," of only two dimensions. On the other hand, consider tactile space, limiting ourselves to the sensations of a single finger, that is in sum the assemblage of positions this finger can occupy. This tactile space that we shall analyse in the following section and which consequently I ask permission not to consider further for the moment, this tactile space, I say, has three dimensions. Why has space properly so called as many dimensions as tactile space and more than simple visual space? It is because touch does not operate at a distance, while vision does operate at a distance. These two assertions have the same meaning and we have just seen what this is.

Now I return to a point over which I passed rapidly in order not to interrupt the discussion. How do we know that the impressions made on our retina by A at the instant α and B at the instant β are transmitted by the same retinal fibre, although these impressions are qualitatively different? I have suggested a simple hypothesis, while adding that other hypotheses, decidedly more complex, would seem to me more probably true. Here then are these hypotheses, of which I have already said a word. How do we know that the impressions produced by the red object A at the instant α, and by the blue object B at the instant β, if these two objects have been imaged on the same point of the retina, have something in common? The simple hypothesis above made may be rejected and we may suppose that these two impressions, qualitatively different, are transmitted by two different though contiguous nervous fibres. What means have I then of knowing that these fibres are contiguous? It is probable that we should have none, if the eye were immovable. It is the movements of the eye which have told us that there is the same relation between the sensation of blue at the point A and the sensation of blue at the point B of the retina as between the sensation of red at the point A and the sensation of red at the point B. They have shown us, in fact, that the same movements, corresponding to the same muscular sensations, carry us from the first to the second, or from the third to the fourth. I do not emphasize these considerations, which belong, as one sees, to the question of local signs raised by Lotze.

III. Tactile Space

Thus I know how to recognize the identity of two points, the point occupied by A at the instant α and the point occupied by B at the instant β, but only *on one condition*, namely, that I have not budged between the instants α and β. That does not suffice for our object. Suppose, therefore, that I have moved in any manner in the interval between these two instants, how shall I know whether the point occupied by A at the instant α is identical with the point occupied by B at the instant β? I suppose that at the instant α, the object A was in contact with my first finger and that in the same way, at the instant β, the object B touches this first finger; but at the same time, my muscular sense has told me that in the interval my body has moved. I have considered above two series of muscular sensations S and S', and I have said it sometimes happens that we are led to consider two such series S and S' as inverse one of the other, because we have often observed that when these two series succeed one another our primitive impressions are re-established.

If then my muscular sense tells me that I have moved between the two instants α and β, but so as to feel successively the two series of muscular sensations S and S' that I consider inverses, I shall still conclude, just as if I had not budged, that the points occupied by A at the instant α and by B at the instant β are identical, if I ascertain that my first finger touches A at the instant α and B at the instant β.

This solution is not yet completely satisfactory, as one will see. Let us see, in fact, how many dimensions it would make us attribute to space. I wish to compare the two points occupied by A and B at the instants α and β, or (what amounts to the same thing since I suppose that my finger touches A at the instant α and B at the instant β) I wish to compare the two points occupied by my finger at the two instants α and β. The sole means I use for this comparison is the series Σ of muscular sensations which have accompanied the movements of my body between these two instants. The different imaginable series Σ form evidently a physical continuum of which the number of dimensions is very great. Let us agree, as I have done, not to consider as distinct the two series Σ and $\Sigma + S + S'$, when S and S' are inverses one of the other in the sense above given to this word; in spite of this agreement, the aggregate of distinct series Σ will still form a physical continuum and the number of dimensions will be less but still very great.

To each of these series Σ corresponds a point of space; to two series Σ and Σ' thus correspond two points M and M'. The means we have hitherto used enable us to recognize that M and M' are not distinct in two cases: (1) if Σ is identical with Σ'; (2) if $\Sigma' = \Sigma + S + S'$, S and S' being inverses one of the other. If in all the other cases we should regard M and M' as distinct, the manifold of points would have as many dimensions as the aggregate of distinct series Σ, that is, much more than three.

For those who already know geometry, the following explanation would be easily comprehensible. Among the imaginable series of muscular sensations, there are those which correspond to series of movements where the finger does not budge. I say that if one does not consider as distinct the series Σ and $\Sigma + \sigma$, where the series σ corresponds to movements where the finger does not budge, the aggregate of series will constitute a continuum of three dimensions, but that if one regards as distinct two series Σ and Σ' unless $\Sigma' = \Sigma + S + S'$, S and S' being inverses, the aggregate of series will constitute a continuum of more than three dimensions.

In fact, let there be in space a surface A, on this surface a line B, on this line a point M. Let C_0 be the aggregate of all series Σ. Let C_1 be the aggregate of all the series Σ, such that at the end of corresponding movements the finger is found upon the surface A, and C_2 or C_3 the aggregate of series Σ such that at the end the finger is found on B, or at M. It is clear, first, that C_1 will constitute a cut which will divide C_0, that C_2 will be a cut which will divide C_1, and C_3 a cut which will divide C_2. Thence it results, in accordance with our definitions, that if C_3 is a continuum of n dimensions, C_0 will be a physical continuum of $n + 3$ dimensions.

Therefore, let Σ and $\Sigma' = \Sigma + \sigma$ be two series forming part of

C_3; for both, at the end of the movements, the finger is found at M; thence results that at the beginning and at the end of the series σ, the finger is at the same point M. This series σ is therefore one of those which correspond to movements where the finger does not budge. If Σ and $\Sigma + \sigma$ are not regarded as distinct, all the series of C_3 blend into one; therefore C_3 will have 0 dimension, and C_0 will have 3, as I wished to prove. If, on the contrary, I do not regard Σ and $\Sigma + \sigma$ as blending (unless $\sigma = S + S'$, S and S' being inverses), it is clear that C_3 will contain a great number of series of distinct sensations; because, without the finger budging, the body may take a multitude of different attitudes. Then C_3 will form a continuum and C_0 will have more than three dimensions, and this also I wished to prove.

We who do not yet know geometry cannot reason in this way; we can only verify. But then a question arises; how, before knowing geometry, have we been led to distinguish from the others these series σ where the finger does not budge? It is, in fact, only after having made this distinction that we could be led to regard Σ and $\Sigma + \sigma$ as identical, and it is on this condition alone, as we have just seen, that we can arrive at space of three dimensions.

We are led to distinguish the series σ, because it often happens that when we have executed the movements which correspond to these series σ of muscular sensations, the tactile sensations which are transmitted to us by the nerve of the finger that we have called the first finger, persist and are not altered by these movements. Experience alone tells us that and it alone could tell us.

If we have distinguished the series of muscular sensations $S + S'$ formed by the union of two inverse series, it is because they preserve the totallity of our impressions; if now we distinguish the series σ, it is because they preserve *certain* of our impressions. (When I say that a series of muscular sensations S "preserves" one of our impressions A, I mean that we ascertain that if we feel the impression A, then the muscular sensations S, we *still* feel the impression A *after* these sensations S.)

I have said above it often happens that the series σ do not alter the tactile impressions felt by our first finger; I said *often*, I did not say *always*. This it is that we express in our ordinary language by saying that the tactile impressions would not be altered if the finger has not moved, *on the condition* that *neither has* the object *A*, which was in contact with this finger, moved. Before knowing geometry, we could not give this explanation; all we could do is to ascertain that the impression often persists, but not always.

But that the impression often continues is enough to make the series σ appear remarkable to us, to lead us to put in the same class the series Σ and Σ + σ, and hence not regard them as distinct. Under these conditions we have seen that they will engender a physical continuum of three dimensions.

Behold then a space of three dimensions engendered by my first finger. Each of my fingers will create one like it. It remains to consider how we are led to regard them as identical with visual space, as identical with geometric space.

But one reflection before going further; according to the foregoing, we know the points of space, or more generally the final situation of our body, only by the series of muscular sensations revealing to us the movements which have carried us from a certain initial situation to this final situation. But it is clear that this final situation will depend, on the one hand, upon these movements and, *on the other hand, upon the initial situation* from which we set out. Now these movements are revealed to us by our muscular sensations; but nothing tells us the initial situation; nothing can distinguish it for us from all the other possible situations. This puts well in evidence the essential relativity of space.

IV. Identity of the Different Spaces

We are therefore led to compare the two continua *C* and *C'* engendered, for instance, one by my first finger *D*, the other by my second finger *D'*. These two physical continua both have three dimensions. To each element of the continuum *C*, or, if you prefer, to each point of the first tactile space, corresponds a series of muscular sensations Σ, which carry me from a certain initial situation to a certain final situation.* Moreover, the same point of this first space will correspond to Σ and to Σ +σ, if σ is a series of which we know that it does not make the finger D move.

Similarly to each element of the continuum C', or to each point of the second tactile space, corresponds a series of sensations Σ', and the same point will correspond to Σ' and to Σ'+σ', if σ' is a series which does not make the finger D' move.

What makes us distinguish the various series designated σ from those called σ' is that the first do not alter the tactile impressions felt by the finger D and the second preserve those the finger D' feels.

Now see what we ascertain: in the beginning my finger D' feels a sensation A'; I make movements which produce muscular sensations S; my finger D feels the impression A; I make movements which produce a series of sensations σ; my finger D continues to feel the impression A, since this is the characteristic property of the series σ; I then make movements which produce the series S' of muscular sensations, *inverse* to S in the sense above given to this word. I ascertain then that my finger D' feels anew the impression A'. (It is of course understood that S has been suitably chosen.)

This means that the series S +σ+ S', preserving the tactile impressions of the finger D', is one of the series I have called σ'. Inversely, if one takes any series, σ', S' +σ' +S will be one of the series that we call σ.

Thus if S is suitably chosen, S +σ+ S' will be a series σ', and by making σ vary in all possible ways, we shall obtain all the possible series σ'.

Not yet knowing geometry, we limit ourselves to verifying all that, but here is how those who know geometry would explain the fact. In the beginning my finger D' is at the point M, in contact with

---

* In place of saying that we refer space to axes rigidly bound to our body, perhaps it would be better to say, in conformity to what precedes, that we refer it to axes rigidly bound to the initial situation of our body.

the object *a,* which makes it feel the impression A'. I make the movements corresponding to the series $S;$ I have said that this series should be suitably chosen, I should so make this choice that these movements carry the finger D to the point originally occupied by the finger D', that is, to the point $M;$ this finger D will thus be in contact with the object *a,* which will make it feel the impression A.

I then make the movements corresponding to the series σ; in these movements, by hypothesis, the position of the finger D does not change, this finger therefore remains in contact with the object *a* and continues to feel the impression A. Finally I make the movements corresponding to the series S'. As S' is inverse to S, these movements carry the finger D' to the point previously occupied by the finger $D,$ that is, to the point M. If, as may be supposed, the object *a* has not budged, this finger D' will be in contact with this object and will feel anew the impression A'.... *Q. E. D.*

Let us see the consequences. I consider a series of muscular sensations Σ. To this series will correspond a point M of the first tactile space. Now take again the two series *s* and *s'*, inverses of one another, of which we have just spoken. To the series $S + \Sigma + S'$ will correspond a point N of the second tactile space, since to any series of muscular sensations corresponds, as we have said, a point, whether in the first space or in the second.

I am going to consider the two points N and $M,$ thus defined, as corresponding. What authorizes me to do so? For this correspondence to be admissible, it is necessary that if two points M and M', corresponding in the first space to two series Σ and Σ', are identical, so also are the two corresponding points of the second space N and N', that is, the two points which correspond to the two series $S + \Sigma + S'$ and $S + \Sigma' + S'$. Now we shall see that this condition is fulfilled.

First a remark. As S and S' are inverses of one another, we shall have $S + S' = 0$, and consequently $S + S' + \Sigma = \Sigma + S + S' = \Sigma$, or again $\Sigma + S + S' + \Sigma' = \Sigma + \Sigma'$; but it does not follow that we have $S + \Sigma + S' = \Sigma$; because, though we have used the addition sign to represent the succession of our sensations, it is clear that the order of this succession is not indifferent: we cannot, therefore, as in ordinary addition, invert the order of the terms; to use abridged language, our operations are associative, but not commutative.

That fixed, in order that Σ and Σ' should correspond to the same point $M = M'$ of the first space, it is necessary and sufficient for us to have $\Sigma' = \Sigma + \sigma$. We shall then have: $S + \Sigma' + S' = S + \Sigma + \sigma + S' = S + \Sigma + S' + S + \sigma + S'$.

But we have just ascertained that $S + \sigma + S'$ was one of the series σ'. We shall therefore have: $S + \Sigma' + S' = S + \Sigma + S' + \sigma$, which means that the series $S + \Sigma' + S'$ and $S + \Sigma + S'$ correspond to the same point $N = N'$ of the second space. *Q. E. D.*

Our two spaces therefore correspond point for point; they can be "transformed" one into the other; they are isomorphic. How are we led to conclude thence that they are identical?

Consider the two series σ and $S + \sigma + S' = \sigma'$. I have said that often, but not always, the series σ preserves the tactile impression A felt by the finger D; and similarly it often happens, but not always, that the series σ' preserves the tactile impression A' felt by the finger D'. Now I ascertain that it happens *very often* (that is, much more often than what I have just called "often") that when the series σ has preserved the impression A of the finger D, the series σ' preserves at the same time the impression A' of the finger D'; and, inversely, that if the first impression is altered, the second is likewise. That happens *very often,* but not always.

We interpret this experimental fact by saying that the unknown object a which gives the impression A to the finger D is identical with the unknown object a' which gives the impression A' to the finger D'. And in fact when the first object moves, which the disappearance of the impression A tells us, the second likewise moves, since the impression A' disappears likewise. When the first object remains motionless, the second remains motionless. If these two objects are identical, as the first is at the point M of the first space and the second at the point N of the second space, these two points are identical. This is how we are led to regard these two spaces as identical; or better this is what we mean when we say that they are identical.

What we have just said of the identity of the two tactile spaces makes unnecessary our discussing the question of the identity of tactile space and visual space, which could be treated in the same way.

V. SPACE AND EMPIRICISM

It seems that I am about to be led to conclusions in conformity with empiristic ideas. I have, in fact, sought to put in evidence the role of experience and to analyse the experimental facts which intervene in the genesis of space of three dimensions. But whatever may be the importance of these facts, there is one thing we must not forget and to which besides I have more than once called attention. These experimental facts are often verified but not always. That evidently does not mean that space has often three dimensions, but not always.

I know well that it is easy to save oneself and that, if the facts do not verify, it will be easily explained by saying that the exterior objects have moved. If experience succeeds, we say that it teaches us about space; if it does not succeed, we hie to exterior objects which we accuse of having moved; in other words, if it does not succeed, it is given a fillip.

These fillips are legitimate; I do not refuse to admit them; but they suffice to tell us that the properties of space are not experimental truths, properly so called. If we had wished to verify other laws, we could have succeeded also, by giving other analogous fillips. Should we not always have been able to justify these fillips by the same reasons? One could at most have said to us: "Your fillips are doubtless legitimate, but you abuse them; why move the exterior objects so often?"

To sum up, experience does not prove to us that space has three dimensions; it only proves to us that it is convenient to attribute three to it, because thus the number of fillips is reduced to a minimum.

I will add that experience bring us into contact only with representative space, which is a physical continuum, never with geometric space, which is a mathematical continuum. At the very most it would appear to tell us that it is convenient to give to geometric space three dimensions, so that it may have as many as representative space.

The empiric question may be put under another form. Is it impossible to conceive physical phenomena, the mechanical phenomena for example, otherwise than in space of three dimensions? We should thus have an objective experimental proof, so to speak, independent of our physiology, of our modes of representation.

But it is not so; I shall not here discuss the question completely, I shall confine myself to recalling the striking example given us by the mechanics of Hertz. You know that the great physicist did not believe in the existence of forces, properly so called; he supposed that visible material points are subjected to certain invisible bonds which join them to other invisible points and that it is the effect of these invisible bonds that we attribute to forces.

But that is only a part of his ideas. Suppose a system formed of n material points, visible or not; that will give in all $3n$ co-ordinates; let us regard them as the co-ordinates of a *single* point in space of $3n$ dimensions. This single point would be constrained to remain upon a surface (of any number of dimensions $<3n$) in virtue of the bonds of which we have just spoken; to go on this surface from one point to another, it would always take the shortest way; this would be the single principle which would sum up all mechanics.

Whatever should be thought of this hypothesis, whether we be allured by its simplicity, or repelled by its artificial character, the simple fact that Hertz was able to conceive it, and to regard it as more convenient than our habitual hypotheses, suffices to prove that our ordinary ideas, and, in particular, the three dimensions of space, are in no ways imposed upon mechanics with an invincible force.

VI. MIND AND SPACE

Experience, therefore, has played only a single role, it has served as occasion. But this role was nonetheless very important; and I have thought it necessary to give it prominence. This role would have been useless if there existed an *à priori* form imposing itself upon our sensitivity, and which was space of three dimensions.

Does this form exist, or, if you choose, can we represent to ourselves space of more than three dimensions? And first what does this question mean? In the true sense of the word, it is clear that we cannot represent to ourselves space of four, nor space of three, dimensions; we cannot first represent them to ourselves empty, and no more can we represent to ourselves an object either in space of four, or in space of three, dimensions: (1) Because these spaces are both infinite and we cannot represent to ourselves a figure *in* space, that is, the part *in* the whole, without representing the whole, and that is impossible, because it is infinite; (2) because these spaces are both mathematical continua and we can represent to ourselves only the physical continuum; (3) because these spaces are both homogenous, and the frames in which we enclose our sensations, being limited, cannot be homogenous.

Thus the question put can only be understood in another manner; is it possible to imagine that, the results of the experiences related above having been different, we might have been led to attribute to space more than three dimensions; to imagine, for instance, that the sensation of accommodation might not be constantly in accord with the sensation of convergence of the eyes; or indeed that the experiences of which we have spoken in paragraph 2 and of which we express the result by saying "that touch does not operate at a distance," might have led us to an inverse conclusion?

And *then evidently yes* that is possible. From the moment one imagines an experience, one imagines just by that the two contrary results it may give. That is possible, but that is difficult, because we have to overcome a multitude of associations of ideas, which are the fruit of a long personal experience and of the still longer experience of the race. Is it these associations (or at least those of them that we have inherited from our ancestors), which constitute this *à priori* form of which it is said that we have pure intuition? Then I do not see why one should declare it refractory to analysis and should deny me the right of investigating its origin.

When it is said that our sensations are "extended" only one thing can be meant, that is that they are always associated with the idea of certain muscular sensations, corresponding to the movements which enable us to reach the object which causes them, which enable us, in other words, to defend ourselves against it. And it is just because this association is useful for the defence of the organism, that it is so old in the history of the species and that it seems to us indestructible. Nevertheless, it is only an association and we can conceive that it may be broken; so that we may not say that sensation cannot enter consciousness without entering in space, but that in fact it does not enter consciousness without entering in space, which means, without being entangled in this association.

No more can I understand one's saying that the idea of time is logically subsequent to space, since we can represent it to ourselves only under the form of a straight line; as well say that time is logically subsequent to the cultivation of the prairies, since it is usually represented armed with a scythe. That one cannot represent to himself simultaneously the different parts of time, goes without saying, since the essential character of these parts is precisely not to be simultaneous. That does not mean that we have not the intuition of time. So far as that goes, no more should we have that of space, because neither can we represent it, in the proper sense of the word, for the reasons I have mentioned. What we represent to ourselves under the name of straight is a crude image which as ill resembles the geometric straight as it does time itself.

Why has it been said that every attempt to give a fourth dimension to space always carries this one back to one of the other three? It is easy to understand. Consider our muscular sensations and the "series" they may form. In consequence of numerous experiences, the ideas of these series are associated together in a very complex woof, our series are *classed*. Allow me, for convenience of language, to express my thought in a way altogether crude and even inexact by saying that our series of muscular sensations are classed in three classes corresponding to the three dimensions of space. Of course this classification is much more complicated than that, but that will suffice to make my reasoning understood. If I wish to imagine a fourth dimension, I shall suppose another series of muscular sensations, making part of a fourth class. But as *all* my muscular sensations have already been classed in one of the three pre-existent classes, I can only represent to myself a series belonging to one of these three classes, so that my fourth dimension is carried back to one of the other three.

What does that prove? This: that it would have been necessary first to destroy the old classification and replace it by a new one in which the series of muscular sensations should have been distributed into four classes. The difficulty would have disappeared.

It is presented sometimes under a more striking form. Suppose I am enclosed in a chamber between the six impassable boundaries formed by the four walls, the floor and the ceiling; it will be impossible for me to get out and to imagine my getting out. Pardon, can you not imagine that the door opens, or that two of these walls separate? But of course, you answer, one must suppose that these walls remain immovable. Yes, but it is evident that I have the right to move; and then the walls that we suppose absolutely at rest will be in motion with regard to me. Yes, but such a relative motion cannot be anything; when objects are at rest, their relative motion with regard to any axes is that of a rigid solid; now, the apparent motions that you imagine are not in conformity with the laws of motion of a rigid solid. Yes, but it is experience which has taught us the laws of motion of a rigid solid; nothing would prevent our *imagining* them different. To sum up, for me to imagine that I get out of my prison, I have only to imagine that the walls seem to open, when I move.

I believe, therefore, that if by space is understood a mathematical continuum of three dimensions, were it otherwise amorphous, it is the mind which constructs it, but it does not construct it out of nothing; it needs materials and models. These materials, like these models, pre-exist within it. But there is not a single model which is imposed upon it; it has *choice;* it may choose, for instance, between space of four and space of three dimensions. What then is the role of experience? It gives the indications following which the choice is made.

Another thing: whence does space get its quantitative character? It comes from the role which the series of muscular sensations play in its genesis. These are series which may *repeat themselves,* and it is from their repetition that that number comes; it is because they can repeat themselves indefinitely that space is infinite. And finally we have seen, at the end of section 3, that it is also because of this that space is relative. So it is repetition which has given to space its essential characteristics; now, repetition supposes time; this is enough to tell that time is logically anterior to space.

VII. Role of the Semi-circular Canals

I have not hitherto spoken of the role of certain organs to which the physiologists attribute with reason a capital importance, I mean the semi-circular canals. Numerous experiments have sufficiently shown that these canals are necessary to our sense of orientation; but the physiologists are not entirely in accord; two opposing theories have been proposed, that of Mach-Delage and that of M. de Cyon.

M. de Cyon is a physiologist who has made his name illustrious by important discoveries on the innervation of the heart; I cannot, however, agree with his ideas on the question before us. Not being a physiologist, I hesitate to criticize the experiments he has directed against the adverse theory of Mach-Delage; it seems to me, however, that they are not convincing, because in many of them the *total* pressure was made to vary in one of the canals, while, physiologically, what varies is the *difference* between the pressures on the two extremities of the canal; in others the organs were subjected to profound lesions, which must alter their functions.

Besides, this is not important; the experiments, if they were irreproachable, might be convincing against the old theory. They would not be convincing *for* the new theory. In fact, if I have rightly understood the theory, my explaining it will be enough for one to understand that it is impossible to conceive of an experiment confirming it.

The three pairs of canals would have as sole function to tell us that space has three dimensions. Japanese mice have only two pairs of canals; they believe, it would seem, that space has only two dimensions, and they manifest this opinion in the strangest way; they put themselves in a circle, and, so ordered, they spin rapidly around. The lampreys, having only one pair of canals, believe that space has only one dimension, but their manifestations are less turbulent.

It is evident that such a theory is inadmissible. The sense-organs are designed to tell us of *changes* which happen in the exterior world. We could not understand why the Creator should have given us organs destined to cry without cease: remember that space has three dimensions, since the number of these three dimensions is not subject to change.

We must, therefore, come back to the theory of Mach-Delage. What the nerves of the canals can tell us is the difference of pressure on the two extremities of the same canal, and thereby: (1) the direction of the vertical with regard to three axes rigidly bound to the head; (2) the three components of the acceleration of translation of the center of gravity of the head; (3) the centrifugal forces developed by the rotation of the head; (4) the acceleration of the motion of rotation of the head.

It follows from the experiments of M. Delage that it is this last indication which is much the most important; doubtless because the nerves are less sensible to the difference of pressure itself than to the brusque variations of this difference. The first three indications may thus be neglected.

Knowing the acceleration of the motion of rotation of the head at each instant, we deduce from it, by an unconscious integration, the final orientation of the head, referred to a certain initial orientation taken as origin. The circular canals contribute, therefore, to inform us of the movements that we have executed, and that on the same ground as the muscular sensations. When, therefore, above we speak of the series S or of the series Σ, we should say, not that these were series of muscular sensations alone, but that they were series at the same time of muscular sensations due to the semi-circular canals. Apart from this addition, we should have nothing to change in what precedes.

In the series S and Σ, these sensations of the semi-circular canals evidently hold a very important place. Yet alone they would not suffice, because they can tell us only of the movements of the head; they tell us nothing of the relative movements of the body, or of the members in regard to the head. And more, it seems that they tell us only of the rotations of the head and not of the translations it may undergo.

PART II

THE
PHYSICAL
SCIENCES

CHAPTER V

ANALYSIS AND PHYSICS

I

You have doubtless often been asked of what good are mathematics and whether these delicate constructions entirely mind-made are not artificial and born of our caprice.

Among those who put this question I should make a distinction; practical people ask of us only the means of money-making. These merit no reply; rather would it be proper to ask of them what is the good of accumulating so much wealth and whether, to get time to acquire it, we are to neglect art and science, which alone give us souls capable of enjoying it, "and for life's sake to sacrifice all reasons for living."

Besides, a science made solely in view of applications is impossible; truths are fecund only if bound together. If we devote ourselves solely to those truths whence we expect an immediate result, the intermediary links are wanting and there will no longer be a chain.

The men most disdainful of theory get from it, without suspecting it, their daily bread; deprived of this food, progress would quickly cease, and we should soon congeal into the immobility of China.

But enough of uncompromising practicians! Besides these, there are those who are only interested in nature and who ask us if we can enable them to know it better.

To answer these, we have only to show them the two monuments already rough-hewn, Celestial Mechanics and Mathematical Physics.

They would doubtless concede that these structures are well worth the trouble they have cost us. But this is not enough. Mathematics have a triple aim. They must furnish an instrument for the study of nature. But that is not all: they have a philosophic aim and, I dare maintain, an esthetic aim. They must aid the philosopher to fathom the notions of number, of space, of time. And above all, their adepts find therein delights analogous to those given by painting and music. They admire the delicate harmony of numbers and forms; they marvel when a new discovery opens to them an unexpected perspective; and has not the joy they thus feel the esthetic character, even though the senses take no part therein? Only a privileged few are called to enjoy it fully, it is true, but is not this the case for all the noblest arts?

This is why I do not hesitate to say that mathematics deserve to be cultivated for their own sake, and the theories inapplicable to physics as well as the others. Even if the physical aim and the esthetic aim were not united, we ought not to sacrifice either.

But more: these two aims are inseparable and the best means of attaining one is to aim at the other, or at least never to lose sight of it. This is what I am about to try to demonstrate in setting forth the nature of the relations between the pure science and its applications.

The mathematician should not be for the physicist a mere purveyor of formulae; there should be between them a more intimate collaboration. Mathematical physics and pure analysis are not merely adjacent powers, maintaining good neighbourly relations; they mutually interpenetrate and their spirit is the same. This will be better understood when I have shown what physics gets from mathematics and what mathematics, in return, borrows from physics.

II

The physicist cannot ask of the analyst to reveal to him a new truth; the latter could at most only aid him to foresee it. It is a long time since one still dreamt of forestalling experiment, or of constructing the entire world on certain premature hypotheses. Since all those constructions in which one yet took a naïve delight it is an age, today only their ruins remain.

All laws are therefore deduced from experiment; but to enunciate them, a special language is needful; ordinary language is too poor, it is besides too vague, to express relations so delicate, so rich, and so precise.

This therefore is one reason why the physicist cannot do without mathematics; it furnishes him the only language he can speak. And a well-made language is no indifferent thing; not to go beyond physics, the unknown man who invented the word *heat* devoted many generations to error. Heat has been treated as a substance, simply because it was designated by a substantive, and it has been thought indestructible.

On the other hand, he who invented the word *electricity* had the unmerited good fortune to implicitly endow physics with a *new* law, that of the conservation of electricity, which, by a pure chance, has been found exact, at least until now.

Well, to continue the simile, the writers who embellish a language, who treat it as an object of art, make of it at the same time a more supple instrument, more apt for rendering shades of thought.

We understand, then, how the analyst, who pursues a purely esthetic aim, helps create, just by that, a language more fit to satisfy the physicist.

But this is not all: law springs from experiment, but not immediately. Experiment is individual, the law deduced from it is general; experiment is only approximate, the law is precise, or at least pretends to be. Experiment is made under conditions always complex, the enunciation of the law eliminates these complications. This is what is called "correcting the systematic errors."

In a word, to get the law from experiment, it is necessary to generalize; this is a necessity imposed upon the most circumspect observer. But how generalize? Every particular truth may evidently be extended in an infinity of ways. Among these thousand routes opening before us, it is necessary to make a choice, at least provisional; in this choice, what shall guide us?

It can only be analogy. But how vague is this word! Primitive man knew only crude analogies, those which strike the senses, those of colours or of sounds. He never would have dreamt of likening light to radiant heat.

What has taught us to know the true, profound analogies, those the eyes do not see but reason divines?

It is the mathematical spirit, which disdains matter to cling only to pure form. This it is which has taught us to give the same name to things differing only in material, to call by the same name, for instance, the multiplication of quaternions and that of whole numbers.

If quaternions, of which I have just spoken, had not been so promptly utilized by the English physicists, many persons would doubtless see in them only a useless fancy, and yet, in teaching us to liken what appearances separate, they would have already rendered us more apt to penetrate the secrets of nature.

Such are the services the physicist should expect of analysis; but for this science to be able to render them, it must be cultivated in the broadest fashion without immediate expectation of utility—the mathematician must have worked as artist.

What we ask of him is to help us to see, to discern our way in the labyrinth which opens before us. Now, he sees best who stands highest. Examples abound, and I limit myself to the most striking.

The first will show us how changing the language suffices to reveal generalizations not before suspected.

When Newton's law has been substituted for Kepler's, we still know only elliptic motion. Now, insofar as concerns this motion, the two laws differ only in form; we pass from one to the other by a simple differentiation. And yet from Newton's law may be deduced by an immediate generalization all the effects of perturbations and the whole of celestial mechanics. If, on the other hand, Kepler's enunciation had been retained, no one would ever have regarded the orbits of the perturbed planets, those complicated curves of which no one has ever written the equation, as the natural generalizations of the ellipse. The progress of observations would only have served to create belief in chaos.

The second example is equally deserving of consideration.

When Maxwell began his work, the laws of electro-dynamics admitted up to his time accounted for all the known facts. It was not a new experiment which came to invalidate them. But in looking at them under a new bias, Maxwell saw that the equations became more symmetrical when a term was added, and besides, this term was too small to produce effects appreciable with the old methods.

You know that Maxwell's *à priori* views awaited for twenty years an experimental confirmation; or if you prefer, Maxwell was twenty years ahead of experiment. How was this triumph obtained?

It was because Maxwell was profoundly steeped in the sense of mathematical symmetry; would he have been so, if others before him had not studied this symmetry for its own beauty?

It was because Maxwell was accustomed to "think in vectors," and yet it was through the theory of imaginaries (neomonics) that vectors were introduced into analysis. And those who invented imaginaries hardly suspected the advantage which would be obtained from them for the study of the real world; of this the name given them is proof sufficient.

In a word, Maxwell was perhaps not an able analyst, but this ability would have been for him only a useless and bothersome baggage. On the other hand, he had in the highest degree the intimate sense of mathematical analogies. Therefore it is that he made good mathematical physics.

Maxwell's example teaches us still another thing.

How should the equations of mathematical physics be treated? Should we simply deduce all the consequences, and regard them as intangible realities? Far from it; what they should teach us above all is what can and what should be changed. It is thus that we get from them something useful.

The third example goes to show us how we may perceive mathematical analogies between phenomena which have physically no relation either apparent or real, so that the laws of one of these phenomena aid us to divine those of the other.

The very same equation, that of Laplace, is met in the theory of Newtonian attraction, in that of the motion of liquids, in that of the electric potential, in that of magnetism, in that of the propagation of heat and in still many others. What is the result? These theories seem images copied one from the other; they are mutually illuminating, borrowing their language from each other; ask electricians if they do not felicitate themselves on having invented the phrase flow of force, suggested by hydrodynamics and the theory of heat.

Thus mathematical analogies not only may make us foresee physical analogies, but besides do not cease to be useful when these latter fail.

To sum up, the aim of mathematical physics is not only to facilitate for the physicist the numerical calculation of certain constants or the integration of certain differential equations. It is besides, it is above all, to reveal to him the hidden harmony of things in making him see them in a new way.

Of all the parts of analysis, the most elevated, the purest, so to speak, will be the most fruitful in the hands of those who know how to use them.

III

Let us now see what analysis owes to physics.

It would be necessary to have completely forgotten the history of science not to remember that the desire to understand nature has had on the development of mathematics the most constant and happiest influence.

In the first place the physicist sets us problems whose solution he expects of us. But in proposing them to us, he has largely paid us in advance for the service we shall render him, if we solve them.

If I may be allowed to continue my comparison with the fine arts, the pure mathematician who should forget the existence of the exterior world would be like a painter who knew how to harmoniously combine colours and forms, but who lacked models. His creative power would soon be exhausted.

The combinations which numbers and symbols may form are an infinite multitude. In this multitude how shall we choose those which are worthy to fix our attention? Shall we let ourselves be guided solely by our caprice? This caprice, which itself would besides soon tire, would doubtless carry us very far apart and we should quickly cease to understand each other.

But this is only the smaller side of the question. Physics will doubtless prevent our straying, but it will also preserve us from a danger much more formidable; it will prevent our ceaselessly going around in the same circle.

History proves that physics has not only forced us to choose among problems which came in a crowd; it has also imposed upon us such as we should without it never have dreamt of. However varied may be the imagination of man, nature is still a thousand times richer. To follow her we must take ways we have neglected, and these paths lead us often to summits whence we discover new countries. What could be more useful!

It is with mathematical symbols as with physical realities; it is in comparing the different aspects of things that we are able to comprehend their inner harmony, which alone is beautiful and consequently worthy of our efforts.

The first example I shall cite is so old we are tempted to forget it; it is nevertheless the most important of all.

The sole natural object of mathematical thought is the whole number. It is the external world which has imposed the continuum upon us, which we doubtless have invented, but which it has forced us to invent. Without it there would be no infinitesimal analysis; all mathematical science would reduce itself to arithmetic or to the theory of substitutions.

On the contrary, we have devoted to the study of the continuum almost all our time and all our strength. Who will regret it; who will think that this time and this strength have been wasted? Analysis unfolds before us infinite perspectives that arithmetic never suspects; it shows us at a glance a majestic assemblage whose array is simple and symmetric; on the contrary, in the theory of numbers, where reigns the unforeseen, the view is, so to speak, arrested at every step.

Doubtless it will be said that outside of the whole number there is no rigour, and consequently no mathematical truth; that the whole number hides everywhere, and that we must strive to render transparent the screens which cloak it, even if to do so we must resign ourselves to interminable repetitions. Let us not be such purists and let us be grateful to the continuum, which, if *all* springs from the whole number, was alone capable of making *so much* proceed therefrom.

Need I also recall that M. Hermite obtained a surprising advantage from the introduction of continuous variables into the theory of numbers? Thus the whole number's own domain is itself invaded, and this invasion has established order where disorder reigned.

See what we owe to the continuum and consequently to physical nature.

Fourier's series is a precious instrument of which analysis makes continual use, it is by this means that it has been able to represent discontinuous functions; Fourier invented it to solve a problem of physics relative to the propagation of heat. If this problem had not come up naturally, we should never have dared to give discontinuity its rights; we should still long have regarded continuous functions as the only true functions.

The notion of function has been thereby considerably extended and has received from some logician-analysts an unforeseen development. These analysts have thus adventured into regions where reigns the purest abstraction and have gone as far away as possible from the real world. Yet it is a problem of physics which has furnished them the occasion.

After Fourier's series, other analogous series have entered the domain of analysis; they have entered by the same door; they have been imagined in view of applications.

The theory of partial differential equations of the second order has an analogous history. It has been developed chiefly by and for physics. But it may take many forms, because such an equation does not suffice to determine the unknown function, it is necessary to adjoin to it complementary conditions which are called conditions at the limits; whence many different problems.

If the analysts had abandoned themselves to their natural tendencies, they would never have known but one, that which Madame Kovalevski has treated in her celebrated memoir. But there are a multitude of others which they would have ignored. Each of the theories of physics, that of electricity, that of heat, presents us these equations under a new aspect. It may therefore be said that without these theories we should not know partial differential equations.

It is needless to multiply examples. I have given enough to be able to conclude: when physicists ask of us the solution of a problem, it is not a duty-service they impose upon us, it is on the contrary we who owe them thanks.

IV

But this is not all; physics not only gives us the occasion to solve problems; it aids us to find the means thereto, and that in two ways. It makes us foresee the solution; it suggests arguments to us.

I have spoken above of Laplace's equation which is met in a multitude of diverse physical theories. It is found again in geometry, in the theory of conformal representation and in pure analysis, in that of imaginaries.

In this way, in the study of functions of complex variables, the analyst, alongside of the geometric image, which is his usual instrument, finds many physical images which he may make use of with the same success. Thanks to these images he can see at a glance what pure deduction would show him only successively. He masses thus the separate elements of the solution, and by a sort of intuition divines before being able to demonstrate.

To divine before demonstrating! Need I recall that thus have been made all the important discoveries? How many are the truths that physical analogies permit us to present and that we are not in condition to establish by rigorous reasoning!

For example, mathematical physics introduces a great number of developments in series. No one doubts that these developments converge; but the mathematical certitude is lacking. These are so many conquests assured for the investigators who shall come after us.

On the other hand, physics furnishes us not alone solutions; it furnishes us besides, in a certain measure, arguments. It will suffice to recall how Felix Klein, in a question relative to Riemann surfaces, has had recourse to the properties of electric currents.

It is true, the arguments of this species are not rigourous, in the sense the analyst attaches to this word. And here a question arises: How can a demonstration not sufficiently rigorous for the analyst suffice for the physicist? It seems there cannot be two rigours, that rigour is or is not, and that, where it is not, there cannot be deduction.

This apparent paradox will be better understood by recalling under what conditions number is applied to natural phenomena. Whence come in general the difficulties encountered in seeking rigour? We strike them almost always in seeking to establish that some quantity tends to some limit, or that some function is continuous, or that it has a derivative.

Now the numbers the physicist measures by experiment are never known except approximately; and besides, any function always differs as little as you choose from a discontinuous function, and at the same time it differs as little as you choose from a continuous function. The physicist may, therefore, at will suppose that the function studied is continuous, or that it is discontinuous; that it has or has not a derivative; and may do so without fear of ever being contradicted, either by present experience or by any future experiment. We see that with such liberty he makes sport of difficulties which stop the analyst. He may always reason as if all the functions which occur in his calculations were entire polynomials.

Thus the sketch which suffices for physics is not the deduction which analysis requires. It does not follow thence that one cannot aid in finding the other. So many physical sketches have already been transformed into rigourous demonstrations that today this transformation is easy. There would be plenty of examples did I not fear in citing them to tire the reader.

I hope I have said enough to show that pure analysis and mathematical physics may serve one another without making any sacrifice one to the other, and that each of these two sciences should rejoice in all which elevates its associate.

CHAPTER VI

ASTRONOMY

Governments and parliaments must find that astronomy is one of the sciences which cost most dear: the least instrument costs hundreds of thousands of dollars, the least observatory costs millions; each eclipse carries with it supplementary appropriations. And all that for stars which are so far away, which are complete strangers to our electoral contests, and in all probability will never take any part in them. It must be that our politicians have retained a remnant of idealism, a vague instinct for what is grand; truly, I think they have been calumniated; they should be encouraged and shown that this instinct does not deceive them, that they are not dupes of that idealism.

We might indeed speak to them of navigation, of which no one can underestimate the importance, and which has need of astronomy. But this would be to take the question by its smaller side.

Astronomy is useful because it raises us above ourselves; it is useful because it is grand; that is what we should say. It shows us how small is man's body, how great his mind, since his intelligence can embrace the whole of this dazzling immensity, where his body is only an obscure point, and enjoy its silent harmony. Thus we attain the consciousness of our power, and this is something which cannot cost too dear, since this consciousness makes us mightier.

But what I should wish before all to show is, to what point astronomy has facilitated the work of the other sciences, more directly useful, since it has given us a soul capable of comprehending nature.

Think how diminished humanity would be if, under heavens constantly overclouded, as Jupiter's must be, it had forever remained ignorant of the stars. Do you think that in such a world we should be what we are? I know well that under this somber vault we should have been deprived of the light of the sun, necessary to organisms like those which inhabit the earth. But if you please, we shall assume that these clouds are phosphorescent and emit a soft and constant light. Since we are making hypotheses, another will cost no more. Well! I repeat my question: Do you think that in such a world we should be what we are?

The stars send us not only that visible and gross light which strikes our bodily eyes, but from them also comes to us a light far more subtle, which illuminates our minds and whose effects I shall try to show you. You know what man was on the earth some thousands of years ago, and what he is today. Isolated amidst a nature where everything was a mystery to him, terrified at each unexpected manifestation of incomprehensible forces, he was incapable of seeing in the conduct of the universe anything but caprice; he attributed all phenomena to the action of a multitude of little genii, fantastic and exacting, and to act on the world he sought to conciliate them by means analogous to those employed to gain the good graces of a minister or a deputy. Even his failures did not enlighten him, any more than today a beggar refused is discouraged to the point of ceasing to beg.

Today we no longer beg of nature; we command her, because we have discovered certain of her secrets and shall discover others each day. We command her in the name of laws she cannot challenge because they are hers; these laws we do not madly ask her to change, we are the first to submit to them. Nature can only be governed by obeying her.

What a change must our souls have undergone to pass from the one state to the other! Does anyone believe that, without the lessons of the stars, under the heavens perpetually overclouded that I have just supposed, they would have changed so quickly? Would the metamorphosis have been possible, or at least would it not have been much slower?

And first of all, astronomy it is which taught that there are laws. The Chaldeans, who were the first to observe the heavens with some attention, saw that this multitude of luminous points is not a confused crowd wandering at random, but rather a disciplined army. Doubtless the rules of this discipline escaped them, but the harmonious spectacle of the starry night sufficed to give them the impression of regularity, and that was in itself already a great thing. Besides, these rules were discerned by Hipparchus, Ptolemy, Copernicus, Kepler, one after another, and finally, it is needless to recall that Newton it was who enunciated the oldest, the most precise, the most simple, the most general of all natural laws.

And then, taught by this example, we have seen our little terrestrial world better and, under the apparent disorder, there also we have found again the harmony that the study of the heavens had revealed to us. It also is regular, it also obeys immutable laws, but they are more complicated, in apparent conflict one with another, and an eye untrained by other sights would have seen there only chaos and the reign of chance or caprice. If we had not known the stars, some bold spirits might perhaps have sought to foresee physical phenomena; but their failures would have been frequent, and they would have excited only the derision of the vulgar; do we not see, that even in our day the meteorologists sometimes deceive themselves, and that certain persons are inclined to laugh at them?

How often would the physicists, disheartened by so many checks, have fallen into discouragement, if they had not had, to sustain their confidence, the brilliant example of the success of the astronomers! This success showed them that nature obeys laws; it only remained to know what laws; for that they only needed patience, and they had the right to demand that the skeptics should give them credit.

This is not all: astronomy has not only taught us that there are laws, but that from these laws there is no escape, that with them there is no possible compromise. How much time should we have needed to comprehend that fact, if we had known only the terrestrial world, where each elemental force would always seem to us in conflict with other forces? Astronomy has taught us that the laws are infinitely precise, and that if those we enunciate are approximative, it is because we do not know them well. Aristotle, the most scientific mind of antiquity, still accorded a part to accident, to chance, and seemed to think that the laws of nature, at least here below, determine only the large features of phenomena. How much has the ever-increasing precision of astronomical predictions contributed to correct such an error, which would have rendered nature unintelligible!

But are these laws not local, varying in different places, like those which men make; does not that which is truth in one corner of the universe, on our globe for instance, or in our little solar system, become error a little farther away? And then could it not be asked whether laws depending on space do not also depend upon time, whether they are not simple habitudes, transitory, therefore, and ephemeral? Again it is astronomy that answers this question. Consider the double stars; all describe conics; thus, as far as the telescope carries, it does not reach the limits of the domain which obeys Newton's law.

Even the simplicity of this law is a lesson for us; how many complicated phenomena are contained in the two lines of its enunciation; persons who do not understand celestial mechanics may form some idea of it at least from the size of the treatises devoted to this science; and then it may be hoped that the complication of physical phenomena likewise hides from us some simple cause still unknown.

It is therefore astronomy which has shown us what are the general characteristics of natural laws; but among these characteristics there is one, the most subtle and the most important of all, which I shall ask leave to stress.

How was the order of the universe understood by the ancients;

for instance, by Pythagoras, Plato or Aristotle? It was either an immutable type fixed once for all, or an ideal to which the world sought to approach. Kepler himself still thought thus when, for instance, he sought whether the distances of the planets from the sun had not some relation to the five regular polyhedrons. This idea contained nothing absurd, but it was sterile, since nature is not so made. Newton has shown us that a law is only a necessary relation between the present state of the world and its immediately subsequent state. All the other laws since discovered are nothing else; they are in sum, differential equations; but it is astronomy which furnished the first model for them, without which we should doubtless long have erred.

Astronomy has also taught us to set at naught appearances. The day Copernicus proved that what was thought the most stable was in motion, that what was thought moving was fixed, he showed us how deceptive could be the infantile reasonings which spring directly from the immediate data of our senses. True, his ideas did not easily triumph, but since this triumph there is no longer a prejudice so inveterate that we cannot shake it off. How can we estimate the value of the new weapon thus won?

The ancients thought everything was made for man, and this illusion must be very tenacious, since it must ever be combated. Yet it is necessary to divest oneself of it; or else one will be only an eternal myope, incapable of seeing the truth. To comprehend nature one must be able to get out of self, so to speak, and to contemplate her from many different points of view; otherwise we never shall know more than one side. Now, to get out of self is what he who refers everything to himself cannot do. Who delivered us from this illusion? It was those who showed us that the earth is only one of the smallest planets of the solar system, and that the solar system itself is only an imperceptible point in the infinite spaces of the stellar universe.

At the same time astronomy taught us not to be afraid of big numbers. This was needful, not only for knowing the heavens, but to know the earth itself; and was not so easy as it seems to us today. Let us try to go back and picture to ourselves what a Greek would have thought if told that red light vibrates four hundred millions of millions of times per second. Without any doubt, such an assertion would have appeared to him pure madness, and he never would have lowered himself to test it. Today an hypothesis will no longer appear absurd to us because it obliges us to imagine objects much larger or smaller than those our senses are capable of showing us, and we no longer comprehend those scruples which arrested our predecessors and prevented them from discovering certain truths simply because they were afraid of them. But why? It is because we have seen the heavens enlarging and enlarging without cease; because we know that the sun is 150 millions of kilometers from the earth and that the distances of the nearest stars are hundreds of thousands of times greater yet. Habituated to the contemplation of the infinitely great, we have become apt to comprehend the infinitely small. Thanks to the education it has received, our imagination, like the eagle's eye that the sun does not dazzle, can look truth in the face.

Was I wrong in saying that it is astronomy which has made us a soul capable of comprehending nature; that under heavens always overcast and starless, the earth itself would have been for us eternally unintelligible; that we should there have seen only caprice and disorder; and that, not knowing the world, we should never have been able to subdue it? What science could have been more useful? And in thus speaking I put myself at the point of view of those who only value practical applications. Certainly, this point of view is not mine; as for me, on the contrary, if I admire the conquests of industry, it is above all because if they free us from material cares, they will one day give to all the leisure to contemplate nature. I do not say: science is useful, because it teaches us to construct machines. I say: Machines are useful, because in working for us, they will some day leave us more time to make science. But finally it is worth remarking that between the two points of view there is no antagonism, and that man having pursued a disinterested aim, all else has been added unto him.

Auguste Comte has said somewhere, that it would be idle to seek to know the composition of the sun, since this knowledge would be of no use to sociology. How could he be so short-sighted? Have we not just seen that it is by astronomy that, to speak his language, humanity has passed from the theological to the positive state? He found an explanation for that because it had happened. But how has he not understood that what remained to do was not less considerable and would be not less profitable? Physical astronomy, which he seems to condemn, has already begun to bear fruit, and it will give us much more, for it only dates from yesterday.

First was discovered the nature of the sun, what the founder of positivism wished to deny us, and there bodies were found which exist on the earth, but had here remained undiscovered; for example, helium, that gas almost as light as hydrogen. That already contradicted Comte. But to the spectroscope we owe a lesson precious in a quite different way; in the most distant stars, it shows us the same substances. It might have been asked whether the terrestrial elements were not due to some chance which had brought together more tenuous atoms to construct of them the more complex edifice that the chemists call atoms; whether, in other regions of the universe, other fortuitous meetings had not engendered edifices entirely different. Now we know that this is not so, that the laws of our chemistry are the general laws of nature, and that they owe nothing to the chance which caused us to be born on the earth.

But, it will be said, astronomy has given to the other sciences all it can give them, and now that the heavens have procured for us the instruments which enable us to study terrestrial nature, they could without danger veil themselves forever. After what we have just said, is there still need to answer this objection? One could have reasoned the same in Ptolemy's time; then also men thought they knew everything, and they still had almost everything to learn.

The stars are majestic laboratories, gigantic crucibles, such as no chemist could dream. There reign temperatures impossible for us to realize. Their only defect is being a little far away; but the telescope will soon bring them near to us, and then we shall see how matter acts there. What good fortune for the physicist and the chemist!

Matter will there exhibit itself to us under a thousand different states, from those rarefied gases which seem to form the nebula and which are luminous with I know not what glimmering of mysterious origin, even to the incandescent stars and to the planets so near and yet so different.

Perchance even, the stars will some day teach us something about life; that seems an insensate dream and I do not at all see how it can be realized; but, a hundred years ago, would not the chemistry of the stars have also appeared a mad dream?

But limiting our views to horizons less distant, there still will remain to us promises less contingent and yet sufficiently seductive. If the past has given us much, we may rest assured that the future will give us still more.

After all, it could scarce be believed how useful belief in astrology has been to humanity. If Kepler and Tycho Brahe made a living, it was because they sold to naïve kings predictions founded on the conjunctions of the stars. If these princes had not been so credulous, we should perhaps continue to believe that nature obeys caprice, and we should still wallow in ignorance.

CHAPTER VII

THE HISTORY OF
MATHEMATICAL PHYSICS

What is the present state of mathematical physics? What are the problems it is led to set itself? What is its future? Is its orientation about to be modified?

Ten years hence will the aim and the methods of this science appear to our immediate successors in the same light as to ourselves; or, on the contrary, are we about to witness a profound transformation? Such are the questions we are forced to raise in entering today upon our investigation.

If it is easy to propound them: to answer is difficult. If we felt tempted to risk a prediction, we should easily resist this temptation, by thinking of all the stupidities the most eminent savants of a hundred years ago would have uttered, if someone had asked them what the science of the nineteenth century would be. They would have thought themselves bold in their predictions, and after the event, how very timid we should have found them. Do not, therefore, expect of me any prophecy.

But if, like all prudent physicians, I shun giving a prognosis, yet I cannot dispense with a little diagnostic; well, yes, there are indications of a serious crisis, as if we might expect an approaching transformation. Still, be not too anxious: we are sure the patient will not die of it, and we may even hope that this crisis will be salutary, for the history of the past seems to guarantee us this. This crisis, in fact, is not the first, and to understand it, it is important to recall those which have preceded. Pardon then a brief historical sketch.

THE PHYSICS OF CENTRAL FORCES

Mathematical physics, as we know, was born of celestial mechanics, which gave birth to it at the end of the eighteenth century, at the moment when it itself attained its complete development. During its first years especially the infant strikingly resembled its mother.

The astronomic universe is formed of masses, very great, no doubt, but separated by intervals so immense that they appear to us only as material points. These points attract each other inversely as the square of the distance, and this attraction is the sole force which influences their movements. But if our senses were sufficiently keen to show us all the details of the bodies which the physicist studies, the spectacle thus disclosed would scarcely differ from the one the astronomer contemplates. There also we should see material points, separated from one another by intervals, enormous in comparison with their dimensions, and describing orbits according to regular laws. These infinitesimal stars are the atoms. Like the stars proper, they attract or repel each other, and this attraction or this repulsion following the straight line which joins them, depends only on the distance. The law according to which this force varies as function of the distance is perhaps not the law of Newton, but it is an analogous law; in place of the exponent -2, we have probably a different exponent, and it is from this change of exponent that arises all the diversity of physical phenomena, the variety of qualities and of sensations, all the world, coloured and sonorous, which surrounds us; in a word, all nature.

Such is the primitive conception in all its purity. It only remains to seek in the different cases what value should be given to this exponent in order to explain all the facts. It is on this model that Laplace, for example, constructed his beautiful theory of capillarity; he regards it only as a particular case of attraction, or, as he says, of universal gravitation, and no one is astonished to find it in the middle of one of the five volumes of the "Mécanique céleste." More recently Briot believes he penetrated the final secret of optics in demonstrating that the atoms of ether attract each other in the inverse ratio of the sixth power of the distance; and Maxwell, Maxwell himself, does he not say somewhere that the atoms of gases repel each other in the inverse ration of the fifth power of the distance? We have the exponent -6, or -5, in place of the exponent -2, but it is always an exponent.

Among the theories of this epoch, one alone is an exception, that of Fourier; in it are indeed atoms acting at a distance one upon the other; they mutually transmit heat, but they do not attract, they never budge. From this point of view, Fourier's theory must have appeared to the eyes of his contemporaries, to those of Fourier himself, as imperfect and provisional.

This conception was not without grandeur; it was seductive, and many among us have not finally renounced it; they know that one will attain the ultimate elements of things only by patiently disentangling the complicated skein that our senses give us; that it is necessary to advance step by step, neglecting no intermediary; that our fathers were wrong in wishing to skip stations; but they believe that when one shall have arrived at these ultimate elements, there again will be found the majestic simplicity of celestial mechanics.

Neither has this conception been useless; it has rendered us an inestimable service, since it has contributed to make precise the fundamental notion of the physical law.

I will explain myself; how did the ancients understand law? It was for them an internal harmony, static, so to say, and immutable; or else it was like a model that nature tried to imitate. For us a law is something quite different; it is a constant relation between the phenomenon of today and that of tomorrow; in a word, it is a differential equation.

Behold the ideal form of physical law; well, it is Newton's law which first clothed it forth. If then one has acclimated this form in physics, it is precisely by copying as far as possible this law of Newton, that is by imitating celestial mechanics. This is, moreover, the idea I have tried to bring out in chapter VI.

THE PHYSICS OF THE PRINCIPLES

Nevertheless, a day arrived when the conception of central forces no longer appeared sufficient, and this is the first of those crises of which I just now spoke.

What was done then? The attempt to penetrate into the detail of the structure of the universe, to isolate the pieces of this vast mechanism, to analyse one by one the forces which put them in motion, was abandoned, and we were content to take as guides certain general principles the express object of which is to spare us this minute study. How so? Suppose we have before us any machine; the initial wheel work and the final wheel work alone are visible, but the transmission, the intermediary machinery by which the movement is communicated from one to the other, is hidden in the interior and escapes our view; we do not know whether the communication is made by gearing or by belts, by connecting-rods or by other contrivances. Do we say that it is impossible for us to understand anything about this machine so long as we are not permitted to take it to pieces? You know well we do not, and that the principle of the conservation of energy suffices to determine for us the most interesting point. We easily ascertain that the final wheel turns ten times less quickly than the initial wheel, since these two wheels are visible; we are able thence to conclude that a couple applied to the one will be balanced by a couple ten times greater applied to the other. For that there is no need to penetrate the mechanism of this equilibrium and to know how the forces compensate each other in the interior of the machine; it suffices to be assured that this compensation cannot fail to occur.

Well, in regard to the universe, the principle of the conservation of energy is able to render us the same service. The universe is also a machine, much more complicated than all those of industry, of which almost all the parts are profoundly hidden from us; but in observing the motion of those that we can see, we are able, by the aid of this principle, to draw conclusions which remain true whatever may be the details of the invisible mechanism which animates them.

The principle of the conservation of energy, or Mayer's principle, is certainly the most important, but it is not the only one; there are others from which we can derive the same advantage. These are:

Carnot's principle, or the principle of the degradation of energy.

Newton's principle, or the principle of the equality of action and reaction.

The principle of relativity, according to which the laws of physical phenomena must be the same for a stationary observer as for an observer carried along in a uniform motion of translation; so that we have not and cannot have any means of discerning whether or not we are carried along in such a motion.

The principle of the conservation of mass, or Lavoisier's principle.

I will add the principle of least action.

The application of these five or six general principles to the different physical phenomena is sufficient for our learning of them all that we could reasonably hope to know of them. The most remarkable example of this new mathematical physics is, beyond question, Maxwell's electro-magnetic theory of light.

We know nothing as to what the ether is, how its molecules are disposed, whether they attract or repel each other; but we know that this medium transmits at the same time the optical perturbations and the electrical perturbations; we know that this transmission must take place in conformity with the general principles of mechanics, and that suffices us for the establishment of the equations of the electro-magnetic field.

These principles are results of experiments boldly generalized; but they seem to derive from their very generality a high degree of certainty. In fact, the more general they are, the more frequent are the opportunities to check them, and the verifications multiplying, taking the most varied, the most unexpected forms, end by no longer leaving place for doubt.

Utility of the Old Physics. Such is the second phase of the history of mathematical physics and we have not yet emerged from it. Shall we say that the first has been useless? that during fifty years science went the wrong way, and that there is nothing left but to forget so many accumulated efforts that a vicious conception condemned in advance to failure? Not the least in the world. Do you think the second phase could have come into existence without the first? The hypothesis of central forces contained all the principles; it involved them as necessary consequences; it involved both the conservation of energy and that of masses, and the equality of action and reaction, and the law of least action, which appeared, it is true, not as experimental truths, but as theorems; the enunciation of which had at the same time something more precise and less general than under their present form.

It is the mathematical physics of our fathers which has familiarized us little by little with these various principles; which has habituated us to recognize them under the different vestments in which they disguise themselves. They have been compared with the data of experience, it has been seen how it was necessary to modify their enunciation to adapt them to these data; thereby they have been extended and consolidated. Thus they came to be regarded as experimental truths; the conception of central forces became then a useless support, or rather an embarrassment, since it made the principles partake of its hypothetical character.

The frames then have not broken, because they are elastic; but they have enlarged; our fathers, who established them, did not labor in vain, and we recognize in the science of today the general traits of the sketch which they traced.

CHAPTER VIII

The Present Crisis of Mathematical Physics

The New Crisis. Are we now about to enter upon a third period? Are we on the eve of a second crisis? These principles on which we have built all, are they about to crumble away in their turn? This has been for some time a pertinent question.

When I speak thus, you no doubt think of radium, that grand revolutionist of the present time, and in fact I shall come back to it presently; but there is something else. It is not alone the conservation of energy which is in question; all the other principles are equally in danger, as we shall see in passing them successively in review.

Carnot's Principle. Let us commence with the principle of Carnot. This is the only one which does not present itself as an immediate consequence of the hypothesis of central forces; more than that, it seems, if not to directly contradict that hypothesis, at least not to be reconciled with it without a certain effort. If physical phenomena were due exclusively to the movements of atoms whose mutual attraction depended only on the distance, it seems that all these phenomena should be reversible; if all the initial velocities were reversed, these atoms, always subjected to the same forces, ought to go over their trajectories in the contrary sense, just as the earth would describe in the retrograde sense this same elliptic orbit which it describes in the direct sense, if the initial conditions of its motion had been reversed. On this account, if a physical phenomenon is possible, the inverse phenomenon should be equally so, and one should be able to reascend the course of time. Now, it is not so in nature, and this is precisely what the principle of Carnot teaches us; heat can pass from the warm body to the cold body; it is impossible afterwards to make it take the inverse route and to reestablish differences of temperature which have been effaced. Motion can be wholly dissipated and transformed into heat by friction; the contrary transformation can never be made except partially.

We have striven to reconcile this apparent contradiction. If the world tends towards uniformity, this is not because its ultimate parts, at first unlike, tend to become less and less different; it is because, shifting at random, they end by blending. For an eye which should distinguish all the elements, the variety would remain always as great; each grain of this dust preserves its originality and does not model itself on its neighbours; but as the blend becomes more and more intimate, our gross senses perceive only the uniformity. This is why, for example, temperatures tend to a level, without the possibility of going backwards.

A drop of wine falls into a glass of water; whatever may be the law of the internal motion of the liquid, we shall soon see it coloured of a uniform rosy tint, and however much from this moment one may shake it afterwards, the wine and the water do not seem capable of again separating. Here we have the type of the irreversible physical phenomenon: to hide a grain of barley in a heap of wheat, this is easy; afterwards to find it again and get it out, this is practically impossible. All this Maxwell and Boltzmann have explained; but the one who has seen it most clearly, in a book too little read because it is a little difficult to read, is Gibbs, in his *Elementary Principles of Statistical Mechanics*.

For those who take this point of view, Carnot's principle is only an imperfect principle, a sort of concession to the infirmity of our senses; it is because our eyes are too gross that we do not distinguish the elements of the blend; it is because our hands are too gross that we cannot force them to separate; the imaginary demon of Maxwell, who is able to sort the molecules one by one, could well constrain the world to return backwards. Can it return of itself? That is not impossible; that is only infinitely improbable. The chances are that we should wait a long time for the concourse of circumstances which would permit a retrogradation; but sooner or later they will occur, after years whose number it would take millions of figures to write. These reservations, however, all remained theoretic; they were not very disquieting, and Carnot's principle retained all its principal value. But here the scene changes. The biologist, armed with his microscope, long ago noticed in his preparations irregular movements of little particles in suspension; this is the Brownian movement. He first thought this was a vital phenomenon, but soon he saw that the inanimate bodies danced with no less ardor than the others; then he turned the matter over to the physicists. Unhappily, the physicists remained long uninterested in this question; one concentrates the light to illuminate the microscopic preparation, thought they; with light goes heat; thence inequalities of temperature and in the liquid interior currents which produce the movements referred to.

It occurred to M. Gouy to look more closely, and he saw, or thought he saw, that this explanation is untenable, that the movements become brisker as the particles are smaller, but that they are not influenced by the mode of illumination. If then these movements never cease, or rather are reborn without cease, without borrowing anything from an external source of energy, what ought we to believe? To be sure, we should not on this account renounce our belief in the conservation of energy, but we see under our eyes now motion transformed into heat by friction, now inversely heat changed into motion, and that without loss since the movement lasts forever. This is the contrary of Carnot's principle. If this be so, to see the world return backwards, we no longer have need of the infinitely keen eye of Maxwell's demon; our microscope suffices. Bodies too large, those, for example, which are a tenth of a millimeter, are hit from all sides by moving atoms, but they do not budge, because these shocks are very numerous and the law of chance makes them compensate each other; but the smaller particles receive too few shocks for this compensation to take place with certainty and are incessantly knocked about. And behold already one of our principles in peril.

The Principle of Relativity. Let us pass to the principle of relativity: this not only is confirmed by daily experience, not only is it a necessary consequence of the hypothesis of central forces, but it is irresistibly imposed upon our good sense, and yet it also is assailed. Consider two electrified bodies; though they seem to us at rest, they are both carried along by the motion of the earth; an electric charge in motion, Rowland has taught us, is equivalent to a current; these two charged bodies are, therefore, equivalent to two parallel currents of the same sense and these two currents should attract each other. In measuring this attraction, we shall measure the velocity of the earth; not its velocity in relation to the sun or the fixed stars, but its absolute velocity.

I well know what will be said: it is not its absolute velocity that is measured, it is its velocity in relation to the ether. How unsatisfactory that is! Is it not evident that from the principle so understood we could no longer infer anything? It could no longer tell us anything just because it would no longer fear any contradiction. If we succeed in measuring anything, we shall always be free to say that this is not the absolute velocity, and if it is not the velocity in relation to the ether, it might always be the velocity in relation to some new unknown fluid with which we might fill space.

Indeed, experiment has taken upon itself to ruin this interpretation of the principle of relativity; all attempts to measure the velocity of the earth in relation to the ether have led to negative results. This time experimental physics has been more faithful to the principle than mathematical physics; the theorists, to put in accord their other general views, would not have spared it; but experiment has been stubborn in confirming it. The means have been varied; finally Michelson pushed precision to its last limits; nothing came of it. It is precisely to explain this obstinacy that the mathematicians are forced today to employ all their ingenuity.

Their task was not easy, and if Lorentz has got through it, it is only by accumulating hypotheses.

The most ingenious idea was that of local time. Imagine two observers who wish to adjust their timepieces by optical signals; they exchange signals, but as they know that the transmission of light is not instantaneous, they are careful to cross them. When station B perceives the signal from station A, its clock should not mark the same hour as that of station A at the moment of sending the signal, but this hour augmented by a constant representing the duration of the transmission. Suppose, for example, that station A sends its signal when its clock marks the hour O, and that station B perceives it when its clock marks the hour t. The clocks are adjusted if the slowness equal to t represents the duration of the transmission, and to verify it, station B sends in its turn a signal when its clock marks O; then station A should perceive it when its clock marks t. The timepieces are then adjusted.

And in fact they mark the same hour at the same physical instant, but on the one condition, that the two stations are fixed. Otherwise the duration of the transmission will not be the same in the two senses, since the station A, for example, moves forward to meet the optical perturbation emanating from B, whereas the station B flees before the perturbation emanating from A. The watches adjusted in that way will not mark, therefore, the true time; they will mark what may be called the *local time,* so that one of them will gain on the other. It matters little, since we have no means of perceiving it. All the phenomena which happen at A, for example, will be late, but all will be equally so, and the observer will not perceive it, since his watch is slow; so, as the principle of relativity would have it, he will have no means of knowing whether he is at rest or in absolute motion.

Unhappily, that does not suffice, and complementary hypotheses are necessary; it is necessary to admit that bodies in motion undergo a uniform contraction in the sense of the motion. One of the diameters of the earth, for example, is shrunk by one two-hundred-millionth in consequence of our planet's motion, while the other diameter retains its normal length. Thus the last little differences are compensated. And then, there is still the hypothesis about forces. Forces, whatever be their origin, gravity as well as elasticity, would be reduced in a certain proportion in a world animated by a uniform translation; or, rather, this would happen for the components perpendicular to the translation; the components parallel would not change. Resume, then, our example of two electrified bodies; these bodies repel each other, but at the same time if all is carried along in a uniform translation, they are equivalent to two parallel currents of the same sense which attract each other. This electro-dynamic attraction diminishes, therefore, the electro-static repulsion, and the total repulsion is feebler than if the two bodies were at rest. But since to measure this repulsion we must balance it by another force, and all these other forces are reduced in the same proportion, we perceive nothing. Thus, all seems arranged, but are all the doubts dissipated? What would happen if one could communicate by non-luminous signals whose velocity of propagation differed from that of light? If, after having adjusted the watches by the optical procedure, we wished to verify the adjustment by the aid of these new signals, we should observe discrepancies which would render evident the common translation of the two stations. And are such signals inconceivable, if we admit with Laplace that universal gravitation is transmitted a million times more rapidly than light?

Thus, the principle of relativity has been valiantly defended in these latter times, but the very energy of the defence proves how serious was the attack.

Newton's Principle. Let us speak now of the principle of Newton, on the equality of action and reaction. This is intimately bound up with the preceding, and it seems indeed that the fall of the one would involve that of the other. Thus we must not be astonished to find here the same difficulties.

Electrical phenomena, according to the theory of Lorentz, are due to the displacements of little charged particles, called electrons, immersed in the medium we call ether. The movements of these electrons produce perturbations in the neighbouring ether; these perturbations propagate themselves in every direction with the velocity of light, and in turn other electrons, originally at rest, are made to vibrate when the perturbation reaches the parts of the ether which touch them. The electrons, therefore, act on one another, but this action is not direct, it is accomplished through the ether as intermediary. Under these conditions can there be compensation between action and reaction, at least for an observer who should take account only of the movements of matter, that is, of the electrons, and who should be ignorant of those of the ether that he could not see? Evidently not. Even if the compensation should be exact, it could not be simultaneous. The perturbation is propagated with a finite velocity; it, therefore, reaches the second electron only when the first has long ago entered upon its rest. This second electron, therefore, will undergo, after a delay, the action of the first, but will certainly not at that moment react upon it, since around this first electron nothing any longer budges.

The analysis of the facts permits us to be still more precise. Imagine, for example, a Hertzian oscillator, like those used in wireless telegraphy; it sends out energy in every direction; but we can provide it with a parabolic mirror, as Hertz did with his smallest oscillators, so as to send all the energy produced in a single direction. What happens then according to the theory? The apparatus recoils, as if it were a cannon and the projected energy a ball; and that is contrary to the principle of Newton, since our projectile here has no mass, it is not matter, it is energy. The case is still the same, moreover, with a beacon light provided with a reflector, since light is nothing but a perturbation of the electro-magnetic field. This beacon light should recoil as if the light it sends out were a projectile. What is the force that should produce this recoil? It is what is called the Maxwell-Bartholi pressure. It is very minute, and it has been difficult to put it in evidence even with the most sensitive radiometers; but it suffices that it exists.

If all the energy issuing from our oscillator falls on a receiver, this will act as if it had received a mechanical shock, which will represent in a sense the compensation of the oscillator's recoil; the reaction will be equal to the action, but it will not be simultaneous; the receiver will move on, but not at the moment when the oscillator recoils. If the energy propagates itself indefinitely without encountering a receiver, the compensation will never occur.

Shall we say that the space which separates the oscillator from the receiver and which the perturbation must pass over in going from the one to the other is not void, that it is full not only of ether, but of air, or even in the interplanetary spaces of some fluid subtle but still ponderable; that this matter undergoes the shock like the receiver at the moment when the energy reaches it, and recoils in its turn when the perturbation quits it? That would save Newton's principle, but that is not true. If energy in its diffusion remained always attached to some material substratum, then matter in motion would carry along light with it, and Fizeau has demonstrated that it does nothing of the sort, at least for air. Michelson and Morley have since confirmed this. It might be supposed also that the movements of matter proper are exactly compensated by those of the ether; but that would lead us to the same reflections as before now. The principle so understood will explain everything, since, whatever might be the visible movements, we always could imagine hypothetical movements which compensate them. But if it is able to explain everything, this is because it does not enable us to foresee anything; it does not enable us to decide between the different possible hypotheses, since it explains everything beforehand. It therefore becomes useless.

And then the suppositions that it would be necessary to make on the movements of the ether are not very satisfactory. If the electric charges double, it would be natural to imagine that the velocities of the diverse atoms of ether double also, and for the compensation, it would be necessary that the mean velocity of the ether quadruple.

This is why I have long thought that these consequences of theory, contrary to Newton's principle, would end some day by being abandoned, and yet the recent experiments on the movements of the electrons issuing from radium seem rather to confirm them.

Lavoisier's Principle. I arrive at the principle of Lavoisier on the conservation of mass. Certainly, this is one not to be touched without unsettling all mechanics. And now certain persons think that it seems true to us only because in mechanics merely moderate velocities are considered, but that it would cease to be true for bodies animated by velocities comparable to that of light. Now these velocities, it is believed at present, have been realized; the cathode rays or those of radium may be formed of very minute particles or of electrons which are displaced with velocities smaller no doubt than that of light, but which might be its one tenth or one third.

These rays can be deflected, whether by an electric field, or by a magnetic field, and we are able, by comparing these deflections, to measure at the same time the velocity of the electrons and their mass (or rather the relation of their mass to their charge). But when it was seen that these velocities approached that of light, it was decided that a correction was necessary. These molecules, being electrified, cannot be displaced without agitating the ether; to put them in motion it is necessary to overcome a double inertia, that of the molecule itself and that of the ether. The total or apparent mass that one measures is composed, therefore, of two parts: the real or mechanical mass of the molecule and the electro-dynamic mass representing the inertia of the ether.

The calculations of Abraham and the experiments of Kaufmann have then shown that the mechanical mass, properly so called, is null, and that the mass of the electrons, or, at least, of the negative electrons, is of exclusively electro-dynamic origin. This is what forces us to change the definition of mass; we cannot any longer distinguish mechanical mass and electro-dynamic mass, since then the first would vanish; there is no mass other than electro-dynamic inertia. But in this case the mass can no longer be constant; it augments with the velocity, and it even depends on the direction, and a body animated by a notable velocity will not oppose the same inertia to the forces which tend to deflect it from its route, as to those which tend to accelerate or to retard its progress.

There is still a resource; the ultimate elements of bodies are electrons, some charged negatively, the others charged positively.

The negative electrons have no mass, this is understood; but the positive electrons, from the little we know of them, seem much greater. Perhaps they have, besides their electro-dynamic mass, a true mechanical mass. The real mass of a body would, then, be the sum of the mechanical masses of its positive electrons, the negative electrons not counting; mass so defined might still be constant.

Alas! this resource also evades us. Recall what we have said of the principle of relativity and of the efforts made to save it. And it is not merely a principle which it is a question of saving, it is the indubitable results of the experiments of Michelson.

Well, as was above seen, Lorentz, to account for these results, was obliged to suppose that all forces, whatever their origin, were reduced in the same proportion in a medium animated by a uniform translation; this is not sufficient; it is not enough that this take place for the real forces, it must also be the same for the forces of inertia; it is therefore necessary, he says, that *the masses of all the particles be influenced by a translation to the same degree as the electro-magnetic masses of the electrons.*

So the mechanical masses must vary in accordance with the same laws as the electrodynamic masses; they cannot, therefore, be constant.

Need I point out that the fall of Lavoisier's principle involves that of Newton's? This latter signifies that the center of gravity of an isolated system moves in a straight line; but if there is no longer a constant mass, there is no longer a center of gravity, we no longer know even what this is. This is why I said above that the experiments on the cathode rays appeared to justify the doubts of Lorentz concerning Newton's principle.

From all these results, if they were confirmed, would arise an entirely new mechanics, which would be, above all, characterized by this fact, that no velocity could surpass that of light,* any more than any temperature can fall below absolute zero.

---

* Because bodies would oppose an increasing inertia to the causes which would tend to accelerate their motion; and this inertia would become infinite when one approached the velocity of light.

No more for an observer, carried along himself in a translation he does not suspect, could any apparent velocity surpass that of light; and this would be then a contradiction, if we did not recall that this observer would not use the same clocks as a fixed observer, but, indeed, clocks marking "local time."

Here we are then facing a question I content myself with stating. If there is no longer any mass, what becomes of Newton's law? Mass has two aspects: it is at the same time a co-efficient of inertia and an attracting mass entering as factor into Newtonian attraction. If the co-efficient of inertia is not constant, can the attracting mass be? That is the question.

Mayer's Principle. At least, the principle of the conservation of energy yet remained to us, and this seemed more solid. Shall I recall to you how it was in its turn thrown into discredit? This event has made more noise than the preceding, and it is in all the memoirs. From the first works of Becquerel, and, above all, when the Curies had discovered radium, it was seen that every radioactive body was an inexhaustible source of radiation. Its activity seemed to subsist without alteration throughout the months and the years. This was in itself a strain on the principles; these radiations were in fact energy, and from the same morsel of radium this issued and forever issued. But these quantities of energy were too slight to be measured; at least that was the belief and we were not much disquieted.

The scene changed when Curie bethought himself to put radium in a calorimeter; it was then seen that the quantity of heat incessantly created was very notable.

The explanations proposed were numerous; but in such case we cannot say, the more the better. Insofar as no one of them has prevailed over the others, we cannot be sure there is a good one among them. Since some time, however, one of those explanations seems to be getting the upper hand and we may reasonably hope that we hold the key to the mystery.

Sir W. Ramsay has striven to show that radium is in process of transformation, that it contains a store of energy enormous but not inexhaustible. The transformation of radium then would produce a million times more heat than all known transformations; radium would wear itself out in 1,250 years; this is quite short, and you see that we are at least certain to have this point settled some hundreds of years from now. While waiting, our doubts remain.

CHAPTER IX

THE FUTURE OF
MATHEMATICAL PHYSICS

The Principles and Experiment. In the midst of so much ruin, what remains standing? The principle of least action is hitherto intact, and Larmor appears to believe that it will long survive the others; in reality, it is still more vague and more general.

In presence of this general collapse of the principles, what attitude will mathematical physics take? And first, before too much excitement, it is proper to ask if all that is really true. All these derogations to the principles are encountered only among infinitesimals; the microscope is necessary to see the Brownian movement; electrons are very light; radium is very rare, and one never has more than some milligrams of it at a time. And, then, it may be asked whether, besides the infinitesimal seen, there was not another infinitesimal unseen counter-poise to the first.

So there is an interlocutory question, and, as it seems, only experiment can solve it. We shall, therefore, only have to hand over the matter to the experimenters, and, while waiting for them to finally decide the debate, not to preoccupy ourselves with these disquieting problems, and to tranquilly continue our work as if the principles were still uncontested. Certes, we have much to do without leaving the domain where they may be applied in all security; we have enough to employ our activity during this period of doubts.

The Role of the Analyst. And as to these doubts, is it indeed true that we can do nothing to disembarrass science of them? It must indeed be said, it is not alone experimental physics that has given birth to them; mathematical physics has as well contributed. It is the experimenters who have seen radium throw out energy, but it is the theorists who have put in evidence all the difficulties raised by the propagation of light across a medium in motion; but for these it is probable we should not have become conscious of them. Well, then, if they have done their best to put us into this embarrassment, it is proper also that they help us to get out of it.

They must subject to critical examination all these new views I have just outlined before you, and abandon the principles only after having made a loyal effort to save them. What can they do in this sense? That is what I will try to explain.

It is a question before all of endeavouring to obtain a more satisfactory theory of the electro-dynamics of bodies in motion. It is there especially, as I have sufficiently shown above, that difficulties accumulate. It is useless to heap up hypotheses, we cannot satisfy all the principles at once; so far, one has succeeded in safeguarding some only on condition of sacrificing the others; but all hope of obtaining better results is not yet lost. Let us take, then, the theory of Lorentz, turn it in all senses, modify it little by little, and perhaps everything will arrange itself.

Thus in place of supposing that bodies in motion undergo a contraction in the sense of the motion, and that this contraction is the same whatever be the nature of these bodies and the forces to which they are otherwise subjected, could we not make a more simple and natural hypothesis? We might imagine, for example, that it is the ether which is modified when it is in relative motion in reference to the material medium which penetrates it, that, when it is thus modified, it no longer transmits perturbations with the same velocity in every direction. It might transmit more rapidly those which are propagated parallel to the motion of the medium, whether in the same sense or in the opposite sense, and less rapidly those which are propagated perpendicularly. The wave surfaces would no longer be spheres, but ellipsoids, and we could dispense with that extraordinary contraction of all bodies.

I cite this only as an example, since the modifications that might be essayed would be evidently susceptible of infinite variation.

Aberration and Astronomy. It is possible also that astronomy may some day furnish us data on this point; she it was in the main who raised the question in making us acquainted with the phenomenon of the aberration of light. If we make crudely the theory of aberration, we reach a very curious result. The apparent positions of the stars differ from their real positions because of the earth's motion, and as this motion is variable, these apparent positions vary. The real position we cannot ascertain, but we can observe the variations of the apparent position. The observations of the aberration show us, therefore, not the earth's motion, but the variations of this motion; they cannot, therefore, give us information about the absolute motion of the earth.

At least this is true in first approximation, but the case would be no longer the same if we could appreciate the thousandths of a second. Then it would be seen that the amplitude of the oscillation depends not alone on the variation of the motion, a variation which is well known, since it is the motion of our globe on its elliptic orbit, but on the mean value of this motion, so that the constant of aberration would not be quite the same for all the stars, and the differences would tell us the absolute motion of the earth in space.

This, then, would be, under another form, the ruin of the principle of relativity. We are far, it is true, from appreciating the thousandth of a second, but, after all, say some, the earth's total absolute velocity is perhaps much greater than its relative velocity with respect to the sun. If, for example, it were 300 kilometres per second in place of 30, this would suffice to make the phenomenon observable.

I believe that in reasoning thus one admits a too simple theory of aberration. Michelson has shown us, I have told you, that the physical procedures are powerless to put in evidence absolute motion; I am persuaded that the same will be true of the astronomic procedures, however far precision be carried.

However that may be, the data astronomy will furnish us in this regard will some day be precious to the physicist. Meanwhile, I believe that the theorists, recalling the experience of Michelson, may anticipate a negative result, and that they would accomplish a useful work in constructing a theory of aberration which would explain this in advance.

Electrons and Spectra. These dynamics of electrons can be approached from many sides, but among the ways leading thither is one which has been somewhat neglected, and yet this is one of those which promise us the most surprises. It is movements of electrons which produce the lines of the emission spectra; this is proved by the Zeeman effect; in an incandescent body what vibrates is sensitive to the magnet, therefore electrified. This is a very important first point, but no one has gone further. Why are the lines of the spectrum distributed in accordance with a regular law? These laws have been studied by the experimenters in their least details; they are very precise and comparatively simple. A first study of these distributions recalls the harmonics encountered in acoustics; but the difference is great. Not only are the numbers of vibrations not the successive multiples of a single number, but we do not even find anything analogous to the roots of those transcendental equations to which we are led by so many problems of mathematical physics: that of the vibrations of an elastic body of any form, that of the Hertzian oscillations in a generator of any form, the problem of Fourier for the cooling of a solid body.

The laws are simpler, but they are of wholly other nature, and to cite only one of these differences, for the harmonics of high order, the number of vibrations tends towards a finite limit, instead of increasing indefinitely.

That has not yet been accounted for, and I believe that there we have one of the most important secrets of nature. A Japanese physicist, M. Nagaoka, has recently proposed an explanation; according to him, atoms are composed of a large positive electron surrounded by a ring formed of a very great number of very small negative electrons. Such is the planet Saturn with its rings. This is a very interesting attempt, but not yet wholly satisfactory; this attempt should be renewed. We will penetrate, so to speak, into the inmost recess of matter. And from the particular point of view which we today occupy, when we know why the vibrations of incandescent bodies differ thus from ordinary elastic vibrations, why the electrons do not behave like the matter which is familiar to us, we shall better comprehend the dynamics of electrons and it will be perhaps more easy for us to reconcile it with the principles.

Conventions Preceding Experiment. Suppose, now, that all these efforts fail, and, after all, I do not believe they will, what must be done? Will it be necessary to seek to mend the broken principles by giving what we French call a *coup de pouce?* That evidently is always possible, and I retract nothing of what I have said above.

Have you not written, you might say if you wished to seek a quarrel with me—have you not written that the principles, though of experimental origin, are now unassailable by experiment because they have become conventions? And now you have just told us that the most recent conquests of experiment put these principles in danger.

Well, formerly I was right and today I am not wrong. Formerly I was right, and what is now happening is a new proof of it. Take, for example, the calorimetric experiment of Curie on radium. Is it possible to reconcile it with the principle of the conservation of energy? This has been attempted in many ways; but there is among them one I should like you to notice; this is not the explanation which tends today to prevail, but it is one of those which have been proposed. It has been conjectured that radium was only an intermediary, that it only stored radiations of unknown nature which flashed through space in every direction, traversing all bodies, save radium, without being altered by this passage and without exercising any action upon them. Radium alone took from them a little of their energy and afterwards gave it out to us in various forms.

What an advantageous explanation, and how convenient! First, it is unverifiable and thus irrefutable. Then again it will serve to account for any derogation whatever to Mayer's principle; it answers in advance not only the objection of Curie, but all the objections that future experimenters might accumulate. This new and unknown energy would serve for everything.

This is just what I said, and therewith we are shown that our principle is unassailable by experiment.

But then, what have we gained by this stroke? The principle is intact, but thenceforth of what use is it? It enabled us to foresee that in such or such circumstance we could count on such a total quantity of energy; it limited us; but now that this indefinite provision of new energy is placed at our disposal, we are no longer limited by anything; and, as I have written in *Science and Hypothesis,* if a principle ceases to be fecund, experiment without contradicting it directly will nevertheless have condemned it.

Future Mathematical Physics. This, therefore, is not what would have to be done; it would be necessary to rebuild anew. If we were reduced to this necessity, we could moreover console ourselves. It would not be necessary thence to conclude that science can weave only a Penelope's web, that it can raise only ephemeral structures, which it is soon forced to demolish from top to bottom with its own hands.

As I have said, we have already passed through a like crisis. I have shown you that in the second mathematical physics, that of the principles, we find traces of the first, that of central forces; it will be just the same if we must know a third. Just so with the animal that exuviates, that breaks its too narrow carapace and makes itself a fresh one, under the new envelope one will recognize the essential traits of the organism which have persisted.

We cannot foresee in what way we are about to expand; perhaps it is the kinetic theory of gases which is about to undergo development and serve as model to the others. Then the facts which first appeared to us as simple thereafter would be merely resultants of a very great number of elementary facts which only the laws of chance would make cooperate for a common end. Physical law would then assume an entirely new aspect; it would no longer be solely a differential equation, it would take the character of a statistical law.

Perhaps, too, we shall have to construct an entirely new mechanics that we only succeed in catching a glimpse of, where, inertia increasing with the velocity, the velocity of light would become an impassable limit. The ordinary mechanics, more simple, would remain a first approximation, since it would be true for velocities not too great, so that the old dynamics would still be found under the new. We should not have to regret having believed in the principles, and even, since velocities too great for the old formulas would always be only exceptional, the surest way in practise would be still to act as if we continued to believe in them. They are so useful, it would be necessary to keep a place for them. To determine to exclude them altogether would be to deprive oneself of a precious weapon. I hasten to say in conclusion that we are not yet there, and as yet nothing proves that the principles will not come forth from out of the fray, victorious and intact.[*]

---

[*] These considerations on mathematical physics are borrowed from my St. Louis address.

THE OBJECTIVE VALUE OF SCIENCE

CHAPTER X

IS SCIENCE ARTIFICIAL?

I. THE PHILOSOPHY OF M. LEROY

There are many reasons for being skeptics; should we push this skepticism to the very end or stop on the way? To go to the end is the most tempting solution, the easiest, and that which many have adopted, despairing of saving anything from the shipwreck.

Among the writings inspired by this tendency it is proper to place in the first rank those of M. LeRoy. This thinker is not only a philosopher and a writer of the greatest merit, but he has acquired a deep knowledge of the exact and physical sciences, and even has shown rare powers of mathematical invention. Let us recapitulate in a few words his doctrine, which has given rise to numerous discussions.

Science consists only of conventions, and to this circumstance solely does it owe its apparent certitude; the facts of science and, *a fortiori,* its laws are the artificial work of the scientist; science therefore can teach us nothing of the truth; it can only serve us as a rule of action.

Here we recognize the philosophic theory known under the

318 · *The Value of Science*

name of nominalism; all is not false in this theory; its legitimate domain must be left it, but out of this it should not be allowed to go.

This is not all; M. LeRoy's doctrine is not only nominalistic; it has besides another characteristic which it doubtless owes to M. Bergson, it is anti-intellectualistic. According to M. LeRoy, the intellect deforms all it touches, and that is still more true of its necessary instrument "discourse." There is reality only in our fugitive and changing impressions, and even this reality, when touched, vanishes.

And yet M. LeRoy is not a skeptic; if he regards the intellect as incurably powerless, it is only to give more scope to other sources of knowledge, to the heart, for instance, to sentiment, to instinct or to faith.

However great my esteem for M. LeRoy's talent, whatever the ingenuity of this thesis, I cannot wholly accept it. Certes, I am in accord on many points with M. LeRoy, and he has even cited, in support of his view, various passages of my writings which I am by no means disposed to reject. I think myself only the more bound to explain why I cannot go with him all the way.

M. LeRoy often complains of being accused of skepticism. He could not help being, though this accusation is probably unjust. Are not appearances against him? Nominalist in doctrine, but realist at heart, he seems to escape absolute nominalism only by a desperate act of faith.

The fact is that anti-intellectualistic philosophy in rejecting analysis and "discourse," just by that condemns itself to being intransmissible, it is a philosophy essentially internal, or, at the very least, only its negations can be transmitted; what wonder then that for an external observer it takes the shape of skepticism?

Therein lies the weak point of this philosophy; if it strives to remain faithful to itself, its energy is spent in a negation and a cry of enthusiasm. Each author may repeat this negation and this cry, may vary their form, but without adding anything.

And yet, would it not be more logical in remaining silent? See, you have written long articles; for that, it was necessary to use words. And therein have you not been much more "discursive" and consequently much further from life and truth than the animal who simply lives without philosophizing? Would not this animal be the true philosopher?

However, because no painter has made a perfect portrait, should we conclude that the best painting is not to paint? When a zoologist dissects an animal, certainly he "alters it." Yes, in dissecting it, he condemns himself to never know all of it; but in not dissecting it, he would condemn himself to never know anything of it and consequently to never see anything of it.

Certes, in man are other forces besides his intellect, no one has ever been mad enough to deny that. The first comer makes these blind forces act or lets them act; the philosopher must *speak* of them; to speak of them, he must know of them the little that can be known, he should therefore *see* them act. How? With what eyes, if not with his intellect? Heart, instinct, may guide it, but not render it useless; they may direct the look, but not replace the eye. It may be granted that the heart is the workman, and the intellect only the instrument. Yet is it an instrument not to be done without, if not for action, at least for philosophizing? Therefore a philosopher really anti-intellectualistic is impossible. Perhaps we shall have to declare for the supremacy of action; always it is our intellect which will thus conclude; in allowing precedence to action it will thus retain the superiority of the thinking reed. This also is a supremacy not to be disdained.

Pardon these brief reflections and pardon also their brevity, scarcely skimming the question. The process of intellectualism is not the subject I wish to treat: I wish to speak of science, and about it there is no doubt; by definition, so to speak, it will be intellectualistic or it will not be at all. Precisely the question is, whether it will be.

II. Science, Rule of Action

For M. LeRoy, science is only a rule of action. We are powerless to know anything and yet we are launched, we must act, and at all hazards we have established rules. It is the aggregate of these rules that is called science.

It is thus that men, desirous of diversion, have instituted rules of play, like those of tric-trac, for instance, which, better than science itself, could rely upon the proof by universal consent. It is thus likewise that, unable to choose, but forced to choose, we toss up a coin, head or tail to win.

The rule of tric-trac is indeed a rule of action like science, but does anyone think the comparison just and not see the difference? The rules of the game are arbitrary conventions, and the contrary convention might have been adopted, *which would have been nonetheless good.* On the contrary, science is a rule of action which is successful, generally at least, and I add, while the contrary rule would not have succeeded.

If I say, to make hydrogen cause an acid to act on zinc, I formulate a rule which succeeds; I could have said, make distilled water act on gold; that also would have been a rule, only it would not have succeeded. If, therefore, scientific "recipes" have a value, as a rule of action, it is because we know they succeed, generally at least. But to know this is to know something and then why tell us we can know nothing?

Science foresees, and it is because it foresees, that it can be useful and serve as a rule of action. I well know that its previsions are often contradicted by the event; that shows that science is imperfect and if I add that it will always remain so, I am certain that this is a prevision which, at least, will never be contradicted. Always the scientist is less often mistaken than a prophet who should predict at random. Besides the progress though slow is continuous, so that scientists, though more and more bold, are less and less misled. This is little, but it is enough.

I well know that M. LeRoy has somewhere said that science was mistaken oftener than one thought, that comets sometimes played tricks on astronomers, that scientists, who apparently are men, did not willingly speak of their failures and that, if they should speak of them, they would have to count more defeats than victories.

That day, M. LeRoy evidently overreached himself. If science did not succeed, it could not serve as a rule of action; whence would it get its value? Because it is "lived," that is, because we love it and believe in it? The alchemists had recipes for making gold, they loved them and had faith in them, and yet our recipes are the good ones, although our faith be less lively, because they succeed.

There is no escape from this dilemma; either science does not enable us to foresee, and then it is valueless as a rule of action; or else it enables us to foresee in a fashion more or less imperfect, and then it is not without value as a means of knowledge.

It should not even be said that action is the goal of science; should we condemn studies of the star Sirius, under pretext that we shall probably never exercise any influence on that star? To my eyes, on the contrary, it is the knowledge which is the end, and the action which is the means. If I felicitate myself on the industrial development, it is not alone because it furnishes a facile argument to the advocates of science; it is above all because it gives to the scientist faith in himself and also because it offers an immense field of experience where clash forces too colossal to be interfered with. Without this ballast, who knows whether it would not quit the earth, seduced by the mirage of some scholastic novelty, or whether it would not despair, believing it had fashioned only a dream?

III. THE CRUDE FACT AND THE SCIENTIFIC FACT

What was most paradoxical in M. LeRoy's thesis was that affirmation that *the scientist creates the fact;* this was at the same time its essential point and it is one of those which have been most discussed.

Perhaps, says he (I well believe that this was a concession), it is not the scientist that creates the fact in the rough; it is at least he who creates the scientific fact.

This distinction between the fact in the rough and the scientific fact does not by itself appear to me illegitimate. But I complain first that the boundary has not been traced either exactly or precisely; and then that the author has seemed to suppose that the crude fact, not being scientific, is outside of science.

Finally, I cannot admit that the scientist creates without restraint the scientific fact since it is the crude fact which imposes it upon him.

The examples given by M. LeRoy have greatly astonished me. The first is taken from the notion of atom. The atom chosen as example of fact! I avow that this choice has so disconcerted me that I prefer to say nothing about it. I have evidently misunderstood the author's thought and I could not fruitfully discuss it.

The second case taken as example is that of an eclipse where the crude phenomenon is a play of light and shadow, but where the astronomer cannot intervene without introducing two foreign elements, to wit, a clock and Newton's law.

Finally, M. LeRoy cites the rotation of the earth; it has been answered: but this is not a fact, and he has replied: it was one for Galileo, who affirmed it, as for the inquisitor, who denied it. It always remains that this is not a fact in the same sense as those just spoken of and that to give them the same name is to expose one's self to many confusions.

Here then are four degrees:

1°. It grows dark, says the clown.
2°. The eclipse happened at nine o'clock, says the astronomer.
3°. The eclipse happened at the time deducible from the tables constructed according to Newton's law, says he again.
4°. That results from the earth's turning around the sun, says Galileo finally.

Where then is the boundary between the fact in the rough and the scientific fact? To read M. LeRoy one would believe that it is between the first and the second stage, but who does not see that there is a greater distance from the second to the third, and still more from the third to the fourth.

Allow me to cite two examples which perhaps will enlighten us a little.

I observe the deviation of a galvanometer by the aid of a movable mirror which projects a luminous image or spot on a divided scale. The crude fact is this: I see the spot displace itself on the scale, and the scientific fact is this: a current passes in the circuit.

Or again: when I make an experiment I should subject the result to certain corrections, because I know I must have made errors. These errors are of two kinds, some are accidental and these I shall correct by taking the mean; the others are systematic and I shall be able to correct those only by a thorough study of their causes. The first result obtained is then the fact in the rough, while the scientific fact is the final result after the finished corrections.

Reflecting on this latter example, we are led to subdivide our second stage, and in place of saying:

2. The eclipse happened at nine o'clock, we shall say:

2a. The eclipse happened when my clock pointed to nine, and

2b. My clock being ten minutes slow, the eclipse happened at ten minutes past nine.

And this is not all: the first stage also should be subdivided, and not between these two subdivisions will be the least distance; it is necessary to distinguish between the impression of obscurity felt by one witnessing an eclipse, and the affirmation; it grows dark, which this impression extorts from him. In a sense it is the first which is the only true fact in the rough, and the second is already a sort of scientific fact.

Now then our scale has six stages, and even though there is no reason for halting at this figure, there we shall stop.

What strikes me at the start is this. At the first of our six stages, the fact, still completely in the rough, is, so to speak, individual, it is completely distinct from all other possible facts. From the second stage, already it is no longer the same. The enunciation of the fact would suit an infinity of other facts. So soon as language intervenes, I have at my command only a finite number of terms to express the shades, in number infinite, that my impressions might cover. When I say: it grows dark, that well expresses the impressions I feel in being present at an eclipse; but even in obscurity a multitude of shades could be imagined, and if, instead of that actually realized, had happened a slightly different shade, yet I should still have enunciated this *other* fact by saying: It grows dark.

Second remark: even at the second stage, the enunciation of a fact can only be *true or false*. This is not so of any proposition; if this proposition is the enunciation of a convention, it cannot be said that this enunciation is *true,* in the proper sense of the word, since it could not be true apart from me and is true only because I wish it to be.

When, for instance, I say the unit for length is the metre, this is a decree that I promulgate, it is not something ascertained which forces itself upon me. It is the same, as I think I have elsewhere shown, when it is a question for example of Euclid's postulate.

When I am asked: Is it growing dark? I always know whether I ought to reply yes or no. Although an infinity of possible facts may be susceptible to this same enunciation: it grows dark, I shall always know whether the fact realized belongs or does not belong among those which answer to this enunciation. Facts are classed in categories, and if I am asked whether the fact that I ascertain belongs or does not belong in such a category, I shall not hesitate.

Doubtless this classification is sufficiently arbitrary to leave a large part to man's freedom or caprice. In a word, this classification is a convention. *This convention being given,* if I am asked: Is such a fact true? I shall always know what to answer, and my reply will be imposed upon me by the witness of my senses.

If, therefore, during an eclipse, it is asked: Is it growing dark? all the world will answer yes. Doubtless those speaking a language where bright was called dark, and dark bright, would answer no. But of what importance is that?

In the same way, in mathematics, *when I have laid down the definitions, and the postulates which are conventions,* a theorem henceforth can only be true or false. But to answer the question: Is this theorem true? it is no longer to the witness of my senses that I shall have recourse, but to reasoning.

A statement of fact is always verifiable, and for the verification we have recourse either to the witness of our senses, or to the memory of this witness. This is properly what characterizes a fact. If you put the question to me: Is such a fact true? I shall begin by asking you, if there is occasion, to state precisely the conventions, by asking you, in other words, what language you have spoken; then once settled on this point, I shall interrogate my senses and shall answer yes or no. But it will be my senses that will have made the answer, it will not be *you* when you say to me: I have spoken to you in English or in French.

Is there something to change in all that when we pass to the following stages? When I observe a galvanometer, as I have just said, if I ask an ignorant visitor: Is the current passing? he looks at the wire to try to see something pass; but if I put the same question to my assistant who understands my language, he will know I mean: Does the spot move? and he will look at the scale.

What difference is there then between the statement of a fact in the rough and the statement of a scientific fact? The same difference as between the statement of the same crude fact in French and in German. The scientific statement is the translation of the crude statement into a language which is distinguished above all from the common German or French, because it is spoken by a very much smaller number of people.

Yet let us not go too fast. To measure a current I may use a very great number of types of galvanometers or besides an electro-dynamometer. And then when I shall say there is running in this circuit a current of so many amperes, that will mean: if I adapt to this circuit such a galvanometer I shall see the spot come to the division *a;* but that will mean equally: if I adapt to this circuit such an electro-dynamometer, I shall see the spot go to the division *b.* And that will mean still many other things, because the current can manifest itself not only by mechanical effects, but by effects chemical, thermal, luminous, etc.

Here then is one same statement which suits a very great number of facts absolutely different. Why? It is because I assume a law according to which, whenever such a mechanical effect shall happen, such a chemical effect will happen also. Previous experiments, very numerous, have never shown this law to fail, and then I have understood that I could express by the same statement two facts so invariably bound one to the other.

When I am asked: Is the current passing? I can understand that that means: Will such a mechanical effect happen? But I can understand also: Will such a chemical effect happen? I shall then verify either the existence of the mechanical effect, or that of the chemical effect; that will be indifferent, since in both cases the answer must be the same.

And if the law should one day be found false? If it was perceived that the concordance of the two effects, mechanical and chemical, is not constant? That day it would be necessary to change the scientific language to free it from a grave ambiguity.

And after that? Is it thought that ordinary language by aid of which are expressed the facts of daily life is exempt from ambiguity?

Shall we thence conclude that the facts of daily life are the work of the grammarians?

You ask me: Is there a current? I try whether the mechanical effect exists, I ascertain it and I answer: yes, there is a current. You understand at once that that means that the mechanical effect exists, and that the chemical effect, that I have not investigated, exists likewise. Imagine now, supposing an impossibility, the law we believe true, not to be, and the chemical effect not to exist. Under this hypothesis there will be two distinct facts, the one directly observed and which is true, the other inferred and which is false. It may strictly be said that we have created the second. So that error is the part of man's personal collaboration in the creation of the scientific fact.

But if we can say that the fact in question is false, is this not just because it is not a free and arbitrary creation of our mind, a disguised convention, in which case it would be neither true nor false? And in fact it was verifiable; I had not made the verification, but I could have made it. If I answered amiss, it was because I chose to reply too quickly, without having asked nature, who alone knew the secret.

When, after an experiment, I correct the accidental and systematic errors to bring out the scientific fact, the case is the same; the scientific fact will never be anything but the crude fact translated into another language. When I shall say: It is such an hour, that will be a short way of saying: there is such a relation between the hour indicated by my clock, and the hour it marked at the moment of the passing of such a star and such another star across the meridian. And this convention of language once adopted, when I shall be asked: Is it such an hour? it will not depend upon me to answer yes or no.

Let us pass to the stage before the last: the eclipse happened at the hour given by the tables deduced from Newton's law. This is still a convention of language which is perfectly clear for those who know celestial mechanics or simply for those who have the tables calculated by the astronomers. I am asked: Did the eclipse happen at the hour predicted? I look in the nautical almanac, I see that the eclipse was announced for nine o'clock and I understand that the question means: Did the eclipse happen at nine o'clock? There still we have nothing to change in our conclusions. *The scientific fact is only the crude fact translated into a convenient language.*

It is true that at the last stage things change. Does the earth rotate? Is this a verifiable fact? Could Galileo and the Grand Inquisitor, to settle the matter, appeal to the witness of their senses? On the contrary, they were in accord about the appearances, and, whatever had been the accumulated experiences, they would have remained in accord with regard to the appearances without ever agreeing on their interpretation. It is just on that account that they were obliged to have recourse to procedures of discussion so unscientific.

This is why I think they did not disagree about a *fact:* we have not the right to give the same name to the rotation of the earth, which was the object of their discussion, and to the facts crude or scientific we have hitherto passed in review.

After what precedes, it seems superfluous to investigate whether the fact in the rough is outside of science, because there can neither be science without scientific fact, nor scientific fact without fact in the rough, since the first is only the translation of the second.

And then, has one the right to say that the scientist creates the scientific fact? First of all, he does not create it from nothing, since he makes it with the fact in the rough. Consequently he does not make it freely and *as he chooses*. However able the worker may be, his freedom is always limited by the properties of the raw material on which he works.

After all, what do you mean when you speak of this free creation of the scientific fact and when you take as an example the astronomer who intervenes actively in the phenomenon of the eclipse by bringing his clock? Do you mean: the eclipse happened at nine o'clock; but if the astronomer had wished it to happen at ten, that depended only on him, he had only to advance his clock an hour?

But the astronomer, in perpetrating that bad joke, would evidently have been guilty of an equivocation. When he tells me: the eclipse happened at nine, I understand that nine is the hour deduced from the crude indication of the pendulum by the usual series of corrections. If he has given me solely that crude indication, or if he has made corrections contrary to the habitual rules, he has changed the language agreed upon without forewarning me. If, on the contrary, he took care to forewarn me, I have nothing to complain of, but then it is always the same fact expressed in another language.

In sum, *all the scientist creates in a fact is the language in which he enunciates it.* If he predicts a fact, he will employ this language, and for all those who can speak and understand it, his prediction is free from ambiguity. Moreover, this prediction once made, it evidently does not depend upon him whether it is fulfilled or not.

What then remains of M. LeRoy's thesis? This remains: the scientist intervenes actively in choosing the facts worth observing. An isolated fact has by itself no interest; it becomes interesting if one has reason to think that it may aid in the prediction of other facts; or better, if, having been predicted, its verification is the confirmation of a law. Who shall choose the facts which, corresponding to these conditions, are worthy the freedom of the city in science? This is the free activity of the scientist.

And that is not all. I have said that the scientific fact is the translation of a crude fact into a certain language; I should add that every scientific fact is formed of many crude facts. This is sufficiently shown by the examples cited above. For instance, for the hour of the eclipse my clock marked the hour α at the instant of the eclipse; it marked the hour β at the moment of the last transit of the meridian of a certain star that we take as origin of right ascensions; it marked the hour γ at the moment of the preceding transit of this same star. There are three distinct facts (still it will be noticed that each of them results itself from two simultaneous facts in the rough; but let us pass this over). In place of that I say: the eclipse happened at the hour 24 $(\alpha-\beta)/(\beta-\gamma)$, and the three facts are combined in a single scientific fact. I have concluded that the three readings α, β, γ made on my clock at three different moments lacked interest and that the only thing interesting was the combination $(\alpha-\beta)/(\beta-\gamma)$ of the three. In this conclusion is found the free activity of my mind.

But I have thus used up my power; I cannot make this combination $(\alpha-\beta)/(\beta-\gamma)$ have such a value and not such another, since I cannot influence either the value of α, or that of β, or that of γ, which are imposed upon me as crude facts.

In sum, facts are facts, and *if it happens that they satisfy a prediction, this is not an effect of our free activity.* There is no precise frontier between the fact in the rough and the scientific fact; it can only be said that such an enunciation of fact is *more crude* or, on the contrary, *more scientific* than such another.

IV. "NOMINALISM" AND "THE UNIVERSAL INVARIANT"

If from facts we pass to laws, it is clear that the part of the free activity of the scientist will become much greater. But did not M. LeRoy make it still too great? This is what we are about to examine.

Recall first the examples he has given. When I say: phosphorus melts at 44°, I think I am enunciating a law; in reality it is just the definition of phosphorus; if one should discover a body which, possessing otherwise all the properties of phosphorus, did not melt at

44°, we should give it another name, that is all, and the law would remain true.

Just so when I say: heavy bodies falling freely pass over spaces proportional to the squares of the times, I only give the definition of free fall. Whenever the condition shall not be fulfilled, I shall say that the fall is not free, so that the law will never be wrong.

It is clear that if laws were reduced to that, they could not serve in prediction; then they would be good for nothing, either as means of knowledge, or as principle of action.

When I say: phosphorus melts at 44°, I mean by that: all bodies possessing such or such a property (to wit, all the properties of phosphorus, save fusing-point) fuse at 44°. So understood, my proposition is indeed a law, and this law may be useful to me, because if I meet a body possessing these properties I shall be able to predict that it will fuse at 44°.

Doubtless the law may be found to be false. Then we shall read in the treatises on chemistry: "There are two bodies which chemists long confounded under the name of phosphorus; these two bodies differ only by their points of fusion." That would evidently not be the first time for chemists to attain to the separation of two bodies they were at first not able to distinguish; such, for example, are neodymium and praseodymium, long confounded under the name of didymium.

I do not think the chemists much fear that a like mischance will ever happen to phosphorus. And if, to suppose the impossible, it should happen, the two bodies would probably not have *identically* the same density, *identically* the same specific heat, etc., so that, after having determined with care the density, for instance, one could still foresee the fusion point.

It is, moreover, unimportant; it suffices to remark that there is a law, and that this law, true or false, does not reduce to a tautology.

Will it be said that if we do not know on the earth a body which does not fuse at 44° while having all the other properties of phosphorus, we cannot know whether it does not exist on other planets? Doubtless that may be maintained, and it would then be inferred that the law in question, which may serve as a rule of action to us who inhabit the earth, has yet no general value from the point of view of knowledge, and owes its interest only to the chance which has placed us on this globe. This is possible, but, if it were so, the law would be valueless, not because it reduced to a convention, but because it would be false.

The same is true in what concerns the fall of bodies. It would do me no good to have given the name of free fall to falls which happen in conformity with Galileo's law, if I did not know that elsewhere, in such circumstances, the fall will be *probably* free or *approximately* free. That then is a law which may be true or false, but which does not reduce to a convention.

Suppose the astronomers discover that the stars do not exactly obey Newton's law. They will have the choice between two attitudes; they may say that gravitation does not vary exactly as the inverse of the square of the distance, or else they may say that gravitation is not the only force which acts on the stars and that there is in addition a different sort of force.

In the second case, Newton's law will be considered as the definition of gravitation. This will be the nominalist attitude. The choice between the two attitudes is free, and is made from considerations of convenience, though these considerations are most often so strong that there remains practically little of this freedom.

We can break up this proposition: (1) The stars obey Newton's law, into two others; (2) gravitation obeys Newton's law; (3) gravitation is the only force acting on the stars. In this case proposition (2) is no longer anything but a definition and is beyond the test of experiment; but then it will be on proposition (3) that this check can be exercised. This is indeed necessary, since the resulting proposition (1) predicts verifiable facts in the rough.

It is thanks to these artifices that by an unconscious nominalism the scientists have elevated above the laws what they call principles. When a law has received a sufficient confirmation from experiment, we may adopt two attitudes: either we may leave this law in the fray; it will then remain subjected to an incessant revision, which without any doubt will end by demonstrating that it is only approximative. Or else we may elevate it into a *principle* by adopting conventions such that the proposition may be certainly true. For that the procedure is always the same. The primitive law enunciated a relation between two facts in the rough, A and B; between these two crude facts is introduced an abstract intermediary C, more or less fictitious (such was in the preceding example the impalpable entity, gravitation). And then we have a relation between A and C that we may suppose rigorous and which is the *principle*; and another between C and B which remains a *law* subject to revision.

The principle, henceforth crystallized, so to speak, is no longer subject to the test of experiment. It is not true or false, it is convenient.

Great advantages have often been found in proceeding in that way, but it is clear that if *all* the laws had been transformed into principles *nothing* would be left of science. Every law may be broken up into a principle and a law, but thereby it is very clear that, however far this partition be pushed, there will always remain laws.

Nominalism has therefore limits, and this is what one might fail to recognize if one took to the very letter M. LeRoy's assertions.

A rapid review of the sciences will make us comprehend better what are these limits. The nominalist attitude is justified only when it is convenient; when is it so?

Experiment teaches us relations between bodies; this is the fact in the rough; these relations are extremely complicated. Instead of envisaging directly the relation of the body A and the body B, we introduce between them an intermediary, which is space, and we envisage three distinct relations: that of the body A with the figure A' of space, that of the body B with the figure B' of space, that of the two figures A' and B' to each other. Why is this detour advantageous? Because the relation of A and B was complicated, but differed little from that of A' and B', which is simple; so that this complicated relation may be replaced by the simple relation between A' and B' and by two other relations which tell us that the differences between A and A', on the one hand, between B and B', on the other hand, are *very small*. For example, if A and B are two natural solid bodies which are displaced with slight deformation, we envisage two movable *rigid* figures A' and B'. The laws of the relative displacements of these figures A' and B' will be very simple; they will be those of geometry. And we shall afterwards add that the body A, which always differs very little from A', dilates from the effect of heat and bends from the effect of elasticity. These dilatations and flexions, just because they are very small, will be for our mind relatively easy to study. Just imagine to what complexities of language it would have been necessary to be resigned if we had wished to comprehend in the same enunciation the displacement of the solid, its dilatation and its flexure?

The relation between A and B was a rough law, and was broken up; we now have two laws which express the relations of A and A', of B and B', and a principle which expresses that of A' with B'. It is the aggregate of these principles that is called geometry.

Two other remarks. We have a relation between two bodies A and B, which we have replaced by a relation between two figures A' and B'; but this same relation between the same two figures A' and B' could just as well have replaced advantageously a relation between two other bodies A'' and B'', entirely different from A and B. And that in many ways. If the principles and geometry had not been invented, after having studied the relation of A and B, it would be necessary to begin again *ab ovo* the study of the relation of A'' and B''. That is why geometry is so precious. A geometrical relation can advantageously replace a relation which, considered in the rough state, should be regarded as mechanical, it can replace another which should be regarded as optical, etc.

Yet let no one say: but that proves geometry an experimental science; in separating its principles from laws whence they have been drawn, you artificially separate it itself from the sciences which have given birth to it. The other sciences have likewise principles, but that does not preclude our having to call them experimental.

It must be recognized that it would have been difficult not to make this separation that is pretended to be artificial. We know the role that the kinematics of solid bodies has played in the genesis of geometry; should it then be said that geometry is only a branch of experimental kinematics? But the laws of the rectilinear propagation of light have also contributed to the formation of its principles. Must geometry be regarded both as a branch of kinematics and as a branch of optics? I recall besides that our Euclidean space which is the proper object of geometry has been chosen, for reasons of convenience, from among a certain number of types which pre-exist in our mind and which are called groups.

If we pass to mechanics, we still see great principles whose origin is analogous, and, as their "radius of action," so to speak, is smaller, there is no longer reason to separate them from mechanics proper and to regard this science as deductive.

In physics, finally, the role of the principles is still more diminished. And in fact they are only introduced when it is of advantage. Now they are advantageous precisely because they are few, since each of them very nearly replaces a great number of laws. Therefore it is not of interest to multiply them. Besides, an outcome is necessary, and for that it is needful to end by leaving abstraction to take hold of reality.

Such are the limits of nominalism, and they are narrow.

M. LeRoy has insisted, however, and he has put the question under another form.

Since the enunciation of our laws may vary with the conventions that we adopt, since these conventions may modify even the natural relations of these laws, is there in the manifold of these laws something independent of these conventions and which may, so to speak, play the role of *universal invariant?* For instance, the fiction has been introduced of beings who, having been educated in a world different from ours, would have been led to create a non-Euclidean geometry. If these beings were afterwards suddenly transported into our world, they would observe the same laws as we, but they would enunciate them in an entirely different way. In truth there would still be something in common between the two enunciations, but this is because these beings do not yet differ enough from us. Beings still more strange may be imagined, and the part common to the two systems of enunciations will shrink more and more. Will it thus shrink in convergence towards zero, or will there remain an irreducible residue which will then be the universal invariant sought?

The question calls for precise statement. Is it desired that this common part of the enunciations be expressible in words? It is clear then that there are not words common to all languages, and we cannot pretend to construct I know not what universal invariant which should be understood both by us and by the fictitious non-Euclidean geometers of whom I have just spoken; no more than we can construct a phrase which can be understood both by Germans who do not understand French and by French who do not understand German. But we have fixed rules which permit us to translate the French enunciations into German, and inversely. It is for that that grammars and dictionaries have been made. There are also fixed rules for translating the Euclidean language into the non-Euclidean language, or, if there are not, they could be made.

And even if there were neither interpreter nor dictionary, if the Germans and the French, after having lived centuries in separate worlds, found themselves all at once in contact, do you think there would be nothing in common between the science of the German books and that of the French books? The French and the Germans would certainly end by understanding each other, as the American Indians ended by understanding the language of their conquerors after the arrival of the Spanish.

But, it will be said, doubtless the French would be capable of understanding the Germans even without having learned German, but this is because there remains between the French and the Germans something in common, since both are men. We should still attain to an understanding with our hypothetical non-Euclideans, though they be not men, because they would still retain something human. But in any case a minimum of humanity is necessary.

This is possible, but I shall observe first that this little humanness which would remain in the non-Euclideans would suffice not only to make possible the translation of *a little* of their language, but to make possible the translation of *all* their language.

Now, that there must be a minimum is what I concede; suppose there exists I know not what fluid which penetrates between the molecules of our matter, without having any action on it and without being subject to any action coming from it. Suppose beings sensible to the influence of this fluid and insensible to that of our matter. It is clear that the science of these beings would differ absolutely from ours and that it would be idle to seek an "invariant" common to these two sciences. Or again, if these beings rejected our logic and did not admit, for instance, the principle of contradiction.

But truly I think it without interest to examine such hypotheses.

And then, if we do not push whimsicality so far, if we introduce only fictitious beings having senses analogous to ours and sensible to the same impressions, and moreover admitting the principles of our logic, we shall then be able to conclude that their language, however different from ours it may be, would always be capable of translation. Now the possibility of translation implies the existence of an invariant. To translate is precisely to disengage this invariant. Thus, to decipher a cryptogram is to seek what in this document remains invariant, when the letters are permuted.

What now is the nature of this invariant it is easy to understand, and a word will suffice us. The invariant laws are the relations between the crude facts, while the relations between the "scientific facts" remain always dependent on certain conventions.

CHAPTER XI

SCIENCE AND REALITY

V. CONTINGENCE AND DETERMINISM

I do not intend to treat here the question of the contingence of the laws of nature, which is evidently insoluble, and on which so much has already been written. I only wish to call attention to what different meanings have been given to this word, *contingence*, and how advantageous it would be to distinguish them.

If we look at any particular law, we may be certain in advance that it can only be approximative. It is, in fact, deduced from experimental verifications, and these verifications were and could be only approximate. We should always expect that more precise measurements will oblige us to add new terms to our formulas; this is what has happened, for instance, in the case of Marriotte's law.

Moreover the statement of any law is necessarily incomplete. This enunciation should comprise the enumeration of *all* the antecedents in virtue of which a given consequent can happen. I should first describe *all* the conditions of the experiment to be made and the law would then be stated: if all the conditions are fulfilled, the phenomenon will happen.

But we shall be sure of not having forgotten *any* of these conditions only when we shall have described the state of the entire universe at the instant *t;* all the parts of this universe may, in fact, exercise an influence more or less great on the phenomenon which must happen at the instant *t + dt.*

Now it is clear that such a description could not be found in the enunciation of the law; besides, if it were made, the law would become incapable of application; if one required so many conditions, there would be very little chance of their ever being all realized at any moment.

Then as one can never be certain of not having forgotten some essential condition, it cannot be said: if such and such conditions are realized, such a phenomenon will occur; it can only be said: if such and such conditions are realized, it is probable that such a phenomenon will occur, very nearly.

Take the law of gravitation, which is the least imperfect of all known laws. It enables us to foresee the motions of the planets. When I use it, for instance, to calculate the orbit of Saturn, I neglect the action of the stars, and in doing so, I am certain of not deceiving myself, because I know that these stars are too far away for their action to be sensible.

I announce, then, with a quasi-certitude that the co-ordinates of Saturn at such an hour will be comprised between such and such limits. Yet is that certitude absolute? Could there not exist in the universe some gigantic mass, much greater than that of all the known stars and whose action could make itself felt at great distances? That mass might be animated by a colossal velocity, and after having circulated from all time at such distances that its influence had remained hitherto insensible to us, it might come all at once to pass near us. Surely it would produce in our solar system enormous perturbations that we could not have foreseen. All that can be said is that such an event is wholly improbable, and then, instead of saying: Saturn will be near such a point of the heavens, we must limit ourselves to saying: Saturn will probably be near such a point of the heavens. Although this probability may be practically equivalent to certainty, it is only a probability.

For all these reasons, no particular law will ever be more than approximate and probable. Scientists have never failed to recognize this truth; only they believe, right or wrong, that every law may be replaced by another closer and more probable, that this new law will itself be only provisional, but that the same movement can continue indefinitely, so that science in progressing will possess laws more and more probable, that the approximation will end by differing as little as you choose from exactitude and the probability from certitude.

If the scientists who think thus were right, must it still be said that *the* laws of nature are contingent, even though *each* law, taken in particular, may be qualified as contingent? Or must one require, before concluding the contingence *of the* natural laws, that this progress have an end, that the scientist finish some day by being arrested in his search for a closer and closer approximation and that, beyond a certain limit, he thereafter meet in nature only caprice?

In the conception of which I have just spoken (and which I shall call the scientific conception), every law is only a statement, imperfect and provisional, but it must one day be replaced by another, a superior law, of which it is only a crude image. No place therefore remains for the intervention of a free will.

It seems to me that the kinetic theory of gases will furnish us a striking example.

You know that in this theory all the properties of gases are explained by a simple hypothesis; it is supposed that all the gaseous molecules move in every direction with great velocities and that they follow rectilineal paths which are disturbed only when one molecule passes very near the sides of the vessel or another molecule. The effects our crude senses enable us to observe are the mean effects, and in these means, the great deviations compensate, or at least it is very improbable that they do not compensate; so that the observable phenomena follow simple laws such as that of Mariotte or of Gay-Lussac. But this compensation of deviations is only probable. The molecules incessantly change place and in these continual displacements the figures they form pass successively through all possible combinations. Singly these combinations are very numerous; almost all are in conformity with Mariotte's law, only a few deviate from it. These also will happen, only it would be necessary to wait a long time for them. If a gas were observed during a sufficiently long time, it would certainly be finally seen to deviate, for a very short time, from Mariotte's law. How long would it be necessary to wait? If it were desired to calculate the probable number of years, it would be found that this number is so great that to write only the number of places of figures employed would still require half a score of places of figures. No matter; enough that it may be done.

I do not care to discuss here the value of this theory. It is evident that if it be adopted, Mariotte's law will thereafter appear only as contingent, since a day will come when it will not be true. And yet, think you the partisans of the kinetic theory are adversaries of determinism? Far from it; they are the most ultra of mechanists. Their molecules follow rigid paths, from which they depart only under the influence of forces which vary with the distance, following a perfectly determinate law. There remains in their system not the smallest place either for freedom, or for an evolutionary factor, properly so-called, or for anything whatever that could be called contingence. I add, to avoid mistake, that neither is there any evolution of Mariotte's law itself; it ceases to be true after I know not how many centuries; but at the end of a fraction of a second it again becomes true and that for an incalculable number of centuries.

And since I have pronounced the word *evolution*, let us clear away another mistake. It is often said: Who knows whether the laws do not evolve and whether we shall not one day discover that they were not at the Carboniferous epoch what they are today? What are we to understand by that? What we think we know about the past state of our globe, we deduce from its present state. And how is this deduction made? It is by means of laws supposed known. The law being a relation between the antecedent and the consequent, enables us equally well to deduce the consequent from the antecedent, that is, to foresee the future, and to deduce the antecedent from the consequent, that is, to conclude from the present to the past. The astronomer who knows the present situation of the stars can from it deduce their future situation by Newton's law, and this is what he does when he constructs ephemerides; and he can equally deduce from it their past situation. The calculations he thus can make cannot teach him that Newton's law will cease to be true in the future, since this law is precisely his point of departure; not more can they tell him it was not true in the past. Still in what concerns the future, his ephemerides can one day be tested and our descendants will perhaps recognize that they were false. But in what concerns the past, the geologic past which had no witnesses, the results of his calculation, like those of all speculations where we seek to deduce the past from the present, escape by their very nature every species of test. So that if the laws of nature were not the same in the Carboniferous age as at the present epoch, we shall never be able to know it, since we can know nothing of this age only what we deduce from the hypothesis of the permanence of these laws.

Perhaps it will be said that this hypothesis might lead to contradictory results and that we shall be obliged to abandon it. Thus, in what concerns the origin of life, we may conclude that there have always been living beings, since the present world shows us always life springing from life; and we may also conclude that there have not always been, since the application of the existent laws of physics to the present state of our globe teaches us that there was a time when this globe was so warm that life on it was impossible. But contradictions of this sort can always be removed in two ways; it may be supposed that the actual laws of nature are not exactly what we have assumed; or else it may be supposed that the laws of nature actually are what we have assumed, but that it has not always been so.

It is evident that the actual laws will never be sufficiently well known for us not to be able to adopt the first of these two solutions and for us to be constrained to infer the evolution of natural laws.

On the other hand, suppose such an evolution; assume, if you wish, that humanity lasts sufficiently long for this evolution to have witnesses. The *same* antecedent shall produce, for instance, different consequents at the Carboniferous epoch and at the Quaternary. That evidently means that the antecedents are closely alike; if all the circumstances were identical, the Carboniferous epoch would be indistinguishable from the Quaternary. Evidently this is not what is supposed. What remains is that such antecedent, accompanied by such accessory circumstance, produces such consequent; and that the same antecedent, accompanied by such other accessory circumstance, produces such other consequent. Time does not enter into the affair.

The law, such as ill-informed science would have stated it, and which would have affirmed that this antecedent always produces this consequent, without taking account of the accessory circumstances, this law, which was only approximate and probable, must be replaced by another law more approximate and more probable, which brings in these accessory circumstances. We always come back, therefore, to that same process which we have analysed above, and if humanity should discover something of this sort, it would not say that it is the laws which have evoluted, but the circumstances which have changed.

Here, therefore, are several different senses of the word *contingence*. M. LeRoy retains them all and he does not sufficiently distinguish them, but he introduces a new one. Experimental laws are only approximate, and if some appear to us as exact, it is because we have artificially transformed them into what I have above called a principle. We have made this transformation freely, and as the caprice which has determined us to make it is something eminently contingent, we have communicated this contingence to the law itself. It is in this sense that we have the right to say that determinism supposes freedom, since it is freely that we become determinists. Perhaps it will be found that this is to give large scope to nominalism and that the introduction of this new sense of the word *contingence* will not help much to solve all those questions which naturally arise and of which we have just been speaking.

I do not at all wish to investigate here the foundations of the principle of induction; I know very well that I shall not succeed; it is as difficult to justify this principle as to get on without it. I only wish to show how scientists apply it and are forced to apply it.

When the same antecedent recurs, the same consequent must likewise recur; such is the ordinary statement. But reduced to these terms this principle could be of no use. For one to be able to say that the same antecedent recurred, it would be necessary for the circumstances *all* to be reproduced, since no one is absolutely indifferent, and for them to be *exactly* reproduced. And, as that will never happen, the principle can have no application.

We should therefore modify the enunciation and say: If an antecedent A has once produced a consequent B, an antecedent A', slightly different from A, will produce a consequent B', slightly different from B. But how shall we recognize that the antecedents A and A' are "slightly different"? If some one of the circumstances can be expressed by a number, and this number has in the two cases values very near together, the sense of the phrase "slightly different" is relatively clear; the principle then signifies that the consequent is a continuous function of the antecedent. And as a practical rule, we reach this conclusion that we have the right to interpolate. This is in fact what scientists do every day, and without interpolation all science would be impossible.

Yet observe one thing. The law sought may be represented by a curve. Experiment has taught us certain points of this curve. In virtue of the principle we have just stated, we believe these points may be connected by a continuous graph. We trace this graph with the eye. New experiments will furnish us new points of the curve. If these points are outside of the graph traced in advance, we shall have to modify our curve, but not to abandon our principle. Through any points, however numerous they may be, a continuous curve may always be passed. Doubtless, if this curve is too capricious, we shall be shocked (and we shall even suspect errors of experiment), but the principle will not be directly put at fault.

Furthermore, among the circumstances of a phenomenon, there are some that we regard as negligible, and we shall consider A and A' as slightly different if they differ only by these accessory circumstances. For instance, I have ascertained that hydrogen unites with oxygen under the influence of the electric spark, and I am certain that these two gases will unite anew, although the longitude of Jupiter may have changed considerably in the interval. We assume, for instance, that the state of distant bodies can have no sensible influence on terrestrial phenomena, and that seems in fact requisite, but there are cases where the choice of these practically indifferent circumstances admits of more arbitrariness or, if you choose, requires more tact.

One more remark: The principle of induction would be inapplicable if there did not exist in nature a great quantity of bodies like one another, or almost alike, and if we could not infer, for instance, from one bit of phosphorus to another bit of phosphorus.

If we reflect on these considerations, the problem of determinism and of contingence will appear to us in a new light.

Suppose we were able to embrace the series of all phenomena of the universe in the whole sequence of time. We could envisage what might be called the *sequences,* I mean relations between antecedent and consequent. I do not wish to speak of constant relations or laws, I envisage separately (individually, so to speak) the different sequences realized.

We should then recognize that among these sequences there are no two altogether alike. But, if the principle of induction, as we have just stated it, is true, there will be those almost alike and that can be classed alongside one another. In other words, it is possible to make a classification of sequences.

It is to the possibility and the legitimacy of such a classification that determinism, in the end, reduces. This is all that the preceding analysis leaves of it. Perhaps under this modest form it will seem less appalling to the moralist.

It will doubtless be said that this is to come back by a detour to M. LeRoy's conclusion which a moment ago we seemed to reject: we are determinists voluntarily. And in fact all classification supposes the active intervention of the classifier. I agree that this may be maintained, but it seems to me that this detour will not have been useless and will have contributed to enlighten us a little.

VI. OBJECTIVITY OF SCIENCE

I arrive at the question set by the title of this article: What is the objective value of science? And first what should we understand by objectivity?

What guarantees the objectivity of the world in which we live is that this world is common to us with other thinking beings. Through the communications that we have with other men, we receive from them ready-made reasonings; we know that these reasonings do not come from us and at the same time we recognize in them the work of reasonable beings like ourselves. And as these reasonings appear to fit the world of our sensations, we think we may infer that these reasonable beings have seen the same thing as we; thus it is we know we have not been dreaming.

Such, therefore, is the first condition of objectivity; what is objective must be common to many minds and consequently transmissible from one to the other, and as this transmission can only come about by that "discourse" which inspires so much distrust in M. LeRoy, we are even forced to conclude: no discourse, no objectivity.

The sensations of others will be for us a world eternally closed. We have no means of verifying that the sensation I call red is the same as that which my neighbour calls red.

Suppose that a cherry and a red poppy produce on me the sensation A and on him the sensation B and that, on the contrary, a leaf produces on me the sensation B and on him the sensation A. It is clear we shall never know anything about it; since I shall call red the sensation A and green the sensation B, while he will call the first green and the second red. In compensation, what we shall be able to ascertain is that, for him as for me, the cherry and the red poppy produce the *same* sensation, since he gives the same name to the sensations he feels and I do the same.

Sensations are therefore intransmissible, or rather all that is pure quality in them is intransmissible and forever impenetrable. But it is not the same with relations between these sensations.

From this point of view, all that is objective is devoid of all quality and is only pure relation. Certes, I shall not go so far as to say that objectivity is only pure quantity (this would be to particularize too far the nature of the relations in question), but we understand how someone could have been carried away into saying that the world is only a differential equation.

With due reserve regarding this paradoxical proposition, we must nevertheless admit that nothing is objective which is not transmissible, and consequently that the relations between the sensations can alone have an objective value.

Perhaps it will be said that the esthetic emotion, which is common to all mankind, is proof that the qualities of our sensations are also the same for all men and hence are objective. But if we think about this, we shall see that the proof is not complete; what is proved is that this emotion is aroused in John as in James by the sensations to which James and John give the same name or by the corresponding combinations of these sensations; either because this emotion is associated in John with the sensation *A,* which John calls red, while parallelly it is associated in James with the sensation *B,* which James calls red; or better because this emotion is aroused, not by the qualities themselves of the sensations, but by the harmonious combination of their relations of which we undergo the unconscious impression.

Such a sensation is beautiful, not because it possesses such a quality, but because it occupies such a place in the woof of our associations of ideas, so that it cannot be excited without putting in motion the "receiver" which is at the other end of the thread and which corresponds to the artistic emotion.

Whether we take the moral, the esthetic or the scientific point of view, it is always the same thing. Nothing is objective except what is identical for all; now we can only speak of such an identity if a comparison is possible, and can be translated into a "money of exchange" capable of transmission from one mind to another. Nothing, therefore, will have objective value except what is transmissible by "discourse," that is, intelligible.

But this is only one side of the question. An absolutely disordered aggregate could not have objective value since it would be unintelligible, but no more can a well-ordered assemblage have it, if it does not correspond to sensations really experienced. It seems to me superfluous to recall this condition, and I should not have dreamt of it, if it had not lately been maintained that physics is not an experimental science. Although this opinion has no chance of being adopted either by physicists or by philosophers, it is good to be warned so as not to let oneself slip over the declivity which would lead thither. Two conditions are therefore to be fulfilled, and if the first separates reality* from the dream, the second distinguishes it from the romance.

Now what is science? I have explained in the preceding article, it is before all a classification, a manner of bringing together facts which appearances separate, though they were bound together by some natural and hidden kinship. Science, in other words, is a system of relations. Now we have just said, it is in the relations alone that objectivity must be sought; it would be vain to seek it in beings considered as isolated from one another.

To say that science cannot have objective value since it teaches us only relations, this is to reason backwards, since, precisely, it is relations alone which can be regarded as objective.

External objects, for instance, for which the word *object* was invented, are really *objects* and not fleeting and fugitive appearances, because they are not only groups of sensations, but groups cemented by a constant bond. It is this bond, and this bond alone, which is the object in itself, and this bond is a relation.

Therefore, when we ask what is the objective value of science, that does not mean: Does science teach us the true nature of things? but it means: Does it teach us the true relations of things?

To the first question, no one would hesitate to reply, no; but I think we may go further; not only science cannot teach us the nature of things; but nothing is capable of teaching it to us and if any god knew it, he could not find words to express it. Not only can we

---

* I here use the word *real* as a synonym of *objective;* I thus conform to common usage; perhaps I am wrong, our dreams are real, but they are not objective.

not divine the response, but if it were given to us, we could understand nothing of it; I ask myself even whether we really understand the question.

When, therefore, a scientific theory pretends to teach us what heat is, or what is electricity, or life, it is condemned beforehand; all it can give us is only a crude image. It is, therefore, provisional and crumbling.

The first question being out of reason, the second remains. Can science teach us the true relations of things? What it joins together should that be put asunder, what it puts asunder should that be joined together?

To understand the meaning of this new question, it is needful to refer to what was said above on the conditions of objectivity. Have these relations an objective value? That means: Are these relations the same for all? Will they still be the same for those who shall come after us?

It is clear that they are not the same for the scientist and the ignorant person. But that is unimportant, because if the ignorant person does not see them all at once, the scientist may succeed in making him see them by a series of experiments and reasonings. The thing essential is that there are points on which all those acquainted with the experiments made can reach accord.

The question is to know whether this accord will be durable and whether it will persist for our successors. It may be asked whether the unions that the science of today makes will be confirmed by the science of tomorrow. To affirm that it will be so we cannot invoke any *à priori* reason; but this is a question of fact, and science has already lived long enough for us to be able to find out by asking its history whether the edifices it builds stand the test of time, or whether they are only ephemeral constructions.

Now what do we see? At the first blush it seems to us that the theories last only a day and that ruins upon ruins accumulate. Today the theories are born, tomorrow they are the fashion, the day after tomorrow they are classic, the fourth day they are superannuated, and the fifth they are forgotten. But if we look more closely, we see that what thus succumb are the theories, properly so called, those which pretend to teach us what things are. But there is in them something which usually survives. If one of them has taught us a true relation, this relation is definitively acquired, and it will be found again under a new disguise in the other theories which will successively come to reign in place of the old.

Take only a single example: the theory of the undulations of the ether taught us that light is a motion; today fashion favors the electro-magnetic theory which teaches us that light is a current. We do not consider whether we could reconcile them and say that light is a current, and that this current is a motion. As it is probable in any case that this motion would not be identical with that which the partisans of the old theory presume, we might think ourselves justified in saying that this old theory is dethroned. And yet something of it remains, since between the hypothetical currents which Maxwell supposes there are the same relations as between the hypothetical motions that Fresnel supposed. There is, therefore, something which remains over and this something is the essential. This it is which explains how we see the present physicists pass without any embarrassment from the language of Fresnel to that of Maxwell. Doubtless many connections that were believed well established have been abandoned, but the greatest number remain and it would seem must remain.

And for these, then, what is the measure of their objectivity? Well, it is precisely the same as for our belief in external objects. These latter are real in this, that the sensations they make us feel appear to us as united to each other by I know not what indestructible cement and not by the hazard of a day. In the same way science reveals to us between phenomena other bonds finer but not less solid; these are threads so slender that they long remained unperceived, but once noticed there remains no way of not seeing them; they are therefore not less real than those which give their reality to external objects; small matter that they are more recently known since neither can perish before the other.

It may be said, for instance, that the ether is no less real than any external body; to say this body exists is to say there is between the colour of this body, its taste, its smell, an intimate bond, solid and persistent; to say the ether exists is to say there is a natural kinship between all the optical phenomena, and neither of the two propositions has less value than the other.

And the scientific syntheses have in a sense even more reality than those of the ordinary senses, since they embrace more terms and tend to absorb in them the partial syntheses.

It will be said that science is only a classification and that a classification cannot be true, but convenient. But it is true that it is convenient, it is true that it is so not only for me, but for all men; it is true that it will remain convenient for our descendants; it is true finally that this cannot be by chance.

In sum, the sole objective reality consists in the relations of things whence results the universal harmony. Doubtless these relations, this harmony, could not be conceived outside of a mind which conceives them. But they are nevertheless objective because they are, will become, or will remain, common to all thinking beings.

This will permit us to revert to the question of the rotation of the earth which will give us at the same time a chance to make clear what precedes by an example.

VII. The Rotation of the Earth

"... Therefore," have I said in *Science and Hypothesis,* "this affirmation, the earth turns round, has no meaning...or rather these two propositions, the earth turns round, and, it is more convenient to suppose that the earth turns round, have one and the same meaning."

These words have given rise to the strangest interpretations. Some have thought they saw in them the rehabilitation of Ptolemy's system, and perhaps the justification of Galileo's condemnation.

Those who had read attentively the whole volume could not, however, delude themselves. This truth, the earth turns round, was put on the same footing as Euclid's postulate, for example. Was that to reject it? But better; in the same language it may very well be said: these two propositions, the external world exists, or, it is more convenient to suppose that it exists, have one and the same meaning. So the hypothesis of the rotation of the earth would have the same degree of certitude as the very existence of external objects.

But after what we have just explained in the fourth part, we may go further. A physical theory, we have said, is by so much the more true, as it puts in evidence more true relations. In the light of this new principle, let us examine the question which occupies us.

No, there is no absolute space; these two contradictory propositions: "The earth turns round" and "The earth does not turn round" are, therefore, neither of them more true than the other. To affirm one while denying the other, *in the kinematic sense,* would be to admit the existence of absolute space.

But if the one reveals true relations that the other hides from us, we can nevertheless regard it as physically more true than the other, since it has a richer content. Now in this regard no doubt is possible.

Behold the apparent diurnal motion of the stars, and the diurnal motion of the other heavenly bodies, and besides, the flattening of the earth, the rotation of Foucault's pendulum, the gyration of cyclones, the trade-winds, what not else? For the Ptolemaist all these phenomena have no bond between them; for the Copernican they are produced by the one same cause. In saying, the earth turns round, I affirm that all these phenomena have an intimate relation, and *that is true,* and that remains true, although there is not and cannot be absolute space.

So much for the rotation of the earth upon itself; what shall we say of its revolution around the sun? Here again, we have three phenomena which for the Ptolemaist are absolutely independent and which for the Copernican are referred back to the same origin; they are the apparent displacements of the planets on the celestial sphere, the aberration of the fixed stars, the parallax of these same stars. Is it by chance that all the planets admit an inequality whose period is a year, and that this period is precisely equal to that of aberration, precisely equal besides to that of parallax? To adopt Ptolemy's system is to answer, yes; to adopt that of Copernicus is to answer, no; this is to affirm that there is a bond between the three phenomena and that also is true although there is no absolute space.

In Ptolemy's system, the motions of the heavenly bodies cannot be explained by the action of central forces, celestial mechanics is impossible. The intimate relations that celestial mechanics reveals to us between all the celestial phenomena are true relations; to affirm the immobility of the earth would be to deny these relations, that would be to fool ourselves.

The truth for which Galileo suffered remains, therefore, the truth, although it has not altogether the same meaning as for the vulgar, and its true meaning is much more subtle, more profound and more rich.

VIII. Science for Its Own Sake

Not against M. LeRoy do I wish to defend science for its own sake; maybe this is what he condemns, but this is what he cultivates, since he loves and seeks truth and could not live without it. But I have some thoughts to express.

We cannot know all facts and it is necessary to choose those which are worthy of being known. According to Tolstoi, scientists make this choice at random, instead of making it, which would be reasonable, with a view to practical applications. On the contrary, scientists think that certain facts are more interesting than others, because they complete an unfinished harmony, or because they make one foresee a great number of other facts. If they are wrong, if this hierarchy of facts that they implicitly postulate is only an idle illusion, there could be no science for its own sake, and consequently there could be no science. As for me, I believe they are right, and, for example, I have shown above what is the high value of astronomical facts, not because they are capable of practical applications, but because they are the most instructive of all.

It is only through science and art that civilisation is of value. Some have wondered at the formula: science for its own sake; and yet it is as good as life for its own sake, if life is only misery; and even as happiness for its own sake, if we do not believe that all pleasures are of the same quality, if we do not wish to admit that the goal of civilisation is to furnish alcohol to people who love to drink.

Every act should have an aim. We must suffer, we must work, we must pay for our place at the game, but this is for seeing's sake; or at the very least that others may one day see.

All that is not thought is pure nothingness; since we can think only thought and all the words we use to speak of things can express only thoughts, to say there is something other than thought, is therefore an affirmation which can have no meaning.

And yet—strange contradiction for those who believe in time— geologic history shows us that life is only a short episode between two eternities of death, and that, even in this episode, conscious thought has lasted and will last only a moment. Thought is only a gleam in the midst of a long night.

But it is this gleam which is everything.

SCIENCE
AND
METHOD

INTRODUCTION

In this work I have collected various studies which are more or less directly concerned with scientific methodology. The scientific method consists in observation and experiment. If the scientist had an infinity of time at his disposal, it would be sufficient to say to him, "Look, and look carefully." But, since he has not time to look at everything, and above all to look carefully, and since it is better not to look at all than to look carelessly, he is forced to make a selection. The first question, then, is to know how to make this selection. This question confronts the physicist as well as the historian; it also confronts the mathematician, and the principles which should guide them all are not very dissimilar. The scientist conforms to them instinctively, and by reflecting on these principles one can foresee the possible future of mathematics.

We shall understand this still better if we observe the scientist at work; and, to begin with, we must have some acquaintance with the psychological mechanism of discovery, more especially that of mathematical discovery. Observation of the mathematician's method of working is specially instructive for the psychologist.

In all sciences depending on observation, we must reckon with errors due to imperfections of our senses and of our instruments.

Happily we may admit that, under certain conditions, there is a partial compensation of these errors, so that they disappear in averages. This compensation is due to chance. But what is chance? It is a notion which is difficult of justification, and even of definition; and yet what I have just said with regard to errors of observation, shows that the scientist cannot get on without it. It is necessary, therefore, to give as accurate a definition as possible of this notion, at once so indispensable and so elusive.

These are generalities which apply in the main to all sciences. For instance, there is no appreciable difference between the mechanism of mathematical discovery and the mechanism of discovery in general. Further on I approach questions more particularly concerned with certain special sciences, beginning with pure mathematics.

In the chapters devoted to them, I am obliged to treat of somewhat more abstract subjects, and, to begin with, I have to speak of the notion of space. Everyone knows that space is relative, or rather everyone says so, but how many people think still as if they considered it absolute? Nevertheless, a little reflection will show to what contradictions they are exposed.

Questions concerning methods of instruction are of importance, firstly, on their own account, and secondly, because one cannot reflect on the best method of imbuing virgin brains with new notions without, at the same time, reflecting on the manner in which these notions have been acquired by our ancestors, and consequently on their true origin—that is, in reality, on their true nature. Why is it that, in most cases, the definitions which satisfy scientists mean nothing at all to children? Why is it necessary to give them other definitions? This is the question I have set myself in the chapter which follows, and its solution might, I think, suggest useful reflections to philosophers interested in the logic of sciences.

On the other hand, there are many geometricians who believe that mathematics can be reduced to the rules of formal logic. Untold efforts have been made in this direction. To attain their object they have not hesitated, for instance, to reverse the historical order of the genesis of our conceptions, and have endeavoured to explain the finite by the infinite. I think I have succeeded in showing, for all who approach the problem with an open mind, that there is in this a deceptive illusion. I trust the reader will understand the importance of the question, and will pardon the aridity of the pages I have been constrained to devote to it.

The last chapters, relating to mechanics and astronomy, will be found easier reading.

Mechanics seem to be on the point of undergoing a complete revolution. The ideas which seemed most firmly established are being shattered by daring innovators. It would certainly be premature to decide in their favour from the start, solely because they are innovators; but it is interesting to state their views, and this is what I have tried to do. As far as possible I have followed the historical order, for the new ideas would appear too surprising if we did not see the manner in which they had come into existence.

Astronomy offers us magnificent spectacles, and raises tremendous problems. We cannot dream of applying the experimental method to them directly; our laboratories are too small. But analogy with the phenomena which these laboratories enable us to reach may nevertheless serve as a guide to the astronomer. The Milky Way, for instance, is an assemblage of suns whose motions appear at first sight capricious. But may not this assemblage be compared with that of the molecules of a gas whose properties we have learnt from the kinetic theory of gases? Thus the method of the physicist may come to the aid of the astronomer by a side-track.

Lastly, I have attempted to sketch in a few lines the history of the development of French geodesy. I have shown at what cost, and by what persevering efforts and often dangers, geodesists have secured for us the few notions we possess about the shape of the earth. Is this really a question of method? Yes, for this history certainly teaches us what precautions must surround any serious scientific operation, and what time and trouble are involved in the conquest of a single new decimal.

PART I

THE SCIENTIST
AND
SCIENCE

CHAPTER I

THE SELECTION OF FACTS

Tolstoi explains somewhere in his writings why, in his opinion, "Science for Science's sake" is an absurd conception. We cannot know all the facts, since they are practically infinite in number. We must make a selection; and that being so, can this selection be governed by the mere caprice of our curiosity? Is it not better to be guided by utility, by our practical, and more especially our moral, necessities? Have we not some better occupation than counting the number of lady-birds in existence on this planet?

It is clear that for him the word *utility* has not the meaning assigned to it by business men, and, after them, by the greater number of our contemporaries. He cares but little for the industrial applications of science, for the marvels of electricity or of automobilism, which he regards rather as hindrances to moral progress. For him the useful is exclusively what is capable of making men better.

It is hardly necessary for me to state that, for my part, I could not be satisfied with either of these ideals. I have no liking either for a greedy and narrow plutocracy, or for a virtuous unaspiring democracy, solely occupied in turning the other cheek, in which we should find good people devoid of curiosity, who, avoiding all excesses, would not die of any disease—save boredom. But it is all a matter of taste, and that is not the point I wish to discuss.

Nonetheless the question remains, and it claims our attention. If our selection is only determined by caprice or by immediate necessity, there can be no science for science's sake, and consequently no science. Is this true? There is no disputing the fact that a selection must be made: however great our activity, facts outstrip us, and we can never overtake them; while the scientist is discovering one fact, millions and millions are produced in every cubic inch of his body. Trying to make science contain nature is like trying to make the part contain the whole.

But scientists believe that there is a hierarchy of facts, and that a judicious selection can be made. They are right, for otherwise there would be no science, and science does exist. One has only to open one's eyes to see that the triumphs of industry, which have enriched so many practical men, would never have seen the light if only these practical men had existed, and if they had not been preceded by disinterested fools who died poor, who never thought of the useful, and yet had a guide that was not their own caprice.

What these fools did, as Mach has said, was to save their successors the trouble of thinking. If they had worked solely in view of an immediate application, they would have left nothing behind them, and in face of a new requirement, all would have had to be done again. Now the majority of men do not like thinking, and this is perhaps a good thing, since instinct guides them, and very often better than reason would guide a pure intelligence, at least whenever they are pursuing an end that is immediate and always the same. But instinct is routine, and if it were not fertilized by thought, it would advance no further with man than with the bee or the ant. It is necessary, therefore, to think for those who do not like thinking, and as they are many, each one of our thoughts must be useful in as many circumstances as possible. For this reason, the more general a law is, the greater is its value.

This shows us how our selection should be made. The most interesting facts are those which can be used several times, those which have a chance of recurring. We have been fortunate enough to be born in a world where there are such facts. Suppose that instead of eighty chemical elements we had eighty millions, and that they were not some common and others rare, but uniformly distributed. Then each time we picked up a new pebble there would be a strong probability that it was composed of some unknown substance. Nothing that we knew of other pebbles would tell us anything about it. Before each new object we should be like a new-born child; like him we could but obey our caprices or our necessities. In such a world there would be no science, perhaps thought and even life would be impossible, since evolution could not have developed the instincts of self-preservation. Providentially it is not so; but this blessing, like all those to which we are accustomed, is not appreciated at its true value. The biologist would be equally embarrassed if there were only individuals and no species, and if heredity did not make children resemble their parents.

Which, then, are the facts that have a chance of recurring? In the first place, simple facts. It is evident that in a complex fact many circumstances are united by chance, and that only a still more improbable chance could ever so unite them again. But are there such things as simple facts? and if there are, how are we to recognize them? Who can tell that what we believe to be simple does not conceal an alarming complexity? All that we can say is that we must prefer facts which appear simple, to those in which our rude vision detects dissimilar elements. Then only two alternatives are possible; either this simplicity is real, or else the elements are so intimately mingled that they do not admit of being distinguished. In the first case we have a chance of meeting the same simple fact again, either in all its purity, or itself entering as an element into some complex whole. In the second case the intimate mixture has similarly a greater chance of being reproduced than a heterogeneous assemblage. Chance can mingle, but it cannot unmingle, and a combination of various elements in a well-ordered edifice in which something can be distinguished, can only be made deliberately. There is, therefore, but little chance that an assemblage in which different things can be distinguished should ever be reproduced. On the other hand, there is great probability that a mixture which appears homogeneous at first sight will be reproduced several times. Accordingly facts which appear simple, even if they are not so in reality, will be more easily brought about again by chance.

It is this that justifies the method instinctively adopted by scientists, and what perhaps justifies it still better is that facts which occur frequently appear to us simple just because we are accustomed to them.

But where is the simple fact? Scientists have tried to find it in the two extremes, in the infinitely great and in the infinitely small. The astronomer has found it because the distances of the stars are immense, so great that each of them appears only as a point and qualitative differences disappear, and because a point is simpler than a body which has shape and qualities. The physicist, on the other hand, has sought the elementary phenomenon in an imaginary division of bodies into infinitely small atoms, because the conditions of the problem, which undergo slow and continuous variations as we pass from one point of the body to another, may be regarded as constant within each of these little atoms. Similarly the biologist has been led instinctively to regard the cell as more interesting than the whole animal, and the event has proved him right, since cells belonging to the most diverse organisms have greater resemblances, for those who can recognize them, than the organisms themselves. The sociologist is in a more embarrassing position. The elements, which for him are men, are too dissimilar, too variable, too capricious, in a word, too complex themselves. Furthermore, history does not repeat itself; how, then, is he to select the interesting fact, the fact which is repeated? Method is precisely the selection of facts, and accordingly our first care must be to devise a method. Many have been devised because none holds the field undisputed. Nearly every sociological thesis proposes a new method, which, however, its author is very careful not to apply, so that sociology is the science with the greatest number of methods and the least results.

It is with regular facts, therefore, that we ought to begin; but as soon as the rule is well established, as soon as it is no longer in doubt, the facts which are in complete conformity with it lose their interest, since they can teach us nothing new. Then it is the exception which becomes important. We cease to look for resemblances, and apply ourselves before all else to differences, and of these differences we select first those that are most accentuated, not only because they are the most striking, but because they will be the most instructive. This will be best explained by a simple example. Suppose we are seeking to determine a curve by observing some of the points on it. The practical man who looked only to immediate utility would merely observe the points he required for some special object; these points would be badly distributed on the curve, they would be crowded together in certain parts and scarce in others, so that it would be impossible to connect them by a continuous line, and they would be useless for any other application. The scientist would proceed in a different manner. Since he wishes to study the curve for itself, he will distribute the points to be observed regularly, and as soon as he knows some of them, he will join them by a regular line, and he will then have the complete curve. But how is he to accomplish this? If he has determined one extreme point on the curve, he will not remain close to this extremity, but will move to the other end. After the two extremities, the central point is the most instructive, and so on.

Thus when a rule has been established, we have first to look for the cases in which the rule stands the best chance of being found in fault. This is one of many reasons for the interest of astronomical facts and of geological ages. By making long excursions in space or in time, we may find our ordinary rules completely upset, and these great upsettings will give us a clearer view and better comprehension of such small changes as may occur nearer us, in the small corner of the world in which we are called to live and move. We shall know this corner better for the journey we have taken into distant lands where we had no concern.

But what we must aim at is not so much to ascertain resemblances and differences, as to discover similarities hidden under apparent discrepancies. The individual rules appear at first discordant, but on looking closer we can generally detect a resemblance; though differing in matter, they approximate in form and in the order of their parts. When we examine them from this point of view, we shall see them widen and tend to embrace everything. This is what gives a value to certain facts that come to complete a whole, and show that it is the faithful image of other known wholes.

I cannot dwell further on this point, but these few words will suffice to show that the scientist does not make a random selection of the facts to be observed. He does not count lady-birds, as Tolstoi says, because the number of these insects, interesting as they are, is subject to capricious variations. He tries to condense a great deal of experience and a great deal of thought into a small volume, and that is why a little book on physics contains so many past experiments, and a thousand times as many possible ones, whose results are known in advance.

But so far we have only considered one side of the question. The scientist does not study nature because it is useful to do so. He studies it because he takes pleasure in it, and he takes pleasure in it because it is beautiful. If nature were not beautiful it would not be worth knowing, and life would not be worth living. I am not speaking, of course, of that beauty which strikes the senses, of the beauty of qualities and appearances. I am far from despising this, but it has nothing to do with science. What I mean is that more intimate beauty which comes from the harmonious order of its parts, and which a pure intelligence can grasp. It is this that gives a body a skeleton, so to speak, to the shimmering visions that flatter our senses, and without this support the beauty of these fleeting dreams would be imperfect, because it would be indefinite and ever elusive. Intellectual beauty, on the contrary, is self-sufficing, and it is for it, more perhaps than for the future good of humanity, that the scientist condemns himself to long and painful labours.

It is, then, the search for this special beauty, the sense of the harmony of the world, that makes us select the facts best suited to contribute to this harmony; just as the artist selects those features of his sitter which complete the portrait and give it character and life. And there is no fear that this instinctive and unacknowledged preoccupation will divert the scientist from the search for truth. We may dream of a harmonious world, but how far it will fall short of the real world! The Greeks, the greatest artists that ever were, constructed a heaven for themselves; how poor a thing it is beside the heaven as we know it!

It is because simplicity and vastness are both beautiful that we seek by preference simple facts and vast facts; that we take delight, now in following the giant courses of the stars, now in scrutinizing with a microscope that prodigious smallness which is also a vastness, and now in seeking in geological ages the traces of a past that attracts us because of its remoteness.

Thus we see that care for the beautiful leads us to the same selection as care for the useful. Similarly economy of thought, that economy of effort which, according to Mach, is the constant tendency of science, is a source of beauty as well as a practical advantage. The buildings we admire are those in which the architect has succeeded in proportioning the means to the end, in which the columns seem to carry the burdens imposed on them lightly and without effort, like the graceful caryatids of the Erechtheum.

Whence comes this concordance? Is it merely that things which seem to us beautiful are those which are best adapted to our intelligence, and that consequently they are at the same time the tools that intelligence knows best how to handle? Or is it due rather to evolution and natural selection? Have the peoples whose ideal conformed best to their own interests, properly understood, exterminated the others and taken their place? One and all pursued their ideal without considering the consequences, but while this pursuit led some to their destruction, it gave empire to others. We are tempted to believe this, for if the Greeks triumphed over the barbarians, and if Europe, heir of the thought of the Greeks, dominates the world, it is due to the fact that the savages loved garish colours and the blatant noise of the drum, which appealed to their senses, while the Greeks loved the intellectual beauty hidden behind sensible beauty, and that it is this beauty which gives certainty and strength to the intelligence.

No doubt Tolstoi would be horrified at such a triumph, and he would refuse to admit that it could be truly useful. But this disinterested pursuit of truth for its own beauty is also wholesome, and can make men better. I know very well there are disappointments, that the thinker does not always find the serenity he should, and even that some scientists have thoroughly bad tempers.

Must we therefore say that science should be abandoned, and morality alone be studied? Does anyone suppose that moralists themselves are entirely above reproach when they have come down from the pulpit?

CHAPTER II

THE FUTURE OF MATHEMATICS

If we wish to foresee the future of mathematics, our proper course is to study the history and present condition of the science.

For us mathematicians, is not this procedure to some extent professional? We are accustomed to *extrapolation*, which is a method of deducing the future from the past and the present; and since we are well aware of its limitations, we run no risk of deluding ourselves as to the scope of the results it gives us.

In the past there have been prophets of ill. They took pleasure in repeating that all problems susceptible of being solved had already been solved, and that after them there would be nothing left but gleanings. Happily we are reassured by the example of the past. Many times already men have thought that they had solved all the problems, or at least that they had made an inventory of all that admit of solution. And then the meaning of the word solution has been extended; the insoluble problems have become the most interesting of all, and other problems hitherto undreamt of have presented themselves. For the Greeks a good solution was one that employed only rule and compass; later it became one obtained by the extraction of radicals, then one in which algebraical functions and radicals alone figured. Thus the pessimists found themselves continually passed over, continually forced to retreat, so that at present I verily believe there are none left.

My intention, therefore, is not to refute them, since they are dead. We know very well that mathematics will continue to develop, but we have to find out in what direction. I shall be told "in all directions," and that is partly true; but if it were altogether true, it would become somewhat alarming. Our riches would soon become embarrassing, and their accumulation would soon produce a mass just as impenetrable as the unknown truth was to the ignorant.

The historian and the physicist himself must make a selection of facts. The scientist's brain, which is only a corner of the universe, will never be able to contain the whole universe; whence it follows that, of the innumerable facts offered by nature, we shall leave some aside and retain others. The same is true, *a fortiori,* in mathematics. The mathematician similarly cannot retain pell-mell all the facts that are presented to him, the more so that it is himself—I was almost going to say his own caprice—that creates these facts. It is he who assembles the elements and constructs a new combination from top to bottom; it is generally not brought to him ready-made by nature.

No doubt it is sometimes the case that a mathematician attacks a problem to satisfy some requirement of physics, that the physicist or the engineer asks him to make a calculation in view of some particular application. Will it be said that we geometricians are to confine ourselves to waiting for orders, and, instead of cultivating our science for our own pleasure, to have no other care but that of accommodating ourselves to our clients' tastes? If the only object of mathematics is to come to the help of those who make a study of nature, it is to them we must look for the word of command. Is this the correct view of the matter? Certainly not; for if we had not cultivated the exact sciences for themselves, we should never have created the mathematical instrument, and when the word of command came from the physicist we should have been found without arms.

Similarly, physicists do not wait to study a phenomenon until some pressing need of material life makes it an absolute necessity, and they are quite right. If the scientists of the eighteenth century had disregarded electricity, because it appeared to them merely a curiosity having no practical interest, we should not have, in the twentieth century, either telegraphy or electro-chemistry or electro-traction. Physicists forced to select are not guided in their selection solely by utility. What method, then, do they pursue in making a selection between the different natural facts? I have explained this in the preceding chapter. The facts that interest them are those that may lead to the discovery of a law, those that have an analogy with many other facts and do not appear to us as isolated, but as closely grouped with others. The isolated fact attracts the attention of all, of the layman as well as the scientist. But what the true scientist alone can see is the link that unites several facts which have a deep but hidden analogy. The anecdote of Newton's apple is probably not true, but it is symbolical, so we will treat it as if it were true. Well, we must suppose that before Newton's day many men had seen apples fall, but none had been able to draw any conclusion. Facts would be barren if there were not minds capable of selecting between them and distinguishing those which have something hidden behind them and recognizing what is hidden—minds which, behind the bare fact, can detect the soul of the fact.

In mathematics we do exactly the same thing. Of the various elements at our disposal we can form millions of different combinations, but any one of these combinations, so long as it is isolated, is absolutely without value; often we have taken great trouble to construct it, but it is of absolutely no use, unless it be, perhaps, to supply a subject for an exercise in secondary schools. It will be quite different as soon as this combination takes its place in a class of analogous combinations whose analogy we have recognized; we shall then be no longer in presence of a fact, but of a law. And then the true discoverer will not be the workman who has patiently built up some of these combinations, but the man who has brought out their relation. The former has only seen the bare fact, the latter alone has detected the soul of the fact. The invention of a new word will often be sufficient to bring out the relation, and the word will be creative. The history of science furnishes us with a host of examples that are familiar to all.

The celebrated Viennese philosopher Mach has said that the part of science is to effect economy of thought, just as a machine effects economy of effort, and this is very true. The savage calculates on his fingers, or by putting together pebbles. By teaching children the multiplication table we save them later on countless operations with pebbles. Someone once recognized, whether by pebbles or otherwise, that 6 times 7 is 42, and had the idea of recording the result, and that is the reason why we do not need to repeat the operation. His time was not wasted even if he was only calculating for his own amusement. His operation only took him two minutes, but it would have taken two million, if a million people had had to repeat it after him.

Thus the importance of a fact is measured by the return it gives—that is, by the amount of thought it enables us to economize.

In physics, the facts which give a large return are those which take their place in a very general law, because they enable us to foresee a very large number of others, and it is exactly the same in mathematics. Suppose I apply myself to a complicated calculation and with much difficulty arrive at a result, I shall have gained nothing by my trouble if it has not enabled me to foresee the results of other analogous calculations, and to direct them with certainty, avoiding the blind groping with which I had to be contented the first time. On the contrary, my time will not have been lost if this very groping has succeeded in revealing to me the profound analogy between the problem just dealt with and a much more extensive class of other problems; if it has shown me at once their resemblances and their differences; if, in a word, it has enabled me to perceive the possibility of a generalization. Then it will not be merely a new result that I have acquired, but a new force.

An algebraic formula which gives us the solution of a type of numerical problem, if we finally replace the letters by numbers, is the simple example which occurs to one's mind at once. Thanks to the formula, a single algebraical calculation saves us the trouble of a constant repetition of numerical calculations. But this is only a rough example: everyone feels that there are analogies which cannot be expressed by a formula, and that they are the most valuable.

If a new result is to have any value, it must unite elements long since known, but till then scattered and seemingly foreign to each other, and suddenly introduce order where the appearance of disorder reigned. Then it enables us to see at a glance each of these elements in the place it occupies in the whole. Not only is the new fact valuable on its own account, but it alone gives a value to the old facts it unites. Our mind is frail as our senses are; it would lose itself in the complexity of the world if that complexity were not harmonious; like the short-sighted, it would only see the details, and would be obliged to forget each of these details before examining the next, because it would be incapable of taking in the whole. The only facts worthy of our attention are those which introduce order into this complexity and so make it accessible to us.

Mathematicians attach a great importance to the elegance of their methods and of their results, and this is not mere dilettantism. What is it that gives us the feeling of elegance in a solution or a demonstration? It is the harmony of the different parts, their symmetry, and their happy adjustment; it is, in a word, all that introduces order, all that gives them unity, that enables us to obtain a clear comprehension of the whole as well as of the parts. But that is also precisely what causes it to give a large return; and in fact the more we see this whole clearly and at a single glance, the better we shall perceive the analogies with other neighbouring objects, and consequently the better chance we shall have of guessing the possible generalizations. Elegance may result from the feeling of surprise caused by the unlooked-for occurrence together of objects not habitually associated. In this, again, it is fruitful, since it thus discloses relations till then unrecognized. It is also fruitful even when it only results from the contrast between the simplicity of the means and the complexity of the problem presented, for it then causes us to reflect on the reason for this contrast, and generally shows us that this reason is not chance, but is to be found in some unsuspected law. Briefly stated, the sentiment of mathematical elegance is nothing but the satisfaction due to some conformity between the solution we wish to discover and the necessities of our mind, and it is on account of this very conformity that the solution can be an instrument for us. This aesthetic satisfaction is consequently connected with the economy of thought. Again the comparison with the Erechtheum occurs to me, but I do not wish to serve it up too often.

It is for the same reason that, when a somewhat lengthy calculation has conducted us to some simple and striking result, we are not satisfied until we have shown that we might have foreseen, if not the whole result, at least its most characteristic features. Why is this? What is it that prevents our being contented with a calculation which has taught us apparently all that we wished to know? The reason is that, in analogous cases, the lengthy calculation might not be able to be used again, while this is not true of the reasoning, often semi-intuitive, which might have enabled us to foresee the result. This reasoning being short, we can see all the parts at a single glance, so that we perceive immediately what must be changed to adapt it to all the problems of a similar nature that may be presented. And since it enables us to foresee whether the solution of these problems will be simple, it shows us at least whether the calculation is worth undertaking.

What I have just said is sufficient to show how vain it would be to attempt to replace the mathematician's free initiative by a mechanical process of any kind. In order to obtain a result having any real value, it is not enough to grind out calculations, or to have a machine for putting things in order: it is not order only, but unexpected order, that has a value. A machine can take hold of the bare fact, but the soul of the fact will always escape it.

Since the middle of last century, mathematicians have become more and more anxious to attain to absolute exactness. They are quite right, and this tendency will become more and more marked. In mathematics, exactness is not everything, but without it there is nothing: a demonstration which lacks exactness is nothing at all. This is a truth that I think no one will dispute, but if it is taken too literally it leads us to the conclusion that before 1820, for instance, there was no such thing as mathematics, and this is clearly an exaggeration. The geometricians of that day were willing to assume what we explain by prolix dissertations. This does not mean that they did not see it at all, but they passed it over too hastily, and, in order to see it clearly, they would have had to take the trouble to state it.

Only, is it always necessary to state it so many times? Those who were the first to pay special attention to exactness have given us reasonings that we may attempt to imitate; but if the demonstrations of the future are to be constructed on this model, mathematical works will become exceedingly long, and if I dread length, it is not only because I am afraid of the congestion of our libraries, but because I fear that as they grow in length our demonstrations will lose that appearance of harmony which plays such a useful part, as I have just explained.

It is economy of thought that we should aim at, and therefore it is not sufficient to give models to be copied. We must enable those that come after us to do without the models, and not to repeat a previous reasoning, but summarize it in a few lines. And this has already been done successfully in certain cases. For instance, there was a whole class of reasonings that resembled each other, and were found everywhere; they were perfectly exact, but they were long. One day someone thought of the term "uniformity of convergence," and this term alone made them useless; it was no longer necessary to repeat them, since they could now be assumed. Thus the hair-splitters can render us a double service, first by teaching us to do as they do if necessary, but more especially by enabling us as often as possible not to do as they do, and yet make no sacrifice of exactness.

One example has just shown us the importance of terms in mathematics; but I could quote many others. It is hardly possible to believe what economy of thought, as Mach used to say, can be effected by a well-chosen term. I think I have already said somewhere that mathematics is the art of giving the same name to different things. It is enough that these things, though differing in matter, should be similar in form, to permit of their being, so to speak, run in the same mould. When language has been well chosen, one is astonished to find that all demonstrations made for a known object apply immediately to many new objects: nothing requires to be changed, not even the terms, since the names have become the same.

A well-chosen term is very often sufficient to remove the exceptions permitted by the rules as stated in the old phraseology. This accounts for the invention of negative quantities, imaginary quantities, decimals to infinity, and I know not what else. And we must never forget that exceptions are pernicious, because they conceal laws.

This is one of the characteristics by which we recognize facts which give a great return: they are the facts which permit of these happy innovations of language. The bare fact, then, has sometimes no great interest: it may have been noted many times without rendering any great service to science; it only acquires a value when some more careful thinker perceives the connexion it brings out, and symbolizes it by a term.

The physicists also proceed in exactly the same way. They have invented the term "energy," and the term has been enormously fruitful, because it also creates a law by eliminating exceptions; because it gives the same name to things which differ in matter, but are similar in form.

Among the terms which have exercised the most happy influence I would note "group" and "invariable." They have enabled us to perceive the essence of many mathematical reasonings, and have shown us in how many cases the old mathematicians were dealing with groups without knowing it, and how, believing themselves far removed from each other, they suddenly found themselves close together without understanding why.

Today we should say that they had been examining isomorphic groups. We now know that, in a group, the matter is of little interest, that the form only is of importance, and that when we are well acquainted with one group, we know by that very fact all the isomorphic groups. Thanks to the terms "group" and "isomorphism," which sum up this subtle rule in a few syllables, and make it readily familiar to all minds, the passage is immediate, and can be made without expending any effort of thinking. The idea of group is, moreover, connected with that of transformation. Why do we attach so much value to the discovery of a new transformation? It is because, from a single theorem, it enables us to draw ten or twenty others. It has the same value as a zero added to the right of a whole number.

This is what has determined the direction of the movement of mathematical science up to the present, and it is also most certainly what will determine it in the future. But the nature of the problems which present themselves contributes to it in an equal degree. We cannot forget what our aim should be, and in my opinion this aim is a double one. Our science borders on both philosophy and physics, and it is for these two neighbours that we must work. And so we have always seen, and we shall still see, mathematicians advancing in two opposite directions.

On the one side, mathematical science must reflect upon itself, and this is useful because reflecting upon itself is reflecting upon the human mind which has created it; the more so because, of all its creations, mathematics is the one for which it has borrowed least from outside. This is the reason for the utility of certain mathematical speculations, such as those which have in view the study of postulates, of unusual geometries, of functions with strange behaviour. The more these speculations depart from the most ordinary conceptions, and, consequently, from nature and applications to natural problems, the better will they show us what the human mind can do when it is more and more withdrawn from the tyranny of the exterior world; the better, consequently, will they make us know this mind itself.

But it is to the opposite side, to the side of nature, that we must direct our main forces.

There we meet the physicist or the engineer, who says, "Will you integrate this differential equation for me; I shall need it within a week for a piece of construction work that has to be completed by a certain date?" "This equation," we answer, "is not included in one of the types that can be integrated, of which you know there are not very many." "Yes, I know; but, then, what good are you?" More often than not a mutual understanding is sufficient. The engineer does not really require the integral in finite terms, he only requires to know the general behaviour of the integral function, or he merely wants a certain figure which would be easily deduced from this integral if we knew it. Ordinarily we do not know it, but we could calculate the figure without it, if we knew just what figure and what degree of exactness the engineer required.

Formerly an equation was not considered to have been solved until the solution had been expressed by means of a finite number of known functions. But this is impossible in about ninety-nine cases out of a hundred. What we can always do, or rather what we should always try to do, is to solve the problem *qualitatively,* so to speak—that is, to try to know approximately the general form of the curve which represents the unknown function.

It then remains to find the *exact* solution of the problem. But if the unknown cannot be determined by a finite calculation, we can always represent it by an infinite converging series which enables us to calculate it. Can this be regarded as a true solution? The story goes that Newton once communicated to Leibnitz an anagram somewhat like the following: *aaaaabbbeeeeii,* etc. Naturally, Leibnitz did not understand it at all, but we who have the key know that the anagram, translated into modern phraseology, means, "I know how to integrate all differential equations," and we are tempted to make the comment that Newton was either exceedingly fortunate or that he had very singular illusions. What he meant to say was simply that he could form (by means of indeterminate co-efficients) a series of powers formally satisfying the equation presented.

Today a similar solution would no longer satisfy us, for two reasons—because the convergence is too slow, and because the terms succeed one another without obeying any law. On the other hand, the series θ appears to us to leave nothing to be desired, first, because it converges very rapidly (this is for the practical man who wants his number as quickly as possible), and secondly, because we perceive at a glance the law of the terms, which satisfies the aesthetic requirements of the theorist.

There are, therefore, no longer some problems solved and others unsolved, there are only problems *more or less* solved, according as this is accomplished by a series of more or less rapid convergence or regulated by a more or less harmonious law. Nevertheless an imperfect solution may happen to lead us towards a better one.

Sometimes the series is of such slow convergence that the calculation is impracticable, and we have only succeeded in demonstrating the possibility of the problem. The engineer considers this absurd, and he is right, since it will not help him to complete his construction within the time allowed. He doesn't trouble himself with the question whether it will be of use to the engineers of the twenty-second century. We think differently, and we are sometimes more pleased at having economized a day's work for our grandchildren than an hour for our contemporaries.

Sometimes by groping, so to speak, empirically, we arrive at a formula that is sufficiently convergent. What more would you have? says the engineer; and yet, in spite of everything, we are not satisfied, for we should have liked to be able to *predict* the convergence. And why? Because if we had known how to predict it in the one case, we should know how to predict it in another. We have been successful, it is true, but that is little in our eyes if we have no real hope of repeating our success.

In proportion as the science develops, it becomes more difficult to take it in in its entirety. Then an attempt is made to cut it in pieces and to be satisfied with one of these pieces—in a word, to specialize. Too great a movement in this direction would constitute a serious obstacle to the progress of the science. As I have said, it is by unexpected concurrences between its different parts that it can make progress. Too much specializing would prohibit these concurrences. Let us hope that congresses, such as those of Heidelberg and Rome, by putting us in touch with each other, will open up a view of our neighbours' territory, and force us to compare it with our own, and so escape in a measure from our own little village. In this way they will be the best remedy against the danger I have just noted.

But I have delayed too long over generalities; it is time to enter into details.

Let us review the different particular sciences which go to make up mathematics; let us see what each of them has done, in what direction it is tending, and what we may expect of it. If the preceding views are correct, we should see that the great progress of the past has been made when two of these sciences have been brought into conjunction, when men have become aware of the similarity of their form in spite of the dissimilarity of their matter, when they have modelled themselves upon each other in such a way that each could profit by the triumphs of the other. At the same time we should look to concurrences of a similar nature for progress in the future.

ARITHMETIC

The progress of arithmetic has been much slower than that of algebra and analysis, and it is easy to understand the reason. The feeling of continuity is a precious guide which fails the arithmetician. Every whole number is separated from the rest, and has, so to speak, its own individuality; each of them is a sort of exception, and that is the reason why general theorems will always be less common in the theory of numbers, and also why those that do exist will be more hidden and will longer escape detection.

If arithmetic is backwards as compared with algebra and analysis, the best thing for it to do is to try to model itself on these sciences, in order to profit by their advance. The arithmetician then should be guided by the analogies with algebra. These analogies are numerous, and if in many cases they have not yet been studied sufficiently closely to become serviceable, they have at least been long foreshadowed, and the very language of the two sciences shows that they have been perceived. Thus we speak of transcendental numbers, and so become aware of the fact that the future classification of these numbers has already a model in the classification of transcendental functions. However, it is not yet very clear how we are to pass from one classification to the other; but if it were clear it would be already done, and would no longer be the work of the future.

The first example that comes to my mind is the theory of congruents, in which we find a perfect parallelism with that of algebraic equations. We shall certainly succeed in completing this parallelism, which must exist, for instance, between the theory of algebraic curves and that of congruents with two variables. When the problems relating to congruents with several variables have been solved, we shall have made the first step towards the solution of many questions of indeterminate analysis.

ALGEBRA

The theory of algebraic equations will long continue to attract the attention of geometricians, the sides by which it may be approached being so numerous and so different.

It must not be supposed that algebra is finished because it furnishes rules for forming all possible combinations; it still remains to find interesting combinations, those that satisfy such and such conditions. Thus there will be built up a kind of indeterminate analysis, in which the unknown quantities will no longer be whole numbers but polynomials. So this time it is algebra that will model itself on arithmetic, being guided by the analogy of the whole number, either with the whole polynomial with indefinite co-efficients, or with the whole polynomial with whole co-efficients.

GEOMETRY

It would seem that geometry can contain nothing that is not already contained in algebra or analysis, and that geometric facts are nothing but the facts of algebra or analysis expressed in another language. It might be supposed, then, that after the review that has just been made, there would be nothing left to say having any special bearing on geometry. But this would imply a failure to recognize the great importance of a well-formed language, or to understand what is added to things themselves by the method of expressing, and consequently of grouping, those things.

To begin with, geometric considerations lead us to set ourselves new problems. These are certainly, if you will, analytical problems, but they are problems we should never have set ourselves on the score of analysis. Analysis, however, profits by them, as it profits by those it is obliged to solve in order to satisfy the requirements of physics.

One great advantage of geometry lies precisely in the fact that the senses can come to the assistance of the intellect, and help to determine the road to be followed, and many minds prefer to reduce the problems of analysis to geometric form. Unfortunately our senses cannot carry us very far, and they leave us in the lurch as soon as we wish to pass outside the three classical dimensions. Does this mean that when we have left this restricted domain in which they would seem to wish to imprison us, we must no longer count on anything but pure analysis, and that all geometry of more than three dimensions is vain and without object? In the generation which preceded ours, the greatest masters would have answered "Yes." Today we are so familiar with this notion that we can speak of it, even in a university course, without exciting too much astonishment.

But of what use can it be? This is easy to see. In the first place it gives us a very convenient language, which expresses in very concise terms what the ordinary language of analysis would state in long-winded phrases. More than that, this language causes us to give the same name to things which resemble one another, and states analogies which it does not allow us to forget. It thus enables us still to find our way in that space which is too great for us, by calling to our mind continually the visible space, which is only an imperfect image of it, no doubt, but still an image. Here again, as in all the preceding examples, it is the analogy with what is simple that enables us to understand what is complex.

This geometry of more than three dimensions is not a simple analytical geometry, it is not purely quantitative, but also qualitative, and it is principally on this ground that it becomes interesting. There is a science called *geometry of position*, which has for its object the study of the relations of position of the different elements of a figure, after eliminating their magnitudes. This geometry is purely qualitative; its theorems would remain true if the figures, instead of being exact, were rudely imitated by a child. We can also construct a geometry of position of more than three dimensions. The importance of geometry of position is immense, and I cannot insist upon it too much; what Riemann, one of its principal creators, has gained from it would be sufficient to demonstrate this. We must succeed in constructing it completely in the higher spaces, and we shall then have an instrument which will enable us really to see into hyperspace and to supplement our senses.

The problems of geometry of position would perhaps not have presented themselves if only the language of analysis had been used. Or rather I am wrong, for they would certainly have presented themselves, since their solution is necessary for a host of questions of analysis, but they would have presented themselves isolated, one after the other, and without our being able to perceive their common link.

CANTORISM

I have spoken above of the need we have of returning continually to the first principles of our science, and of the advantage of this process to the study of the human mind. It is this need which has inspired two attempts which have held a very great place in the most recent history of mathematics. The first is Cantorism, and the services it has rendered to the science are well known. Cantor introduced into the science a new method of considering mathematical infinity, and I shall have occasion to speak of it again in Part II, chapter III. One of the characteristic features of Cantorism is that, instead of rising to the general by erecting more and more complicated constructions, and defining by construction, it starts with the *genus supremum* and only defines, as the scholastics would have said, *per genus proximum et differentiam specificam.* Hence the horror he has sometimes inspired in certain minds, such as Hermite's, whose favourite idea was to compare the mathematical with the natural sciences. For the greater number of us these prejudices had been dissipated, but it has come about that we have run against certain paradoxes and apparent contradictions, which would have rejoiced the heart of Zeno of Elea and the school of Megara. Then began the business of searching for a remedy, each man his own way. For my part I think, and I am not alone in so thinking, that the important thing is never to introduce any entities but such as can be completely defined in a finite number of words. Whatever be the remedy adopted, we can promise ourselves the joy of the doctor called in to follow a fine pathological case.

THE SEARCH FOR POSTULATES

Attempts have been made, from another point of view, to enumerate the axioms and postulates more or less concealed which form the foundation of the different mathematical theories, and in this direction Mr. Hilbert has obtained the most brilliant results. It seems at first that this domain must be strictly limited, and that there will be nothing more to do when the inventory has been completed, which cannot be long. But when everything has been enumerated, there will be many ways of classifying it all. A good librarian always finds work to do, and each new classification will be instructive for the philosopher.

I here close this review, which I cannot dream of making complete. I think that these examples will have been sufficient to show the mechanism by which the mathematical sciences have progressed in the past, and the direction in which they must advance in the future.

CHAPTER III

MATHEMATICAL DISCOVERY

The genesis of mathematical discovery is a problem which must inspire the psychologist with the keenest interest. For this is the process in which the human mind seems to borrow least from the exterior world, in which it acts, or appears to act, only by itself and on itself, so that by studying the process of geometric thought we may hope to arrive at what is most essential in the human mind.

This has long been understood, and a few months ago a review called *L'Enseignement mathématique*, edited by MM. Laisant and Fehr, instituted an enquiry into the habits of mind and methods of work of different mathematicians. I had outlined the principal features of this article when the results of the enquiry were published, so that I have hardly been able to make any use of them, and I will content myself with saying that the majority of the evidence confirms my conclusions. I do not say there is unanimity, for on an appeal to universal suffrage we cannot hope to obtain unanimity.

One first fact must astonish us, or rather would astonish us if we were not too much accustomed to it. How does it happen that there are people who do not understand mathematics? If the science invokes only the rules of logic, those accepted by all well-formed minds, if its evidence is founded on principles that are common to all men, and that none but a madman would attempt to deny, how does it happen that there are so many people who are entirely impervious to it?

There is nothing mysterious in the fact that everyone is not capable of discovery. That everyone should not be able to retain a demonstration he has once learnt is still comprehensible. But what does seem most surprising, when we consider it, is that anyone should be unable to understand a mathematical argument at the very moment it is stated to him. And yet those who can only follow the argument with difficulty are in a majority; this is incontestable, and the experience of teachers of secondary education will certainly not contradict me.

And still further, how is error possible in mathematics? A healthy intellect should not be guilty of any error in logic, and yet there are very keen minds which will not make a false step in a short argument such as those we have to make in the ordinary actions of life, which yet are incapable of following or repeating without error the demonstrations of mathematics which are longer, but which are, after all, only accumulations of short arguments exactly analogous to those they make so easily. Is it necessary to add that mathematicians themselves are not infallible?

The answer appears to me obvious. Imagine a long series of syllogisms in which the conclusions of those that precede form the premises of those that follow. We shall be capable of grasping each of the syllogisms, and it is not in the passage from premises to conclusion that we are in danger of going astray. But between the moment when we meet a proposition for the first time as the conclusion of one syllogism, and the moment when we find it once more as the premise of another syllogism, much time will sometimes have elapsed, and we shall have unfolded many links of the chain; accordingly it may well happen that we shall have forgotten it, or, what is more serious, forgotten its meaning. So we may chance to replace it by a somewhat different proposition, or to preserve the same statement but give it a slightly different meaning, and thus we are in danger of falling into error.

A mathematician must often use a rule, and, naturally, he begins by demonstrating the rule. At the moment the demonstration is quite fresh in his memory he understands perfectly its meaning and significance, and he is in no danger of changing it. But later on he commits it to memory, and only applies it in a mechanical way, and then, if his memory fails him, he may apply it wrongly. It is thus, to take a simple and almost vulgar example, that we sometimes make mistakes in calculation, because we have forgotten our multiplication table.

On this view special aptitude for mathematics would be due to nothing but a very certain memory or a tremendous power of attention. It would be a quality analogous to that of the whist player who can remember the cards played, or, to rise a step higher, to that of the chess player who can picture a very great number of combinations and retain them in his memory. Every good mathematician should also be a good chess player and *vice versa,* and similarly he should be a good numerical calculator. Certainly this sometimes happens, and thus Gauss was at once a geometrician of genius and a very precocious and very certain calculator.

But there are exceptions, or rather I am wrong, for I cannot call them exceptions, otherwise the exceptions would be more numerous than the cases of conformity with the rule. On the contrary, it was Gauss who was an exception. As for myself, I must confess I am absolutely incapable of doing an addition sum without a mistake. Similarly I should be a very bad chess player. I could easily calculate that by playing in a certain way I should be exposed to such and such a danger; I should then review many other moves, which I should reject for other reasons, and I should end by making the move I first examined, having forgotten in the interval the danger I had foreseen.

In a word, my memory is not bad, but it would be insufficient to make me a good chess player. Why, then, does it not fail me in a difficult mathematical argument in which the majority of chess players would be lost? Clearly because it is guided by the general trend of the argument. A mathematical demonstration is not a simple juxtaposition of syllogisms; it consists of syllogisms *placed in a certain order,* and the order in which these elements are placed is much more important than the elements themselves. If I have the feeling, so to speak the intuition, of this order, so that I can perceive the whole of the argument at a glance, I need no longer be afraid of forgetting one of the elements; each of them will place itself naturally in the position prepared for it, without my having to make any effort of memory.

It seems to me, then, as I repeat an argument I have learnt, that I could have discovered it. This is often only an illusion; but even then, even if I am not clever enough to create for myself, I rediscover it myself as I repeat it.

We can understand that this feeling, this intuition of mathematical order, which enables us to guess hidden harmonies and relations, cannot belong to everyone. Some have neither this delicate feeling that is difficult to define, nor a power of memory and attention above the common, and so they are absolutely incapable of understanding even the first steps of higher mathematics. This applies to the majority of people. Others have the feeling only in a slight degree, but they are gifted with an uncommon memory and a great capacity for attention. They learn the details one after the other by heart, they can understand mathematics and sometimes apply them, but they are not in a condition to create. Lastly, others possess the special intuition I have spoken of more or less highly developed, and they cannot only understand mathematics, even though their memory is in no way extraordinary, but they can become creators, and seek to make discovery with more or less chance of success, according as their intuition is more or less developed.

What, in fact, is mathematical discovery? It does not consist in making new combinations with mathematical entities that are already known. That can be done by anyone, and the combinations that could be so formed would be infinite in number, and the greater part of them would be absolutely devoid of interest. Discovery consists precisely in not constructing useless combinations, but in constructing those that are useful, which are an infinitely small minority. Discovery is discernment, selection.

How this selection is to be made I have explained above. Mathe matical facts worthy of being studied are those which, by their analogy with other facts, are capable of conducting us to the knowledge of a mathematical law, in the same way that experimental facts conduct us to the knowledge of a physical law. They are those which reveal unsuspected relations between other facts, long since known, but wrongly believed to be unrelated to each other.

Among the combinations we choose, the most fruitful are often those which are formed of elements borrowed from widely separated domains. I do not mean to say that for discovery it is sufficient to bring together objects that are as incongruous as possible. The greater part of the combinations so formed would be entirely fruitless, but some among them, though very rare, are the most fruitful of all.

Discovery, as I have said, is selection. But this is perhaps not quite the right word. It suggests a purchaser who has been shown a large number of samples, and examines them one after the other in order to make his selection. In our case the samples would be so numerous that a whole life would not give sufficient time to examine them. Things do not happen in this way. Unfruitful combinations do not so much as present themselves to the mind of the discoverer. In the field of his consciousness there never appear any but really useful combinations, and some that he rejects, which, however, partake to some extent of the character of useful combinations. Everything happens as if the discoverer were a secondary examiner who had only to interrogate candidates declared eligible after passing a preliminary test.

But what I have said up to now is only what can be observed or inferred by reading the works of geometricians, provided they are read with some reflection.

It is time to penetrate further, and to see what happens in the very soul of the mathematician. For this purpose I think I cannot do better than recount my personal recollections. Only I am going to confine myself to relating how I wrote my first treatise on Fuchsian functions. I must apologize, for I am going to introduce some technical expressions, but they need not alarm the reader, for he has no need to understand them. I shall say, for instance, that I found the demonstration of such and such a theorem under such and such circumstances; the theorem will have a barbarous name that many will not know, but that is of no importance. What is interesting for the psychologist is not the theorem but the circumstances.

For a fortnight I had been attempting to prove that there could not be any function analogous to what I have since called Fuchsian functions. I was at that time very ignorant. Every day I sat down at my table and spent an hour or two trying a great number of combinations, and I arrived at no result. One night I took some black coffee, contrary to my custom, and was unable to sleep. A host of ideas kept surging in my head; I could almost feel them jostling one another, until two of them coalesced, so to speak, to form a stable combination. When morning came, I had established the existence of one class of Fuchsian functions, those that are derived from the hyper-geometric series. I had only to verify the results, which only took a few hours.

Then I wished to represent these functions by the quotient of two series. This idea was perfectly conscious and deliberate; I was guided by the analogy with elliptical functions. I asked myself what must be the properties of these series, if they existed, and I succeeded without difficulty in forming the series that I have called Theta-Fuchsian.

At this moment I left Caen, where I was then living, to take part in a geological conference arranged by the School of Mines. The incidents of the journey made me forget my mathematical work. When we arrived at Coutances, we got into a break to go for a drive, and, just as I put my foot on the step, the idea came to me, though nothing in my former thoughts seemed to have prepared me for it, that the transformations I had used to define Fuchsian functions were identical with those of non-Euclidian geometry. I made no verification, and had no time to do so, since I took up the conversation again as soon as I had sat down in the break, but I felt absolute certainty at once. When I got back to Caen I verified the result at my leisure to satisfy my conscience.

I then began to study arithmetical questions without any great apparent result, and without suspecting that they could have the least connexion with my previous researches. Disgusted at my want of success, I went away to spend a few days at the seaside, and thought of entirely different things. One day, as I was walking on the cliff, the idea came to me, again with the same characteristics of conciseness, suddenness, and immediate certainty, that arithmetical transformations of indefinite ternary quadratic forms are identical with those of non-Euclidian geometry.

Returning to Caen, I reflected on this result and deduced its consequences. The example of quadratic forms showed me that there are Fuchsian groups other than those which correspond with the hyper-geometric series; I saw that I could apply to them the theory of the Theta-Fuchsian series, and that, consequently, there are Fuchsian functions other than those which are derived from the hyper-geometric series, the only ones I knew up to that time. Naturally, I proposed to form all these functions. I laid siege to them systematically and captured all the outworks one after the other. There was one, however, which still held out, whose fall would carry with it that of the central fortress. But all my efforts were of no avail at first, except to make me better understand the difficulty, which was already something. All this work was perfectly conscious.

Thereupon I left for Mont-Valérien, where I had to serve my time in the army, and so my mind was preoccupied with very different matters. One day, as I was crossing the street, the solution of the difficulty which had brought me to a standstill came to me all at once. I did not try to fathom it immediately, and it was only after my service was finished that I returned to the question. I had all the elements, and had only to assemble and arrange them. Accordingly I composed my definitive treatise at a sitting and without any difficulty.

It is useless to multiply examples, and I will content myself with this one alone. As regards my other researches, the accounts I should give would be exactly similar, and the observations related by other mathematicians in the enquiry of *L'Enseignement mathématique* would only confirm them.

One is at once struck by these appearances of sudden illumination, obvious indications of a long course of previous unconscious work. The part played by this unconscious work in mathematical discovery seems to me indisputable, and we shall find traces of it in other cases where it is less evident. Often when a man is working at a difficult question, he accomplishes nothing the first time he sets to work. Then he takes more or less of a rest, and sits down again at his table. During the first half-hour he still finds nothing, and then all at once the decisive idea presents itself to his mind. We might say that the conscious work proved more fruitful because it was interrupted and the rest restored force and freshness to the mind. But it is more probable that the rest was occupied with unconscious work, and that the result of this work was afterwards revealed to the geometrician exactly as in the cases I have quoted, except that the revelation, instead of coming to light during a walk or a journey, came during a period of conscious work, but independently of that work, which at most only performs the unlocking process, as if it were the spur that excited into conscious form the results already acquired during the rest, which till then remained unconscious.

There is another remark to be made regarding the conditions of this unconscious work, which is, that it is not possible, or in any case not fruitful, unless it is first preceded and then followed by a period of conscious work. These sudden inspirations are never produced (and this is sufficiently proved already by the examples I have quoted) except after some days of voluntary efforts which appeared absolutely fruitless, in which one thought one had accomplished nothing, and seemed to be on a totally wrong track. These efforts, however, were not as barren as one thought; they set the unconscious machine in motion, and without them it would not have worked at all, and would not have produced anything.

The necessity for the second period of conscious work can be even more readily understood. It is necessary to work out the results of the inspiration, to deduce the immediate consequences and put them in order and to set out the demonstrations; but, above all, it is necessary to verify them. I have spoken of the feeling of absolute certainty which accompanies the inspiration; in the cases quoted this feeling was not deceptive, and more often than not this will be the case. But we must beware of thinking that this is a rule without exceptions. Often the feeling deceives us without being any less distinct on that account, and we only detect it when we attempt to establish the demonstration. I have observed this fact most notably with regard to ideas that have come to me in the morning or at night when I have been in bed in a semi-somnolent condition.

Such are the facts of the case, and they suggest the following reflections. The result of all that precedes is to show that the unconscious ego, or, as it is called, the subliminal ego, plays a most important part in mathematical discovery. But the subliminal ego is generally thought of as purely automatic. Now we have seen that mathematical work is not a simple mechanical work, and that it could not be entrusted to any machine, whatever the degree of perfection we suppose it to have been brought to. It is not merely a question of applying certain rules, of manufacturing as many combinations as possible according to certain fixed laws. The combinations so obtained would be extremely numerous, useless, and encumbering. The real work of the discoverer consists in choosing between these combinations with a view to eliminating those that are useless, or rather not giving himself the trouble of making them at all. The rules which must guide this choice are extremely subtle and delicate, and it is practically impossible to state them in precise language; they must be felt rather than formulated. Under these conditions, how can we imagine a sieve capable of applying them mechanically?

The following, then, presents itself as a first hypothesis. The subliminal ego is in no way inferior to the conscious ego; it is not purely automatic; it is capable of discernment; it has tact and lightness of touch; it can select, and it can divine. More than that, it can divine better than the conscious ego, since it succeeds where the latter fails. In a word, is not the subliminal ego superior to the conscious ego? The importance of this question will be readily understood. In a recent lecture, M. Boutroux showed how it had arisen on entirely different occasions, and what consequences would be involved by an answer in the affirmative. (See also the same author's *Science et religion,* pp. 313 *et seq.*)

Are we forced to give this affirmative answer by the facts I have just stated? I confess that, for my part, I should be loth to accept it. Let us, then, return to the facts, and see if they do not admit of some other explanation.

It is certain that the combinations which present themselves to the mind in a kind of sudden illumination after a somewhat prolonged period of unconscious work are generally useful and fruitful combinations, which appear to be the result of a preliminary sifting. Does it follow from this that the subliminal ego, having divined by a delicate intuition that these combinations could be useful, has formed none but these, or has it formed a great many others which were devoid of interest, and remained unconscious?

Under this second aspect, all the combinations are formed as a result of the automatic action of the subliminal ego, but those only which are interesting find their way into the field of consciousness. This, too, is most mysterious. How can we explain the fact that, of the thousand products of our unconscious activity, some are invited to cross the threshold, while others remain outside? Is it mere chance that gives them this privilege? Evidently not. For instance, of all the excitements of our senses, it is only the most intense that retain our attention, unless it has been directed upon them by other causes. More commonly the privileged unconscious phenomena, those that are capable of becoming conscious, are those which, directly or indirectly, most deeply affect our sensibility.

It may appear surprising that sensibility should be introduced in connexion with mathematical demonstrations, which, it would seem, can only interest the intellect. But not if we bear in mind the feeling of mathematical beauty, of the harmony of numbers and forms and of geometric elegance. It is a real aesthetic feeling that all true mathematicians recognize, and this is truly sensibility.

Now, what are the mathematical entities to which we attribute this character of beauty and elegance, which are capable of developing in us a kind of aesthetic emotion? Those whose elements are harmoniously arranged so that the mind can, without effort, take in the whole without neglecting the details. This harmony is at once a satisfaction to our aesthetic requirements, and an assistance to the mind which it supports and guides. At the same time, by setting before our eyes a well-ordered whole, it gives us a presentiment of a mathematical law. Now, as I have said above, the only mathematical facts worthy of retaining our attention and capable of being useful are those which can make us acquainted with a mathematical law. Accordingly we arrive at the following conclusion. The useful combinations are precisely the most beautiful, I mean those that can most charm that special sensibility that all mathematicians know, but of which laymen are so ignorant that they are often tempted to smile at it.

What follows, then? Of the very large number of combinations which the subliminal ego blindly forms, almost all are without interest and without utility. But, for that very reason, they are without action on the aesthetic sensibility; the consciousness will never know them. A few only are harmonious, and consequently at once useful and beautiful, and they will be capable of affecting the geometrician's special sensibility I have been speaking of; which, once aroused, will direct our attention upon them, and will thus give them the opportunity of becoming conscious.

This is only a hypothesis, and yet there is an observation which tends to confirm it. When a sudden illumination invades the mathematician's mind, it most frequently happens that it does not mislead him. But it also happens sometimes, as I have said, that it will not stand the test of verification. Well, it is to be observed almost always that this false idea, if it had been correct, would have flattered our natural instinct for mathematical elegance.

Thus it is this special aesthetic sensibility that plays the part of the delicate sieve of which I spoke above, and this makes it sufficiently clear why the man who has it not will never be a real discoverer.

All the difficulties, however, have not disappeared. The conscious ego is strictly limited, but as regards the subliminal ego, we do not know its limitations, and that is why we are not too loth to suppose that in a brief space of time it can form more different combinations than could be comprised in the whole life of a conscient being. These limitations do exist, however. Is it conceivable that it can form all the possible combinations, whose number staggers the imagination? Nevertheless this would seem to be necessary, for if it produces only a small portion of the combinations, and that by chance, there will be very small likelihood of the *right* one, the one that must be selected, being found among them.

Perhaps we must look for the explanation in that period of preliminary conscious work which always precedes all fruitful unconscious work. If I may be permitted a crude comparison, let us represent the future elements of our combinations as something resembling Epicurus's hooked atoms. When the mind is in complete repose these atoms are immovable; they are, so to speak, attached to the wall. This complete repose may continue indefinitely without the atoms meeting, and, consequently, without the possibility of the formation of any combination.

On the other hand, during a period of apparent repose, but of unconscious work, some of them are detached from the wall and set in motion. They plough through space in all directions, like a swarm of gnats, for instance, or, if we prefer a more learned comparison, like the gaseous molecules in the kinetic theory of gases. Their mutual collisions may then produce new combinations.

What is the part to be played by the preliminary conscious work? Clearly it is to liberate some of these atoms, to detach them from the wall and set them in motion. We think we have accomplished nothing, when we have stirred up the elements in a thousand different ways to try to arrange them, and have not succeeded in finding a satisfactory arrangement. But after this agitation imparted to them by our will, they do not return to their original repose, but continue to circulate freely.

Now our will did not select them at random, but in pursuit of a perfectly definite aim. Those it has liberated are not, therefore, chance atoms; they are those from which we may reasonably expect the desired solution. The liberated atoms will then experience collisions, either with each other, or with the atoms that have remained stationary, which they will run against in their course. I apologize once more. My comparison is very crude, but I cannot well see how I could explain my thought in any other way.

However it be, the only combinations that have any chance of being formed are those in which one at least of the elements is one of the atoms deliberately selected by our will. Now it is evidently among these that what I called just now the *right* combination is to be found. Perhaps there is here a means of modifying what was paradoxical in the original hypothesis.

Yet another observation. It never happens that unconscious work supplies *ready-made* the result of a lengthy calculation in which we have only to apply fixed rules. It might be supposed that the subliminal ego, purely automatic as it is, was peculiarly fitted for this kind of work, which is, in a sense, exclusively mechanical. It would seem that, by thinking overnight of the factors of a multiplication sum, we might hope to find the product ready-made for us on waking; or, again, that an algebraic calculation, for instance, or a verification could be made unconsciously. Observation proves that such is by no means the case. All that we can hope from these inspirations, which are the fruits of unconscious work, is to obtain points of departure for such calculations. As for the calculations themselves, they must be made in the second period of conscious work which follows the inspiration, and in which the results of the inspiration are verified and the consequences deduced. The rules of these calculations are strict and complicated; they demand discipline, attention, will, and consequently consciousness. In the subliminal ego, on the contrary, there reigns what I would call liberty, if one could give this name to the mere absence of discipline and to disorder born of chance. Only, this very disorder permits of unexpected couplings.

I will make one last remark. When I related above some personal observations, I spoke of a night of excitement, on which I worked as though in spite of myself. The cases of this are frequent, and it is not necessary that the abnormal cerebral activity should be caused by a physical stimulant, as in the case quoted. Well, it appears that, in these cases, we are ourselves assisting at our own unconscious work, which becomes partly perceptible to the overexcited consciousness, but does not on that account change its nature. We then become vaguely aware of what distinguishes the two mechanisms, or, if you will, of the methods of working of the two egos The psychological observations I have thus succeeded in making appear to me, in their general characteristics, to confirm the views I have been enunciating.

Truly there is great need of this, for in spite of everything they are and remain largely hypothetical. The interest of the question is so great that I do not regret having submitted them to the reader.

CHAPTER IV

CHANCE

I

"How can we venture to speak of the laws of chance? Is not chance the antithesis of all law?" It is thus that Bertrand expresses himself at the beginning of his "Calculus of Probabilities." Probability is the opposite of certainty; it is thus what we are ignorant of, and consequently it would seem to be what we cannot calculate. There is here at least an apparent contradiction, and one on which much has already been written.

To begin with, what is chance? The ancients distinguished between the phenomena which seemed to obey harmonious laws, established once and for all, and those that they attributed to chance, which were those that could not be predicted because they were not subject to any law. In each domain the precise laws did not decide everything, they only marked the limits within which chance was allowed to move. In this conception, the word *chance* had a precise, objective meaning; what was chance for one was also chance for the other and even for the gods.

But this conception is not ours. We have become complete determinists, and even those who wish to reserve the right of human free will at least allow determinism to reign undisputed in the inorganic world. Every phenomenon, however trifling it be, has a cause, and a mind infinitely powerful and infinitely well informed concerning the laws of nature could have foreseen it from the beginning of the ages. If a being with such a mind existed, we could play no game of chance with him; we should always lose.

For him, in fact, the word *chance* would have no meaning, or rather there would be no such thing as chance. That there is for us is only on account of our frailty and our ignorance. And even without going beyond our frail humanity, what is chance for the ignorant is no longer chance for the learned. Chance is only the measure of our ignorance. Fortuitous phenomena are, by definition, those whose laws we are ignorant of.

But is this definition very satisfactory? When the first Chaldean shepherds followed with their eyes the movements of the stars, they did not yet know the laws of astronomy, but would they have dreamt of saying that the stars move by chance? If a modern physicist is studying a new phenomenon, and if he discovers its law on Tuesday, would he have said on Monday that the phenomenon was fortuitous? But more than this, do we not often invoke what Bertrand calls the laws of chance in order to predict a phenomenon? For instance, in the kinetic theory of gases, we find the well-known laws of Mariotte and of Gay-Lussac, thanks to the hypothesis that the velocities of the gaseous molecules vary irregularly, that is to say, by chance. The observable laws would be much less simple, say all the physicists, if the velocities were regulated by some simple elementary law, if the molecules were, as they say, *organized,* if they were subject to some discipline. It is thanks to chance—that is to say, thanks to our ignorance, that we can arrive at conclusions. Then if the word *chance* is merely synonymous with ignorance, what does this mean? Must we translate as follows?—

"You ask me to predict the phenomena that will be produced. If I had the misfortune to know the laws of these phenomena, I could not succeed except by inextricable calculations, and I should have to give up the attempt to answer you; but since I am fortunate enough to be ignorant of them, I will give you an answer at once. And, what is more extraordinary still, my answer will be right."

Chance, then, must be something more than the name we give to our ignorance. Among the phenomena whose causes we are ignorant of, we must distinguish between fortuitous phenomena, about which the calculation of probabilities will give us provisional information, and those that are not fortuitous, about which we can say nothing, so long as we have not determined the laws that govern them. And as regards the fortuitous phenomena themselves, it is clear that the information that the calculation of probabilities supplies will not cease to be true when the phenomena are better known.

The manager of a life insurance company does not know when each of the assured will die, but he relies upon the calculation of probabilities and on the law of large numbers, and he does not make a mistake, since he is able to pay dividends to his shareholders. These dividends would not vanish if a very far-sighted and very indiscreet doctor came, when once the policies were signed, and gave the manager information on the chances of life of the assured. The doctor would dissipate the ignorance of the manager, but he would have no effect upon the dividends, which are evidently not a result of that ignorance.

II

In order to find the best definition of chance, we must examine some of the facts which it is agreed to regard as fortuitous, to which the calculation of probabilities seems to apply. We will then try to find their common characteristics.

We will select unstable equilibrium as our first example. If a cone is balanced on its point, we know very well that it will fall, but we do not know to which side; it seems that chance alone will decide. If the cone were perfectly symmetrical, if its axis were perfectly vertical, if it were subject to no other force but gravity, it would not fall at all. But the slightest defect of symmetry will make it lean slightly to one side or other, and as soon as it leans, be it ever so little, it will fall altogether to that side. Even if the symmetry is perfect, a very slight trepidation, or a breath of air, may make it incline a few seconds of arc, and that will be enough to determine its fall and even the direction of its fall, which will be that of the original inclination.

A very small cause which escapes our notice determines a considerable effect that we cannot fail to see, and then we say that that effect is due to chance. If we knew exactly the laws of nature and the situation of the universe at the initial moment, we could predict exactly the situation of that same universe at a succeeding moment. But, even if it were the case that the natural laws had no longer any secret for us, we could still only know the initial situation *approximately.* If that enabled us to predict the succeeding situation *with the same approximation,* that is all we require, and we should say that the phenomenon had been predicted, that it is governed by laws. But it is not always so; it may happen that small differences in the initial conditions produce very great ones in the final phenomena. A small error in the former will produce an enormous error in the latter. Prediction becomes impossible, and we have the fortuitous phenomenon.

Our second example will be very much like our first, and we will borrow it from meteorology. Why have meteorologists such difficulty in predicting the weather with any certainty? Why is it that showers and even storms seem to come by chance, so that many people think it quite natural to pray for rain or fine weather, though they would consider it ridiculous to ask for an eclipse by prayer? We see that great disturbances are generally produced in regions where the atmosphere is in unstable equilibrium. The meteorologists see very well that the equilibrium is unstable, that a cyclone will be formed somewhere, but exactly where they are not in a position to say; a tenth of a degree more or less at any given point, and the cyclone will burst here and not there, and extend its ravages over districts it would otherwise have spared. If they had been aware of this tenth of a degree, they could have known it beforehand, but the observations were neither sufficiently comprehensive nor sufficiently precise, and that is the reason why it all seems due to the intervention of chance. Here, again, we find the same contrast between a very trifling cause that is inappreciable to the observer, and considerable effects, that are sometimes terrible disasters.

Let us pass to another example, the distribution of the minor planets on the Zodiac. Their initial longitudes may have had some definite order, but their mean motions were different and they have been revolving for so long that we may say that practically they are distributed *by chance* throughout the Zodiac. Very small initial differences in their distances from the sun, or, what amounts to the same thing, in their mean motions, have resulted in enormous differences in their actual longitudes. A difference of a thousandth part of a second in the mean daily motion will have the effect of a second in three years, a degree in ten thousand years, a whole circumference in three or four millions of years, and what is that beside the time that has elapsed since the minor planets became detached from Laplace's nebula? Here, again, we have a small cause and a great effect, or better, small differences in the cause and great differences in the effect.

The game of roulette does not take us so far as it might appear from the preceding example. Imagine a needle that can be turned about a pivot on a dial divided into a hundred alternate red and black sections. If the needle stops at a red section we win; if not, we lose. Clearly, all depends on the initial impulse we give to the needle. I assume that the needle will make ten or twenty revolutions, but it will stop earlier or later according to the strength of the spin I have given it. Only a variation of a thousandth or a two-thousandth in the impulse is sufficient to determine whether my needle will stop at a black section or at the following section, which is red. These are differences that the muscular sense cannot appreciate, which would escape even more delicate instruments. It is, accordingly, impossible for me to predict what the needle I have just spun will do, and that is why my heart beats and I hope for everything from chance. The difference in the cause is imperceptible, and the difference in the effect is for me of the highest importance, since it affects my whole stake.

III

In this connexion I wish to make a reflection that is somewhat foreign to my subject. Some years ago a certain philosopher said that the future was determined by the past, but not the past by the future; or, in other words, that from the knowledge of the present we could deduce that of the future but not that of the past; because, he said, one cause can produce only one effect, while the same effect can be produced by several different causes. It is obvious that no scientist can accept this conclusion. The laws of nature link the antecedent to the consequent in such a way that the antecedent is determined by the consequent just as much as the consequent is by the antecedent. But what can have been the origin of the philosopher's error? We know that, in virtue of Carnot's principle, physical phenomena are irreversible and that the world is tending towards uniformity. When two bodies of different temperatures are in conjunction, the warmer gives up heat to the colder, and accordingly we can predict that the temperatures will become equal. But once the temperatures have become equal, if we are asked about the previous state, what can we answer? We can certainly say that one of the bodies was hot and the other cold, but we cannot guess which of the two was formerly the warmer.

And yet in reality the temperatures never arrive at perfect equality. The difference between the temperatures only tends towards zero asymptotically. Accordingly there comes a moment when our thermometers are powerless to disclose it. But if we had thermometers a thousand or a hundred thousand times more sensitive, we should recognize that there is still a small difference, and that one of the bodies has remained a little warmer than the other, and then we should be able to state that this is the one which was formerly very much hotter than the other.

So we have, then, the reverse of what we found in the preceding examples, great differences in the cause and small differences in the effect. Flammarion once imagined an observer moving away from the earth at a velocity greater than that of light. For him time would have its sign changed, history would be reversed, and Waterloo would come before Austerlitz. Well, for this observer effects and causes would be inverted, unstable equilibrium would no longer be the exception; on account of the universal irreversibility, everything would seem to him to come out of a kind of chaos in unstable equilibrium, and the whole of nature would appear to him to be given up to chance.

IV

We come now to other arguments, in which we shall see somewhat different characteristics appearing, and first let us take the kinetic theory of gases. How are we to picture a receptacle full of gas? Innumerable molecules, animated with great velocities, course through the receptacle in all directions; every moment they collide with the sides or else with one another, and these collisions take place under the most varied conditions. What strikes us most in this case is not the smallness of the causes, but their complexity. And yet the former element is still found here, and plays an important part. If a molecule deviated from its trajectory to left or right in a very small degree as compared with the radius of action of the gaseous molecules, it would avoid a collision, or would suffer it under different conditions, and that would alter the direction of its velocity after the collision perhaps by 90 or 180 degrees.

That is not all. It is enough, as we have just seen, that the molecule should deviate before the collision in an infinitely small degree, to make it deviate after the collision in a finite degree. Then, if the molecule suffers two successive collisions, it is enough that it should deviate before the first collision in a degree of infinite smallness of the second order, to make it deviate after the first collision in a degree of infinite smallness of the first order, and after the second collision in a finite degree. And the molecule will not suffer two collisions only, but a great number each second. So that if the first collision multiplied the deviation by a very large number A, after n collisions it will be multiplied by A^n. It will, therefore, have become very great, not only because A is large—that is to say, because small causes produce great effects—but because the exponent *n* is large, that is to say, because the collisions are very numerous and the causes very complex.

Let us pass to a second example. Why is it that in a shower the drops of rain appear to us to be distributed by chance? It is again because of the complexity of the causes which determine their formation. Ions have been distributed through the atmosphere; for a long time they have been subjected to constantly changing air currents; they have been involved in whirlwinds of very small dimensions, so that their final distribution has no longer any relation to their original distribution. Suddenly the temperature falls, the vapour condenses, and each of these ions becomes the centre of a raindrop. In order to know how these drops will be distributed and how many will fall on each stone of the pavement, it is not enough to know the original position of the ions, but we must calculate the effect of a thousand minute and capricious air currents.

It is the same thing again if we take grains of dust in suspension in water. The vessel is permeated by currents whose law we know nothing of except that it is very complicated. After a certain length of time the grains will be distributed by chance, that is to say, uniformly, throughout the vessel, and this is entirely due to the complication of the currents. If they obeyed some simple law—if, for instance, the vessel were revolving and the currents revolved in circles about its axis—the case would be altered, for each grain would retain its original height and its original distance from the axis.

We should arrive at the same result by picturing the mixing of two liquids or of two fine powders. To take a rougher example, it is also what happens when a pack of cards is shuffled. At each shuffle the cards undergo a permutation similar to that studied in the theory of substitutions. What will be the resulting permutation? The probability that it will be any particular permutation (for instance, that which brings the card occupying the position $\phi(n)$ before the permutation into the position *n*), this probability, I say, depends on the habits of the player. But if the player shuffles the cards long enough, there will be a great number of successive permutations, and the final order which results will no longer be governed by anything but chance; I mean that all the possible orders will be equally probable. This result is due to the great number of successive permutations, that is to say, to the complexity of the phenomenon.

A final word on the theory of errors. It is a case in which the causes have complexity and multiplicity. How numerous are the traps to which the observer is exposed, even with the best instrument? He must take pains to look out for and avoid the most flagrant, those which give birth to systematic errors. But when he has eliminated these, admitting that he succeeds in so doing, there still remain many which, though small, may become dangerous by the accumulation of their effects. It is from these that accidental errors arise, and we attribute them to chance, because their causes are too complicated and too numerous. Here again we have only small causes, but each of them would produce only a small effect; it is by their union and their number that their effects become formidable.

V

There is yet a third point of view, which is less important than the two former, on which I will not lay so much stress. When we are attempting to predict a fact and making an examination of the antecedents, we endeavour to enquire into the anterior situation. But we cannot do this for every part of the universe, and we are content with knowing what is going on in the neighbourhood of the place where the fact will occur, or what appears to have some connexion with the fact. Our enquiry cannot be complete, and we must know how to select. But we may happen to overlook circumstances which, at first sight, seemed completely foreign to the anticipated fact, to which we should never have dreamt of attributing any influence, which nevertheless, contrary to all anticipation, come to play an important part.

A man passes in the street on the way to his business. Someone familiar with his business could say what reason he had for starting at such an hour and why he went by such a street. On the roof a slater is at work. The contractor who employs him could, to a certain extent, predict what he will do. But the man has no thought for the slater, nor the slater for him; they seem to belong to two worlds completely foreign to one another. Nevertheless the slater drops a tile which kills the man, and we should have no hesitation in saying that this was chance.

Our frailty does not permit us to take in the whole universe, but forces us to cut it up in slices. We attempt to make this as little artificial as possible, and yet it happens, from time to time, that two of these slices react upon each other, and then the effects of this mutual action appear to us to be due to chance.

Is this a third way of conceiving of chance? Not always; in fact, in the majority of cases, we come back to the first or second. Each time that two worlds, generally foreign to one another, thus come to act upon each other, the laws of this reaction cannot fail to be very complex, and moreover a very small change in the initial conditions of the two worlds would have been enough to prevent the reaction from taking place. How very little it would have taken to make the man pass a moment later, or the slater drop his tile a moment earlier!

VI

Nothing that has been said so far explains why chance is obedient to laws. Is the fact that the causes are small, or that they are complex, sufficient to enable us to predict, if not what the effects will be *in each case*, at least what they will be *on the average*? In order to answer this question, it will be best to return to some of the examples quoted above.

I will begin with that of roulette. I said that the point where the needle stops will depend on the initial impulse given it. What is the probability that this impulse will be of any particular strength? I do not know, but it is difficult not to admit that this probability is represented by a continuous analytical function. The probability that the impulse will be comprised between α and $\alpha+\epsilon$ will, then, clearly be equal to the probability that it will be comprised between $\alpha + \epsilon$ and $\alpha + 2\epsilon$, *provided that ϵ is very small.* This is a property common to all analytical functions. Small variations of the function are proportional to small variations of the variable.

But we have assumed that a very small variation in the impulse is sufficient to change the colour of the section opposite which the needle finally stops. From α to $\alpha + \epsilon$ is red, from $\alpha + \epsilon$ to $\alpha + 2\epsilon$ is black. The probability of each red section is accordingly the same as that of the succeeding black section, and consequently the total probability of red is equal to the total probability of black.

The datum in the case is the analytical function which represents the probability of a particular initial impulse. But the theorem remains true, whatever this datum may be, because it depends on a property common to all analytical functions. From this it results finally that we have no longer any need of the datum.

What has just been said of the case of roulette applies also to the example of the minor planets. The Zodiac may be regarded as an immense roulette board on which the Creator has thrown a very great number of small balls, to which he has imparted different initial impulses, varying, however, according to some sort of law. Their actual distribution is uniform and independent of that law, for the same reason as in the preceding case. Thus we see why phenomena obey the laws of chance when small differences in the causes are sufficient to produce great differences in the effects. The probabilities of these small differences can then be regarded as proportional to the differences themselves, just because these differences are small, and small increases of a continuous function are proportional to those of the variable.

Let us pass to a totally different example, in which the complexity of the causes is the principal factor. I imagine a card-player shuffling a pack of cards. At each shuffle he changes the order of the cards, and he may change it in various ways. Let us take three cards only in order to simplify the explanation. The cards which, before the shuffle, occupied the positions 1 2 3 respectively may, after the shuffle, occupy the positions
        1 2 3, 2 3 1, 3 1 2, 3 2 1, 1 3 2, 2 1 3.

Each of these six hypotheses is possible, and their probabilities are respectively

$$P_1, P_2, P_3, P_4, P_5, P_6.$$

The sum of these six numbers is equal to 1, but that is all we know about them. The six probabilities naturally depend upon the player's habits, which we do not know.

At the second shuffle the process is repeated, and under the same conditions. I mean, for instance, that p_4 always represents the probability that the three cards which occupied the positions 1 2 3 after the nth shuffle and before the $n+1^{th}$, will occupy the positions 3 2 1 after the $n+1^{th}$ shuffle. And this remains true, whatever the number n may be, since the player's habits and his method of shuffling remain the same.

But if the number of shuffles is very large, the cards which occupied the positions 1 2 3 before the first shuffle may, after the last shuffle, occupy the positions

$$1\ 2\ 3, 2\ 3\ 1, 3\ 1\ 2, 3\ 2\ 1, 1\ 3\ 2, 2\ 1\ 3,$$

and the probability of each of these six hypotheses is clearly the same and equal to $\frac{1}{6}$; and this is true whatever be the numbers $p_1 \ldots p_6$, which we do not know. The great number of shuffles, that is to say, the complexity of the causes, has produced uniformity.

This would apply without change if there were more than three cards, but even with three the demonstration would be complicated, so I will content myself with giving it for two cards only. We have now only two hypotheses

$$1\ 2, 2\ 1,$$

with the probabilities p_1 and $p_2 = 1 - p_1$. Assume that there are n shuffles, and that I win a shilling if the cards are finally in the initial order, and that I lose one if they are finally reversed. Then my mathematical expectation will be

$$(p_1 - p_2)^n$$

The difference $p_1 - p_2$ is certainly smaller than 1, so that if n is very large, the value of my expectation will be nothing, and we do not require to know p_1 and p_2 to know that the game is fair.

Nevertheless there would be an exception if one of the numbers

p_1 and p_2 was equal to 1 and the other to nothing. *It would then hold good no longer, because our original hypotheses would be too simple.*

What we have just seen applies not only to the mixing of cards, but to all mixing, to that of powders and liquids, and even to that of the gaseous molecules in the kinetic theory of gases. To return to this theory, let us imagine for a moment a gas whose molecules cannot collide mutually, but can be deviated by collisions with the sides of the vessel in which the gas is enclosed. If the form of the vessel is sufficiently complicated, it will not be long before the distribution of the molecules and that of their velocities become uniform. This will not happen if the vessel is spherical, or if it has the form of a rectangular parallelepiped. And why not? Because in the former case the distance of any particular trajectory from the centre remains constant, and in the latter case we have the absolute value of the angle of each trajectory with the sides of the parallelepiped.

Thus we see what we must understand by conditions that are *too simple*. They are conditions which preserve something of the original state as an invariable. Are the differential equations of the problem too simple to enable us to apply the laws of chance? This question appears at first sight devoid of any precise meaning, but we know now what it means. They are too simple if something is preserved, if they admit a uniform integral. If something of the initial conditions remains unchanged, it is clear that the final situation can no longer be independent of the initial situation.

We come, lastly, to the theory of errors. We are ignorant of what accidental errors are due to, and it is just because of this ignorance that we know they will obey Gauss's law. Such is the paradox. It is explained in somewhat the same way as the preceding cases. We only need to know one thing—that the errors are very numerous, that they are very small, and that each of them can be equally well negative or positive. What is the curve of probability of each of them? We do not know, but only assume that it is symmetrical. We can then show that the resultant error will follow Gauss's law, and this resultant law is independent of the particular laws which we do not know. Here again the simplicity of the result actually owes its existence to the complication of the data.

VII

But we have not come to the end of paradoxes. I recalled just above Flammarion's fiction of the man who travels faster than light, for whom time has its sign changed. I said that for him all phenomena would seem to be due to chance. This is true from a certain point of view, and yet, at any given moment, all these phenomena would not be distributed in conformity with the laws of chance, since they would be just as they are for us, who, seeing them unfolded harmoniously and not emerging from a primitive chaos, do not look upon them as governed by chance.

What does this mean? For Flammarion's imaginary Lumen, small causes seem to produce great effects; why, then, do things not happen as they do for us when we think we see great effects due to small causes? Is not the same reasoning applicable to his case?

Let us return to this reasoning. When small differences in the causes produce great differences in the effects, why are the effects distributed according to the laws of chance? Suppose a difference of an inch in the cause produces a difference of a mile in the effect. If I am to win in case the effect corresponds with a mile bearing an even number, my probability of winning will be $\frac{1}{2}$. Why is this? Because, in order that it should be so, the cause must correspond with an inch bearing an even number. Now, according to all appearance, the probability that the cause will vary between certain limits is proportional to the distance of those limits, provided that distance is very small. If this hypothesis be not admitted, there would no longer be any means of representing the probability by a continuous function.

Now what will happen when great causes produce small effects? This is the case in which we shall not attribute the phenomenon to chance, and in which Lumen, on the contrary, would attribute it to chance. A difference of a mile in the cause corresponds to a difference of an inch in the effect. Will the probability that the cause will be comprised between two limits *n* miles apart still be proportional to *n*? We have no reason to suppose it, since this distance of *n* miles is great. But the probability that the effect will be comprised between two limits *n* inches apart will be precisely the same, and accordingly it will not be proportional to *n*, and that notwithstanding the fact that this distance of *n* inches is small. There is, then, no means of representing the law of probability of the effects by a continuous curve. I do not mean to say that the curve may not remain continuous in the *analytical* sense of the word. To *infinitely small* variations of the abscissa there will correspond infinitely small variations of the ordinate. But *practically* it would not be continuous, since to *very small* variations of the abscissa there would not correspond very small variations of the ordinate. It would become impossible to trace the curve with an ordinary pencil: that is what I mean.

What conclusion are we then to draw? Lumen has no right to say that the probability of the cause (that of *his* cause, which is our effect) must necessarily be represented by a continuous function. But if that be so, why have we the right? It is because that state of unstable equilibrium that I spoke of just now as initial, is itself only the termination of a long anterior history. In the course of this history complex causes have been at work, and they have been at work for a long time. They have contributed to bring about the mixture of the elements, and they have tended to make everything uniform, at least in a small space. They have rounded off the corners, levelled the mountains, and filled up the valleys. However capricious and irregular the original curve they have been given, they have worked so much to regularize it that they will finally give us a continuous curve, and that is why we can quite confidently admit its continuity.

Lumen would not have the same reasons for drawing this conclusion. For him complex causes would not appear as agents of regularity and of levelling; on the contrary, they would only create differentiation and inequality. He would see a more and more varied world emerge from a sort of primitive chaos. The changes he would observe would be for him unforeseen and impossible to foresee. They would seem to him due to some caprice, but that caprice would not be at all the same as our chance, since it would not be amenable to any law, while our chance has its own laws. All these points would require a much longer development, which would help us perhaps to a better comprehension of the irreversibility of the universe.

VIII

We have attempted to define chance, and it would be good now to ask ourselves a question. Has chance, thus defined so far as it can be, an objective character?

We may well ask it. I have spoken of very small or very complex causes, but may not what is very small for one be great for another, and may not what seems very complex to one appear simple to another? I have already given a partial answer, since I stated above most precisely the case in which differential equations become too simple for the laws of chance to remain applicable. But it would be good to examine the thing somewhat more closely, for there are still other points of view we may take.

What is the meaning of the word *small*? To understand it, we have only to refer to what has been said above. A difference is very small, an interval is small, when within the limits of that interval the probability remains appreciably constant. Why can that probability be regarded as constant in a small interval? It is because we admit that the law of probability is represented by a continuous curve, not only continuous in the analytical sense of the word, but *practically* continuous, as I explained above. This means not only that it will present no absolute hiatus, but also that it will have no projections or depressions too acute or too much accentuated.

What gives us the right to make this hypothesis? As I said above, it is because, from the beginning of the ages, there are complex causes that never cease to operate in the same direction, which cause the world to tend constantly towards uniformity without the possibility of ever going back. It is these causes which, little by little, have levelled the projections and filled up the depressions, and it is for this reason that our curves of probability present none but gentle undulations. In millions and millions of centuries we shall have progressed another step towards uniformity, and these undulations will be ten times more gentle still. The radius of mean curvature of our curve will have become ten times longer. And then a length that today does not seem to us very small, because an arc of such a length cannot be regarded as rectilineal, will at that period be properly qualified as very small, since the curvature will have become ten times less, and an arc of such a length will not differ appreciably from a straight line.

Thus the word *small* remains relative, but it is not relative to this man or that, it is relative to the actual state of the world. It will change its meaning when the world becomes more uniform and all things are still more mixed. But then, no doubt, men will no longer be able to live, but will have to make way for other beings, shall I say much smaller or much larger? So that our criterion, remaining true for all men, retains an objective meaning.

And, further, what is the meaning of the words *very complex*? I have already given one solution, that which I referred to again at the beginning of this section; but there are others. Complex causes, I have said, produce a more and more intimate mixture, but how long will it be before this mixture satisfies us? When shall we have accumulated enough complications? When will the cards be sufficiently shuffled? If we mix two powders, one blue and the other white, there comes a time when the colour of the mixture appears uniform. This is on account of the infirmity of our senses; it would be uniform for the long-sighted, obliged to look at it from a distance, when it would not yet be so for the short-sighted. Even when it had become uniform for all sights, we could still set back the limit by employing instruments. There is no possibility that any man will ever distinguish the infinite variety that is hidden under the uniform appearance of a gas, if the kinetic theory is true. Nevertheless, if we adopt Gouy's ideas on the Brownian movement, does not the microscope seem to be on the point of showing us something analogous?

This new criterion is thus relative like the first, and if it preserves an objective character, it is because all men have about the same senses, the power of their instruments is limited, and, moreover, they only make use of them occasionally.

IX

It is the same in the moral sciences, and particularly in history. The historian is obliged to make a selection of the events in the period he is studying, and he only recounts those that seem to him the most important. Thus he contents himself with relating the most considerable events of the sixteenth century, for instance, and similarly the most remarkable facts of the seventeenth century. If the former are sufficient to explain the latter, we say that these latter conform to the laws of history. But if a great event of the seventeenth century owes its cause to a small fact of the sixteenth century that no history reports and that everyone has neglected, then we say that this event is due to chance, and so the word has the same sense as in the physical sciences; it means that small causes have produced great effects.

The greatest chance is the birth of a great man. It is only by chance that the meeting occurs of two genital cells of different sex that contain precisely, each on its side, the mysterious elements whose mutual reaction is destined to produce genius. It will be readily admitted that these elements must be rare, and that their meeting is still rarer. How little it would have taken to make the spermatozoid which carried them deviate from its course. It would have been enough to deflect it a hundredth part of an inch, and Napoleon would not have been born and the destinies of a continent would have been changed. No example can give a better comprehension of the true character of chance.

One word more about the paradoxes to which the application of the calculation of probabilities to the moral sciences has given rise. It has been demonstrated that no parliament would ever contain a single member of the opposition, or at least that such an event would be so improbable that it would be quite safe to bet against it, and to bet a million to one. Condorcet attempted to calculate how many jurymen it would require to make a miscarriage of justice practically impossible. If we used the results of this calculation, we should certainly be exposed to the same disillusionment as by betting on the strength of the calculation that the opposition would never have a single representative.

The laws of chance do not apply to these questions. If justice does not always decide on good grounds, it does not make so much use as is generally supposed of Bridoye's method. This is perhaps unfortunate, since, if it did, Condorcet's method would protect us against miscarriages.

What does this mean? We are tempted to attribute facts of this nature to chance because their causes are obscure, but this is not true chance. The causes are unknown to us, it is true, and they are even complex; but they are not sufficiently complex, since they preserve something, and we have seen that this is the distinguishing mark of "too simple" causes. When men are brought together, they no longer decide by chance and independently of each other, but react upon one another. Many causes come into action, they trouble the men and draw them this way and that, but there is one thing they cannot destroy, the habits they have of Panurge's sheep. And it is this that is preserved.

X

The application of the calculation of probabilities to the exact sciences also involves many difficulties. Why are the decimals of a table of logarithms or of the number π distributed in accordance with the laws of chance? I have elsewhere studied the question in regard to logarithms, and there it is easy. It is clear that a small difference in the argument will give a small difference in the logarithm, but a great difference in the sixth decimal of the logarithm. We still find the same criterion.

But as regards the number π the question presents more difficulties, and for the moment I have no satisfactory explanation to give.

There are many other questions that might be raised, if I wished to attack them before answering the one I have more especially set myself. When we arrive at a simple result, when, for instance, we find a round number, we say that such a result cannot be due to chance, and we seek for a non-fortuitous cause to explain it. And in fact there is only a very slight likelihood that, out of 10,000 numbers, chance will give us a round number, the number 10,000 for instance; there is only one chance in 10,000. But neither is there more than one chance in 10,000 that it will give us any other particular number, and yet this result does not astonish us, and we feel no hesitation about attributing it to chance, and that merely because it is less striking.

Is this a simple illusion on our part, or are there cases in which this view is legitimate? We must hope so, for otherwise all science would be impossible. When we wish to check a hypothesis, what do we do? We cannot verify all its consequences, since they are infinite in number. We content ourselves with verifying a few, and, if we succeed, we declare that the hypothesis is confirmed, for so much success could not be due to chance. It is always at bottom the same reasoning.

I cannot justify it here completely, it would take me too long, but I can say at least this. We find ourselves faced by two hypotheses, either a simple cause or else that assemblage of complex causes we call chance. We find it natural to admit that the former must produce a simple result, and then, if we arrive at this simple result, the round number for instance, it appears to us more reasonable to attribute it to the simple cause, which was almost certain to give it us, than to chance, which could only give it us once in 10,000 times. It will not be the same if we arrive at a result that is not simple. It is true that chance also will not give it more than once in 10,000 times, but the simple cause has no greater chance of producing it.

PART II

MATHEMATICAL

REASONING

CHAPTER I

THE RELATIVITY OF SPACE

I

It is impossible to picture empty space. All our efforts to imagine pure space from which the changing images of material objects are excluded can only result in a representation in which highly coloured surfaces, for instance, are replaced by lines of slight colouration, and if we continued in this direction to the end, everything would disappear and end in nothing. Hence arises the irreducible relativity of space.

Whoever speaks of absolute space uses a word devoid of meaning. This is a truth that has been long proclaimed by all who have reflected on the question, but one which we are too often inclined to forget.

If I am at a definite point in Paris, at the Place du Panthéon, for instance, and I say, "I will come back *here* tomorrow"; if I am asked, "Do you mean that you will come back to the same point in space?" I should be tempted to answer yes. Yet I should be wrong, since between now and tomorrow the earth will have moved, carrying with it the Place du Panthéon, which will have travelled more than a million miles. And if I wished to speak more accurately, I should gain nothing, since this million of miles has been covered by our globe in its motion in relation to the sun, and the sun in its turn moves in relation to the Milky Way, and the Milky Way itself is no doubt in motion without our being able to recognize its velocity. So that we are, and shall always be, completely ignorant how far the Place du Panthéon moves in a day. In fact, what I meant to say was, "Tomorrow I shall see once more the dome and pediment of the Panthéon," and if there was no Panthéon my sentence would have no meaning and space would disappear.

This is one of the most commonplace forms of the principle of the relativity of space, but there is another on which Delbeuf has laid particular stress. Suppose that in one night all the dimensions of the universe became a thousand times larger. The world will remain *similar* to itself, if we give the word *similitude* the meaning it has in the third book of Euclid. Only, what was formerly a metre long will now measure a kilometre, and what was a millimetre long will become a metre. The bed in which I went to sleep and my body itself will have grown in the same proportion. When I wake in the morning what will be my feeling in face of such an astonishing transformation? Well, I shall not notice anything at all. The most exact measures will be incapable of revealing anything of this tremendous change, since the yard-measures I shall use will have varied in exactly the same proportions as the objects I shall attempt to measure. In reality the change only exists for those who argue as if space were absolute. If I have argued for a moment as they do, it was only in order to make it clearer that their view implies a contradiction. In reality it would be better to say that as space is relative, nothing at all has happened, and that it is for that reason that we have noticed nothing.

Have we any right, therefore, to say that we know the distance between two points? No, since that distance could undergo enormous variations without our being able to perceive it, provided other distances varied in the same proportions. We saw just now that when I say I shall be here tomorrow, that does not mean that tomorrow I shall be at the point in space where I am today, but that tomorrow I shall be at the same distance from the Panthéon as

I am today. And already this statement is not sufficient, and I ought to say that tomorrow and today my distance from the Panthéon will be equal to the same number of times the length of my body.

But that is not all. I imagined the dimensions of the world changing, but at least the world remaining always similar to itself. We can go much further than that, and one of the most surprising theories of modern physicists will furnish the occasion. According to a hypothesis of Lorentz and Fitzgerald,* all bodies carried forward in the earth's motion undergo a deformation. This deformation is, in truth, very slight, since all dimensions parallel with the earth's motion are diminished by a hundred-millionth, while dimensions perpendicular to this motion are not altered. But it matters little that it is slight; it is enough that it should exist for the conclusion I am soon going to draw from it. Besides, though I said that it is slight, I really know nothing about it. I have myself fallen a victim to the tenacious illusion that makes us believe that we think of an absolute space. I was thinking of the earth's motion on its elliptical orbit round the sun, and I allowed 18 miles a second for its velocity. But its true velocity (I mean this time, not its absolute velocity, which has no sense, but its velocity in relation to the ether), this I do not know and have no means of knowing. It is, perhaps, 10 or 100 times as high, and then the deformation will be 100 or 10,000 times as great.

It is evident that we cannot demonstrate this deformation. Take a cube with sides a yard long. It is deformed on account of the earth's velocity; one of its sides, that parallel with the motion, becomes smaller, the others do not vary. If I wish to assure myself of this with the help of a yard-measure, I shall measure first one of the sides perpendicular to the motion, and satisfy myself that my measure fits this side exactly; and indeed neither one nor the other of these lengths is altered, since they are both perpendicular to the motion. I then wish to measure the other side, that parallel with the motion; for this purpose I change the position of my measure, and turn it so as to apply it to this side. But the yard-measure, having

* *Vide infra*, Book III, Chap. II.

changed its direction and having become parallel with the motion, has in its turn undergone the deformation, so that, though the side is no longer a yard long, it will still fit it exactly, and I shall be aware of nothing.

What, then, I shall be asked, is the use of the hypothesis of Lorentz and Fitzgerald if no experiment can enable us to verify it? The fact is that my statement has been incomplete. I have only spoken of measurements that can be made with a yard-measure, but we can also measure a distance by the time that light takes to traverse it, on condition that we admit that the velocity of light is constant, and independent of its direction. Lorentz could have accounted for the facts by supposing that the velocity of light is greater in the direction of the earth's motion than in the perpendicular direction. He preferred to admit that the velocity is the same in the two directions, but that bodies are smaller in the former than in the latter. If the surfaces of the waves of light had undergone the same deformations as material bodies, we should never have perceived the Lorentz-Fitzgerald deformation.

In the one case as in the other, there can be no question of absolute magnitude, but of the measurement of that magnitude by means of some instrument. This instrument may be a yard-measure or the path traversed by light. It is only the relation of the magnitude to the instrument that we measure, and if this relation is altered, we have no means of knowing whether it is the magnitude or the instrument that has changed.

But what I wish to make clear is, that in this deformation the world has not remained similar to itself. Squares have become rectangles or parallelograms, circles ellipses, and spheres ellipsoids. And yet we have no means of knowing whether this deformation is real.

It is clear that we might go much further. Instead of the Lorentz-Fitzgerald deformation, with its extremely simple laws, we might imagine a deformation of any kind whatever; bodies might be deformed in accordance with any laws, as complicated as we liked, and we should not perceive it, provided all bodies without exception were deformed in accordance with the same laws. When I say all bodies without exception, I include, of course, our own bodies and the rays of light emanating from the different objects.

If we look at the world in one of those mirrors of complicated form which deform objects in an odd way, the mutual relations of the different parts of the world are not altered; if, in fact, two real objects touch, their images likewise appear to touch. In truth, when we look in such a mirror we readily perceive the deformation, but it is because the real world exists beside its deformed image. And even if this real world were hidden from us, there is something which cannot be hidden, and that is ourselves. We cannot help seeing, or at least feeling, our body and our members which have not been deformed, and continue to act as measuring instruments. But if we imagine our body itself deformed, and in the same way as if it were seen in the mirror, these measuring instruments will fail us in their turn, and the deformation will no longer be able to be ascertained.

Imagine, in the same way, two universes which are the image one of the other. With each object P in the universe A, there corresponds, in the universe B, an object P^1 which is its image. The coordinates of this image P^1 are determinate functions of those of the object P; moreover, these functions may be of any kind whatever—I assume only that they are chosen once and for all. Between the position of P and that of P^1 there is a constant relation; it matters little what that relation may be, it is enough that it should be constant.

Well, these two universes will be indistinguishable. I mean to say that the former will be for its inhabitants what the second is for its own. This would be true so long as the two universes remained foreign to one another. Suppose we are inhabitants of the universe A; we have constructed our science and particularly our geometry. During this time the inhabitants of the universe B have constructed a science, and as their world is the image of ours, their geometry will also be the image of ours, or, more accurately, it will be the same. But if one day a window were to open for us upon the universe B, we should feel contempt for them, and we should say, "These wretched people imagine that they have made a geometry, but what they so name is only a grotesque image of ours; their straight lines are all twisted, their circles are hunchbacked, and their spheres have capricious inequalities." We should have no suspicion that they were saying the same of us, and that no one will ever know which is right.

We see in how large a sense we must understand the relativity of space. Space is in reality amorphous, and it is only the things that are in it that give it a form. What are we to think, then, of that direct intuition we have of a straight line or of distance? We have so little the intuition of distance in itself that, in a single night, as we have said, a distance could become a thousand times greater without our being able to perceive it, if all other distances had undergone the same alteration. And in a night the universe B might even be substituted for the universe A without our having any means of knowing it, and then the straight lines of yesterday would have ceased to be straight, and we should not be aware of anything.

One part of space is not by itself and in the absolute sense of the word equal to another part of space, for if it is so for us, it will not be so for the inhabitants of the universe B, and they have precisely as much right to reject our opinion as we have to condemn theirs.

I have shown elsewhere what are the consequences of these facts from the point of view of the idea that we should construct non-Euclidian and other analogous geometries. I do not wish to return to this, and I will take a somewhat different point of view.

II

If this intuition of distance, of direction, of the straight line, if, in a word, this direct intuition of space does not exist, whence comes it that we imagine we have it? If this is only an illusion, whence comes it that the illusion is so tenacious? This is what we must examine. There is no direct intuition of magnitude, as we have said, and we can only arrive at the relation of the magnitude to our measuring instruments. Accordingly we could not have constructed space if we had not had an instrument for measuring it. Well, that instrument to which we refer everything, which we use instinctively, is our own body. It is in reference to our own body that we locate exterior objects, and the only special relations of these objects that we can picture to ourselves are their relations with our body. It is our body that serves us, so to speak, as a system of axes of co-ordinates.

For instance, at a moment α the presence of an object A is revealed to me by the sense of sight; at another moment β the presence of another object B is revealed by another sense, that, for instance, of hearing or of touch. I judge that this object B occupies the same place as the object A. What does this mean? To begin with, it does not imply that these two objects occupy, at two different moments, the same point in an absolute space, which, even if it existed, would escape our knowledge, since between the moments α and β the solar system has been displaced and we cannot know what this displacement is. It means that these two objects occupy the same relative position in reference to our body.

But what is meant even by this? The impressions that have come to us from these objects have followed absolutely different paths—the optic nerve for the object A, and the acoustic nerve for the object B; they have nothing in common from the qualitative point of view. The representations we can form of these two objects are absolutely heterogeneous and irreducible one to the other. Only I know that, in order to reach the object A, I have only to extend my right arm in a certain way; even though I refrain from doing it, I represent to myself the muscular and other analogous sensations which accompany that extension, and that representation is associated with that of the object A.

Now I know equally that I can reach the object B by extending my right arm in the same way, an extension accompanied by the same train of muscular sensations. And I mean nothing else but this when I say that these two objects occupy the same position.

I know also that I could have reached the object A by another appropriate movement of the left arm, and I represent to myself the muscular sensations that would have accompanied the movement. And by the same movement of the left arm, accompanied by the same sensations, I could equally have reached the object B.

And this is very important, since it is in this way that I could defend myself against the dangers with which the object A or the object B might threaten me. With each of the blows that may strike us, nature has associated one or several parries which enable us to protect ourselves against them. The same parry may answer to several blows. It is thus, for instance, that the same movement of the right arm would have enabled us to defend ourselves at the moment α against the object A, and at the moment β against the object B. Similarly, the same blow may be parried in several ways, and we have said, for instance, that we could reach the object A equally well either by a certain movement of the right arm, or by a certain movement of the left.

All these parries have nothing in common with one another, except that they enable us to avoid the same blow, and it is that, and nothing but that, we mean when we say that they are movements ending in the same point in space. Similarly, these objects, of which we say that they occupy the same point in space, have nothing in common, except that the same parry can enable us to defend ourselves against them.

Or, if we prefer it, let us imagine innumerable telegraph wires, some centripetal and others centrifugal. The centripetal wires warn us of accidents that occur outside, the centrifugal wires have to provide the remedy. Connexions are established in such a way that when one of the centripetal wires is traversed by a current, this current acts on a central exchange, and so excites a current in one of the centrifugal wires, and matters are so arranged that several centripetal wires can act on the same centrifugal wire, if the same remedy is applicable to several evils, and that one centripetal wire can disturb several centrifugal wires, either simultaneously or one in default of the other, every time that the same evil can be cured by several remedies.

It is this complex system of associations, it is this distribution board, so to speak, that is our whole geometry, or, if you will, all that is distinctive in our geometry. What we call our intuition of a straight line or of distance is the consciousness we have of these associations and of their imperious character.

Whence this imperious character itself comes, it is easy to understand. The older an association is, the more indestructible it will appear to us. But these associations are not, for the most part, conquests made by the individual, since we see traces of them in the newly born infant; they are conquests made by the race. The more necessary these conquests were, the more quickly they must have been brought about by natural selection.

On this account those we have been speaking of must have been among the earliest, since without them the defence of the organism would have been impossible. As soon as the cells were no longer merely in juxtaposition, as soon as they were called upon to give mutual assistance to each other, some such mechanism as we have been describing must necessarily have been organized in order that the assistance should meet the danger without miscarrying.

When a frog's head has been cut off, and a drop of acid is placed at some point on its skin, it tries to rub off the acid with the nearest foot; and if that foot is cut off, it removes it with the other foot. Here we have, clearly, that double parry I spoke of just now, making it possible to oppose an evil by a second remedy if the first fails. It is this multiplicity of parries, and the resulting co-ordination, that is space.

We see to what depths of unconsciousness we have to descend to find the first traces of these spacial associations, since the lowest parts of the nervous system alone come into play. Once we have realized this, how can we be astonished at the resistance we oppose to any attempt to dissociate what has been so long associated? Now, it is this very resistance that we call the evidence of the truths of geometry. This evidence is nothing else than the repugnance we feel at breaking with very old habits with which we have always got on very well.

III

The space thus created is only a small space that does not extend beyond what my arm can reach, and the intervention of memory is necessary to set back its limits. There are points that will always remain out of my reach, whatever effort I may make to stretch out my hand to them. If I were attached to the ground, like a sea-polyp, for instance, which can only extend its tentacles, all these points would be outside space, since the sensations we might experience from the action of bodies placed there would not be associated with the idea of any movement enabling us to reach them, or with any appropriate parry. These sensations would not seem to us to have any spacial character, and we should not attempt to locate them.

But we are not fixed to the ground like the inferior animals. If the enemy is too far off, we can advance upon him first and extend our hand when we are near enough. This is still a parry, but a long-distance parry. Moreover, it is a complex parry, and into the representation we make of it there enter the representation of the muscular sensations caused by the movement of the legs, that of the muscular sensations caused by the final movement of the arm, that of the sensations of the semi-circular canals, etc. Besides, we have to make a representation, not of a complexus of simultaneous sensations, but of a complexus of successive sensations, following one another in a determined order, and it is for this reason that I said just now that the intervention of memory is necessary.

We must further observe that, to reach the same point, I can approach nearer the object to be attained, in order not to have to extend my hand so far. And how much more might be said? It is not one only, but a thousand parries I can oppose to the same danger. All these parries are formed of sensations that may have nothing in common, and yet we regard them as defining the same point in space, because they can answer to the same danger and are one and all of them associated with the notion of that danger. It is the possibility of parrying the same blow which makes the unity of these different parries, just as it is the possibility of being parried in the same way which makes the unity of the blows of such different kinds that can threaten us from the same point in space. It is this double unity that makes the individuality of each point in space, and in the notion of such a point there is nothing else but this.

The space I pictured in the preceding section, which I might call *restricted space*, was referred to axes of co-ordinates attached to my body. These axes were fixed, since my body did not move, and it was only my limbs that changed their position. What are the axes to which the *extended space* is naturally referred—that is to say, the new space I have just defined? We define a point by the succession of movements we require to make to reach it, starting from a certain initial position of the body. The axes are accordingly attached to this initial position of the body.

But the position I call initial may be arbitrarily chosen from among all the positions my body has successively occupied. If a more or less unconscious memory of these successive positions is necessary for the genesis of the notion of space, this memory can go back more or less into the past. Hence results a certain indeterminateness in the very definition of space, and it is precisely this indeterminateness which constitutes its relativity.

Absolute space exists no longer; there is only space relative to a certain initial position of the body. For a conscious being, fixed to the ground like the inferior animals, who would consequently only know restricted space, space would still be relative, since it would be referred to his body, but this being would not be conscious of the relativity, because the axes to which he referred this restricted space would not change. No doubt the rock to which he was chained would not be motionless, since it would be involved in the motion of our planet; for us, consequently, these axes would change every moment, but for him they would not change. We have the faculty of referring our extended space at one time to the position A of our body considered as initial, at another to the position B which it occupied some moments later, which we are free to consider in its turn as initial, and, accordingly, we make unconscious changes in the co-ordinates every moment. This faculty would fail our imaginary being, and, through not having travelled, he would think space absolute. Every moment his system of axes would be imposed on him; this system might change to any extent in reality, for him it would be always the same, since it would always be the *unique* system. It is not the same for us who possess, each moment, several systems between which we can choose at will, and on condition of going back by memory more or less into the past.

That is not all, for the restricted space would not be homogeneous. The different points of this space could not be regarded as equivalent, since some could only be reached at the cost of the greatest efforts, while others could be reached with ease. On the contrary, our extended space appears to us homogeneous, and we say that all its points are equivalent. What does this mean?

If we start from a certain position A, we can, starting from that position, effect certain movements M, characterized by a certain complexus of muscular sensations. But, starting from another position B, we can execute movements M^1 which will be characterized by the same muscular sensations. Then let α be the situation of a certain point in the body, the tip of the forefinger of the right hand, for instance, in the initial position A, and let b be the position of this same forefinger when, starting from that position A, we have executed the movements M. Then let a^1 be the situation of the forefinger in the position B, and b^1 its situation when, starting from the position B, we have executed the movements M^1.

Well, I am in the habit of saying that the points a and b are, in relation to each other, as the points a^1 and b^1, and that means simply that the two series of movements M and M^1 are accompanied by the same muscular sensations. And as I am conscious that, in passing from the position A to the position B, my body has remained capable of the same movements, I know that there is a point in space which is to the point a^1 what some point b is to the point a, so that the two points a and a^1 are equivalent. It is this that is called the homogeneity of space, and at the same time it is for this reason that space is relative, since its properties remain the same whether they are referred to the axes A or to the axes B. So that the relativity of space and its homogeneity are one and the same thing.

Now, if I wish to pass to the great space, which is no longer to serve for my individual use only, but in which I can lodge the universe, I shall arrive at it by an act of imagination. I shall imagine what a giant would experience who could reach the planets in a few steps, or, if we prefer, what I should feel myself in presence of a world in miniature, in which these planets would be replaced by little balls, while on one of these little balls there would move a Lilliputian that I should call myself. But this act of imagination would be impossible for me if I had not previously constructed my restricted space and my extended space for my personal use.

IV

Now we come to the question why all these spaces have three dimensions. Let us refer to the "distribution board" spoken of above. We have, on the one side, a list of the different possible dangers—let us designate them as A1, A2, etc.—and, on the other side, the list of the different remedies, which I will call in the same way B1, B2, etc. Then we have connexions between the contact studs of the first list and those of the second in such a way that when, for instance, the alarm for danger A3 works, it sets in motion or may set in motion the relay corresponding to the parry B4.

As I spoke above of centripetal or centrifugal wires, I am afraid that all I have said may be taken, not as a simple comparison, but as a description of the nervous system. Such is not my thought, and that for several reasons. Firstly, I should not presume to pronounce an opinion on the structure of the nervous system which I do not know, while those who have studied it only do so with circumspection. Secondly, because, in spite of my incompetence, I fully realize that this scheme would be far too simple. And lastly, because, on my list of parries, there appear some that are very complex, which may even, in the case of extended space, as we have seen above, consist of several steps followed by a movement of the arm. It is not a question, then, of physical connexion between two real conductors, but of psychological association between two series of sensations.

If A1 and A2, for instance, are both of them associated with the parry B1, and if A1 is similarly associated with B2, it will generally be the case that A2 and B2 will also be associated. If this fundamental law were not generally true, there would only be an immense confusion, and there would be nothing that could bear any resemblance to a conception of space or to a geometry. How, indeed, have we defined a point in space? We defined it in two ways: on the one hand, it is the whole of the alarms A which are in connexion with the same parry B; on the other, it is the whole of the parries B which are in connexion with the same alarm A. If our law were not true, we should be obliged to say that A1 and A2 correspond with the same point, since they are both in connexion with B1; but we should be equally obliged to say that they do not correspond with the same point, since A1 would be in connexion with B2, and this would not be true of A2—which would be a contradiction.

But from another aspect, if the law were rigorously and invariably true, space would be quite different from what it is. We should have well-defined categories, among which would be apportioned the alarms A on the one side and the parries B on the other. These categories would be exceedingly numerous, but they would be entirely separated one from the other. Space would be formed of points, very numerous but discrete; it would be *discontinuous.* There would be no reason for arranging these points in one order rather than another, nor, consequently, for attributing three dimensions to space.

But this is not the case. May I be permitted for a moment to use the language of those who know geometry already? It is necessary that I should do so, since it is the language best understood by those to whom I wish to make myself clear. When I wish to parry the blow, I try to reach the point whence the blow comes, but it is enough if I come fairly near it. Then the parry B1 may answer to A1, and to A2 if the point which corresponds with B1 is sufficiently close both to that which corresponds with A1 and to that which corresponds with A2. But it may happen that the point which corresponds with another parry B2 is near enough to the point corresponding with A1, and not near enough to the point corresponding with A2. And so the parry B2 may answer to A1 and not be able to answer to A2.

For those who do not yet know geometry, this may be translated simply by a modification of the law enunciated above. Then what happens is as follows. Two parries, B1 and B2, are associated with one alarm A1, and with a very great number of alarms that we will place in the same category as A1, and make to correspond with the same point in space. But we may find alarms A2 which are associated with B2 and not with B1, but on the other hand are associated with B3, which are not with A1, and so on in succession, so that we may write the sequence

B1, A1, B2, A2, B3, A3, B4, A4,

in which each term is associated with the succeeding and preceding terms, but not with those that are several places removed.

It is unnecessary to add that each of the terms of these sequences is not isolated, but forms part of a very numerous category of other alarms or other parries which has the same connexions as it, and may be regarded as belonging to the same point in space. Thus the fundamental law, though admitting of exceptions, remains almost always true. Only, in consequence of these exceptions, these categories, instead of being entirely separate, partially encroach upon each other and mutually overlap to a certain extent, so that space becomes continuous.

Furthermore, the order in which these categories must be arranged is no longer arbitrary, and a reference to the preceding sequence will make it clear that B2 must be placed between A1 and A2, and, consequently, between B1 and B3, and that it could not be placed, for instance, between B3 and B4.

Accordingly there is an order in which our categories range themselves naturally which corresponds with the points in space, and experience teaches us that this order presents itself in the form of a three-circuit distribution board, and it is for this reason that space has three dimensions.

V

Thus the characteristic property of space, that of having three dimensions, is only a property of our distribution board, a property residing, so to speak, in the human intelligence. The destruction of some of these connexions, that is to say, of these associations of ideas, would be sufficient to give us a different distribution board, and that might be enough to endow space with a fourth dimension.

Some people will be astonished at such a result. The exterior world, they think, must surely count for something. If the number of dimensions comes from the way in which we are made, there might be thinking beings living in our world, but made differently from us, who would think that space has more or less than three dimensions. Has not M. de Cyon said that Japanese mice, having only two pairs of semi-circular canals, think that space has two dimensions? Then will not this thinking being, if he is capable of constructing a physical system, make a system of two or four dimensions, which yet, in a sense, will be the same as ours, since it will be the description of the same world in another language?

It quite seems, indeed, that it would be possible to translate our physics into the language of geometry of four dimensions. Attempting such a translation would be giving oneself a great deal of trouble for little profit, and I will content myself with mentioning Hertz's mechanics, in which something of the kind may be seen. Yet it seems that the translation would always be less simple than the text, and that it would never lose the appearance of a translation, for the language of three dimensions seems the best suited to the description of our world, even though that description may be made, in case of necessity, in another idiom.

Besides, it is not by chance that our distribution board has been formed. There is a connexion between the alarm A1 and the parry B1, that is, a property residing in our intelligence. But why is there this connexion? It is because the parry B1 enables us effectively to defend ourselves against the danger A1, and that is a fact exterior to us, a property of the exterior world. Our distribution board, then, is only the translation of an assemblage of exterior facts; if it has three dimensions, it is because it has adapted itself to a world having certain properties, and the most important of these properties is that there exist natural solids which are clearly displaced in accordance with the laws we call laws of motion of unvarying solids. If, then, the language of three dimensions is that which enables us most easily to describe our world, we must not be surprised. This language is founded on our distribution board, and it is in order to enable us to live in this world that this board has been established.

I have said that we could conceive of thinking beings, living in our world, whose distribution board would have four dimensions, who would, consequently, think in hyperspace. It is not certain, however, that such beings, admitting that they were born, would be able to live and defend themselves against the thousand dangers by which they would be assailed.

VI

A few remarks in conclusion. There is a striking contrast between the roughness of this primitive geometry which is reduced to what I call a distribution board, and the infinite precision of the geometry of geometricians. And yet the latter is the child of the former, but not of it alone; it required to be fertilized by the faculty we have of constructing mathematical concepts, such, for instance, as that of the group. It was necessary to find among these pure concepts the one that was best adapted to this rough space, whose genesis I have tried to explain in the preceding pages, the space which is common to us and the higher animals.

The evidence of certain geometrical postulates is only, as I have said, our unwillingness to give up very old habits. But these postulates are infinitely precise, while the habits have about them something essentially fluid. As soon as we wish to think, we are bound to have infinitely precise postulates, since this is the only means of avoiding contradiction. But among all the possible systems of postulates, there are some that we shall be unwilling to choose, because they do not accord sufficiently with our habits. However fluid and elastic these may be, they have a limit of elasticity.

It will be seen that though geometry is not an experimental science, it is a science born in connexion with experience; that we have created the space it studies, but adapting it to the world in which we live. We have chosen the most convenient space, but experience guided our choice. As the choice was unconscious, it appears to be imposed upon us. Some say that it is imposed by experience, and others that we are born with our space ready-made. After the preceding considerations, it will be seen what proportion of truth and of error there is in these two opinions.

In this progressive education which has resulted in the construction of space, it is very difficult to determine what is the share of the individual and what of the race. To what extent could one of us, transported from his birth into an entirely different world, where, for instance, there existed bodies displaced in accordance with the laws of motion of non-Euclidian solids—to what extent, I say, would he be able to give up the ancestral space in order to build up an entirely new space?

The share of the race seems to preponderate largely, and yet if it is to it that we owe the rough space, the fluid space of which I spoke just now, the space of the higher animals, is it not to the unconscious experience of the individual that we owe the infinitely precise space of the geometrician? This is a question that is not easy of solution. I would mention, however, a fact which shows that the space bequeathed to us by our ancestors still preserves a certain plasticity. Certain hunters learn to shoot fish under the water, although the image of these fish is raised by refraction; and, moreover, they do it instinctively. Accordingly they have learnt to modify their ancient instinct of direction, or, if you will, to substitute for the association A1, B1, another association A1, B2, because experience has shown them that the former does not succeed.

CHAPTER II

MATHEMATICAL DEFINITIONS
AND EDUCATION

1. I have to speak here of general definitions in mathematics. At least that is what the title of the chapter says, but it will be impossible for me to confine myself to the subject as strictly as the rule of unity of action demands. I shall not be able to treat it without speaking to some extent of other allied questions, and I must ask your kind forgiveness if I am thus obliged from time to time to walk among the flower-beds to right or left.

What is a good definition? For the philosopher or the scientist, it is a definition which applies to all the objects to be defined, and applies only to them; it is that which satisfies the rules of logic. But in education it is not that; it is one that can be understood by the pupils.

How is it that there are so many minds that are incapable of understanding mathematics? Is there not something paradoxical in this? Here is a science which appeals only to the fundamental principles of logic, to the principle of contradiction, for instance, to what forms, so to speak, the skeleton of our understanding, to what we could not be deprived of without ceasing to think, and yet there are people who find it obscure, and actually they are the majority. That they should be incapable of discovery we can understand, but that they should fail to understand the demonstrations expounded to them, that they should remain blind when they are shown a light that seems to us to shine with a pure brilliance, it is this that is altogether miraculous.

And yet one need have no great experience of examinations to know that these blind people are by no means exceptional beings. We have here a problem that is not easy of solution, but yet must engage the attention of all who wish to devote themselves to education.

What is understanding? Has the word the same meaning for everybody? Does understanding the demonstration of a theorem consist in examining each of the syllogisms of which it is composed in succession, and being convinced that it is correct and conforms to the rules of the game? In the same way, does understanding a definition consist simply in recognizing that the meaning of all the terms employed is already known, and being convinced that it involves no contradiction?

Yes, for some it is; when they have arrived at the conviction, they will say, I understand. But not for the majority. Almost all are more exacting; they want to know not only whether all the syllogisms of a demonstration are correct, but why they are linked together in one order rather than in another. As long as they appear to them engendered by caprice, and not by an intelligence constantly conscious of the end to be attained, they do not think they have understood.

No doubt they are not themselves fully aware of what they require and could not formulate their desire, but if they do not obtain satisfaction, they feel vaguely that something is wanting. Then what happens? At first they still perceive the evidences that are placed before their eyes, but, as they are connected by too attenuated a thread with those that precede and those that follow, they pass without leaving a trace in their brains, and are immediately forgotten; illuminated for a moment, they relapse at once into an eternal night. As they advance further, they will no longer see even this ephemeral light, because the theorems depend one upon another, and those they require have been forgotten. Thus it is that they become incapable of understanding mathematics.

It is not always the fault of their instructor. Often their intellect, which requires to perceive the connecting thread, is too sluggish to seek it and find it. But in order to come to their assistance, we must first of all thoroughly understand what it is that stops them.

Others will always ask themselves what use it is. They will not have understood, unless they find around them, in practice or in nature, the object of such and such a mathematical notion. Under each word they wish to put a sensible image; the definition must call up this image, and at each stage of the demonstration they must see it being transformed and evolved. On this condition only will they understand and retain what they have understood. These often deceive themselves: they do not listen to the reasoning, they look at the figures; they imagine that they have understood when they have only seen.

2. What different tendencies we have here! Are we to oppose them, or are we to make use of them? And if we wish to oppose them, which are we to favour? Are we to show those who content themselves with the pure logic that they have only seen one side of the matter, or must we tell those who are not so easily satisfied that what they demand is not necessary?

In other words, should we constrain young people to change the nature of their minds? Such an attempt would be useless; we do not possess the philosopher's stone that would enable us to transmute the metals entrusted to us one into the other. All that we can do is to work them, accommodating ourselves to their properties.

Many children are incapable of becoming mathematicians who must nonetheless be taught mathematics; and mathematicians themselves are not all cast in the same mould. We have only to read their works to distinguish among them two kinds of minds—logicians like Weierstrass, for instance, and intuitionists like Riemann. There is the same difference among our students. Some prefer to treat their problems "by analysis," as they say, others "by geometry."

It is quite useless to seek to change anything in this, and besides, it would not be desirable. It is good that there should be logicians and that there should be intuitionists. Who would venture to say whether he would prefer that Weierstrass had never written or that there had never been a Riemann? And so we must resign ourselves to the diversity of minds, or rather we must be glad of it.

3. Since the word *understand* has several meanings, the definitions that will be best understood by some are not those that will be best suited to others. We have those who seek to create an image, and those who restrict themselves to combining empty forms, perfectly intelligible, but purely intelligible, and deprived by abstraction of all matter.

I do not know whether it is necessary to quote any examples, but I will quote some nevertheless, and, first, the definition of fractions will furnish us with an extreme example. In the primary schools, when they want to define a fraction, they cut up an apple or a pie. Of course this is done only in imagination and not in reality, for I do not suppose the budget of primary education would allow such an extravagance. In the higher normal school, on the contrary, or in the universities, they say: a fraction is the combination of two whole numbers separated by a horizontal line. By conventions they define the operations that these symbols can undergo; they demonstrate that the rules of these operations are the same as in the calculation of whole numbers; and, lastly, they establish that multiplication of the fraction by the denominator, in accordance with these rules, gives the numerator. This is very good, because it is addressed to young people long since familiarized with the notion of fractions by dint of cutting up apples and other objects, so that their mind, refined by a considerable mathematical education, has, little by little, come to desire a purely logical definition. But what would be the consternation of the beginner to whom we attempted to offer it?

Such, also, are the definitions to be found in a book that has been justly admired and has received several awards of merit—Hilbert's *Grundlagen der Geometrie.* Let us see how he begins. "Imagine three systems of THINGS, which we will call points, straight lines, and planes." What these "things" are we do not know, and we do not need to know—it would even be unfortunate that we should seek to know; all that we have the right to know about them is that we should learn their axioms, this one, for instance: "Two different points always determine a straight line," which is followed by this commentary: "Instead of determine we may say that the straight line passes through these two points, or that it joins these two points, or that the two points are situated on the straight line." Thus "being situated on a straight line" is simply defined as synonymous with "determining a straight line." Here is a book of which I think very highly, but which I should not recommend to a schoolboy. For the matter of that I might do it without fear; he would not carry his reading very far.

I have taken extreme examples, and no instructor would dream of going so far. But, even though he comes nowhere near such models, is he not still exposed to the same danger?

We are in a class of the fourth grade. The teacher is dictating: "A circle is the position of the points in a plane which are the same distance from an interior point called the centre." The good pupil writes this phrase in his copy-book and the bad pupil draws faces, but neither of them understands. Then the teacher takes the chalk and draws a circle on the board. "Ah," think the pupils, "why didn't he say at once, a circle is a round, and we should have understood." No doubt it is the teacher who is right. The pupils' definition would have been of no value, because it could not have been used for any demonstration, and chiefly because it could not have given them the salutary habit of analyzing their conceptions. But they should be made to see that they do not understand what they think they understand, and brought to realize the roughness of their primitive concept, and to be anxious themselves that it should be purified and refined.

4. I shall return to these examples; I only wished to show the two opposite conceptions. There is a violent contrast between them, and this contrast is explained by the history of the science. If we read a book written fifty years ago, the greater part of the arguments appear to us devoid of exactness.

At that period they assumed that a continuous function cannot change its sign without passing through zero, but today we prove it. They assumed that the ordinary rules of calculus are applicable to incommensurable numbers; today we prove it. They assumed many other things that were sometimes untrue.

They trusted to intuition, but intuition cannot give us exactness, nor even certainty, and this has been recognized more and more. It teaches us, for instance, that every curve has a tangent—that is to say, that every continuous function has a derivative—and that is untrue. As certainty was required, it has been necessary to give less and less place to intuition.

How has this necessary evolution come about? It was not long before it was recognized that exactness cannot be established in the arguments unless it is first introduced into the definitions.

For a long time the objects that occupied the attention of mathematicians were badly defined. They thought they knew them because they represented them by their senses or their imagination, but they had only a rough image, and not a precise idea such as reasoning can take hold of.

It is to this that the logicians have had to apply their efforts, and similarly for incommensurable numbers.

The vague idea of continuity which we owe to intuition has resolved itself into a complicated system of inequalities bearing on whole numbers. Thus it is that all those difficulties which terrified our ancestors when they reflected upon the foundations of the infinitesimal calculus have finally vanished.

In analysis today there is no longer anything but whole numbers, or finite or infinite systems of whole numbers, bound together by a network of equalities and inequalities. Mathematics, as it has been said, has been arithmetized.

5. But we must not imagine that the science of mathematics has attained to absolute exactness without making any sacrifice. What it has gained in exactness it has lost in objectivity. It is by withdrawing from reality that it has acquired this perfect purity. We can now move freely over its whole domain, which formerly bristled

with obstacles. But these obstacles have not disappeared; they have only been removed to the frontier, and will have to be conquered again if we wish to cross the frontier and penetrate into the realms of practice.

We used to possess a vague notion, formed of incongruous elements, some *à priori* and others derived from more or less digested experiences, and we imagined we knew its principal properties by intuition. Today we reject the empirical element and preserve only the *à priori* ones. One of the properties serves as definition, and all the others are deduced from it by exact reasoning. This is very good, but it still remains to prove that this property, which has become a definition, belongs to the real objects taught us by experience, from which we had drawn our vague intuitive notion. In order to prove it we shall certainly have to appeal to experience or make an effort of intuition; and if we cannot prove it, our theorems will be perfectly exact but perfectly useless.

Logic sometimes breeds monsters. For half a century there has been springing up a host of weird functions, which seem to strive to have as little resemblance as possible to honest functions that are of some use. No more continuity, or else continuity but no derivatives, etc. More than this, from the point of view of logic, it is these strange functions that are the most general; those that are met without being looked for no longer appear as more than a particular case, and they have only quite a little corner left them.

Formerly, when a new function was invented, it was in view of some practical end. Today they are invented on purpose to show our ancestors' reasonings at fault, and we shall never get anything more than that out of them.

If logic were the teacher's only guide, he would have to begin with the most general, that is to say, with the most weird, functions. He would have to set the beginner to wrestle with this collection of monstrosities. If you don't do so, the logicians might say, you will only reach exactness by stages.

6. Possibly this may be true, but we cannot take such poor account of reality, and I do not mean merely the reality of the sensi-

ble world, which has its value nevertheless, since it is for battling with it that nine-tenths of our pupils are asking for arms. There is a more subtle reality which constitutes the life of mathematical entities, and is something more than logic.

Our body is composed of cells, and the cells of atoms, but are these cells and atoms the whole reality of the human body? Is not the manner in which these cells are adjusted, from which results the unity of the individual, also a reality, and of much greater interest?

Would a naturalist imagine that he had an adequate knowledge of the elephant if he had never studied the animal except through a microscope?

It is the same in mathematics. When the logician has resolved each demonstration into a host of elementary operations, all of them correct, he will not yet be in possession of the whole reality; that indefinable something that constitutes the unity of the demonstration will still escape him completely.

What good is it to admire the mason's work in the edifices erected by great architects, if we cannot understand the general plan of the master? Now pure logic cannot give us this view of the whole; it is to intuition we must look for it.

Take, for instance, the idea of the continuous function. To begin with, it is only a perceptible image, a line drawn with chalk on a blackboard. Little by little it is purified; it is used for constructing a complicated system of inequalities which reproduces all the lines of the original image; when the work is quite finished, the *centering* is removed, as it is after the construction of an arch; this crude representation is henceforth a useless support, and disappears, and there remains only the edifice itself, irreproachable in the eyes of the logician. And yet, if the instructor did not recall the original image, if he did not replace the *centering* for a moment, how would the pupil guess by what caprice all these inequalities had been scaffolded in this way one upon another? The definition would be logically correct, but it would not show him the true reality.

7. And so we are obliged to make a step backwards. No doubt it is hard for a master to teach what does not satisfy him entirely, but the satisfaction of the master is not the sole object of education. We

have first to concern ourselves with the pupil's state of mind, and what we want it to become.

Zoologists declare that the embryonic development of an animal repeats in a very short period of time the whole history of its ancestors of the geological ages. It seems to be the same with the development of minds. The educator must make the child pass through all that his fathers have passed through, more rapidly, but without missing a stage. On this account, the history of any science must be our first guide.

Our fathers imagined they knew what a fraction was, or continuity, or the area of a curved surface; it is we who have realized that they did not. In the same way our pupils imagine that they know it when they begin to study mathematics seriously. If, without any other preparation, I come and say to them: "No, you do not know it; you do not understand what you imagine you understand; I must demonstrate to you what appears to you evident;" and if, in the demonstration, I rely on premisses that seem to them less evident than the conclusion, what will the wretched pupils think? They will think that the science of mathematics is nothing but an arbitrary aggregation of useless subtleties; or they will lose their taste for it; or else they will look upon it as an amusing game, and arrive at a state of mind analogous to that of the Greek sophists.

Later on, on the contrary, when the pupil's mind has been familiarized with mathematical reasoning and ripened by this long intimacy, doubts will spring up of their own accord, and then your demonstration will be welcome. It will arouse new doubts, and questions will present themselves successively to the child, as they presented themselves successively to our fathers, until they reach a point when only perfect exactness will satisfy them. It is not enough to feel doubts about everything; we must know why we doubt.

8. The principal aim of mathematical education is to develop certain faculties of the mind, and among these intuition is not the least precious. It is through it that the mathematical world remains in touch with the real world, and even if pure mathematics could do without it, we should still have to have recourse to it to fill up

the gulf that separates the symbol from reality. The practitioner will always need it, and for every pure geometrician there must be a hundred practitioners.

The engineer must receive a complete mathematical training, but of what use is it to be to him, except to enable him to see the different aspects of things and to see them quickly? He has no time to split hairs. In the complex physical objects that present themselves to him, he must promptly recognize the point where he can apply the mathematical instruments we have put in his hands. How could he do this if we left between the former and the latter that deep gulf dug by the logicians?

9. Besides the future engineers are other less numerous pupils, destined in their turn to become teachers, and so they must go to the very root of the matter; a profound and exact knowledge of first principles is above all indispensable for them. But that is no reason for not cultivating their intuition, for they would form a wrong idea of the science if they never looked at it on more than one side, and, besides, they could not develop in their pupils a quality they did not possess themselves.

For the pure geometrician himself this faculty is necessary: it is by logic that we prove, but by intuition that we discover. To know how to criticize is good, but to know how to create is better. You know how to recognize whether a combination is correct, but much use this will be if you do not possess the art of selecting among all the possible combinations. Logic teaches us that on such and such a road we are sure of not meeting an obstacle; it does not tell us which is the road that leads to the desired end. For this it is necessary to see the end from afar, and the faculty which teaches us to see is intuition. Without it, the geometrician would be like a writer well up in grammar but destitute of ideas. Now how is this faculty to develop, if, as soon as it shows itself, it is hounded out and proscribed, if we learn to distrust it before we know what good can be got from it?

And here let me insert a parenthesis to insist on the importance of written exercises. Compositions in writing are perhaps not given sufficient prominence in certain examinations. In the École Polytechnique, for instance, I am told that insistence on such compositions would close the door to very good pupils who know their subject and understand it very well, and yet are incapable of applying it in the smallest degree. I said just above that the word *understand* has several meanings. Such pupils only understand in the first sense of the word, and we have just seen that this is not sufficient to make either an engineer or a geometrician. Well, since we have to make a choice, I prefer to choose those who understand thoroughly.

10. But is not the art of exact reasoning also a precious quality that the teacher of mathematics should cultivate above all else? I am in no danger of forgetting it: we must give it attention, and that from the beginning. I should be distressed to see geometry degenerate into some sort of low-grade tachymetrics, and I do not by any means subscribe to the extreme doctrines of certain German professors. But we have sufficient opportunity of training pupils in correct reasoning in those parts of mathematics in which the disadvantages I have mentioned do not occur. We have long series of theorems in which absolute logic has ruled from the very start and, so to speak, naturally, in which the first geometricians have given us models that we must continually imitate and admire.

It is in expounding the first principles that we must avoid too much subtlety, for there it would be too disheartening, and useless besides. We cannot prove everything, we cannot define everything, and it will always be necessary to draw upon intuition. What does it matter whether we do this a little sooner or a little later, and even whether we ask for a little more or a little less, provided that, making a correct use of the premisses it gives us, we learn to reason accurately?

11. Is it possible to satisfy so many opposite conditions? Is it possible especially when it is a question of giving a definition? How are we to find a statement that will at the same time satisfy the inexorable laws of logic and our desire to understand the new notion's place in the general scheme of the science, our need of thinking in images? More often than not we shall not find it, and that is why the statement of a definition is not enough; it must be prepared and it must be justified.

What do I mean by this? You know that it has often been said that every definition implies an axiom, since it asserts the existence of the object defined. The definition, then, will not be justified, from the purely logical point of view, until we have *proved* that it involves no contradiction either in its terms or with the truths previously admitted.

But that is not enough. A definition is stated as a convention, but the majority of minds will revolt if you try to impose it upon them as an *arbitrary* convention. They will have no rest until you have answered a great number of questions.

Mathematical definitions are most frequently, as M. Liard has shown, actual constructions built up throughout of simpler notions. But why should these elements have been assembled in this manner, when a thousand other assemblages were possible? Is it simply caprice? If not, why had this combination more right to existence than any of the others? What need does it fill? How was it foreseen that it would play an important part in the development of the science, that it would shorten our reasoning and our calculations? Is there any familiar object in nature that is, so to speak, its indistinct and rough image?

That is not all. If you give a satisfactory answer to all these questions, we shall realize that the newcomer had the right to be baptized. But the choice of a name is not arbitrary either; we must explain what analogies have guided us, and that if we have given analogous names to different things, these things at least differ only in matter, and have some resemblance in form, that their properties are analogous and, so to speak, parallel.

It is on these terms that we shall satisfy all propensities. If the statement is sufficiently exact to please the logician, the justification will satisfy the intuitionist. But we can do better still. Whenever it is possible, the justification will precede the statement and prepare it. The general statement will be led up to by the study of some particular examples.

One word more. The aim of each part of the statement of a definition is to distinguish the object to be defined from a class of other neighbouring objects. The definition will not be understood until you have shown not only the object defined, but the neighbouring objects from which it has to be distinguished, until you have made it possible to grasp the difference, and have added explicitly your reason for saying this or that in stating the definition.

But it is time to leave generalities and to enquire how the somewhat abstract principles I have been expounding can be applied in arithmetic, in geometry, in analysis, and in mechanics.

ARITHMETIC

12. We do not have to define the whole number. On the other hand, operations on whole numbers are generally defined, and I think the pupils learn these definitions by heart and attach no meaning to them. For this there are two reasons: first, they are taught them too early, while their mind still feels no need of them; and then these definitions are not satisfactory from the logical point of view. For addition, we cannot find a good one, simply because we must stop somewhere, and cannot define everything. The definition of addition is to say that it consists in adding. All that we can do is to start with a certain number of concrete examples and say, the operation that has just been performed is called addition.

For subtraction it is another matter. It can be defined logically as the inverse operation of addition. But is that how we should begin? Here, again, we should start with examples, and show by these examples the relation of the two operations. Thus the definition will be prepared and justified.

In the same way for multiplication. We shall take a particular problem; we shall show that it can be solved by adding several equal numbers together; we shall then point out that we arrive at the result quicker by multiplication, the operation the pupils perform already by rote, and the logical definition will spring from this quite naturally.

We shall define division as the inverse operation of multiplication; but we shall begin with an example drawn from the familiar notion of sharing, and we shall show by this example that multiplication reproduces the dividend.

There remain the operations on fractions. There is no difficulty except in the case of multiplication. The best way is first to expound the theory of proportions, as it is from it alone that the logical definition can spring. But, in order to gain acceptance for the definitions that are met with at the start in this theory, we must prepare them by numerous examples drawn from classical problems of the rule of three, and we shall be careful to introduce fractional data. We shall not hesitate, either, to familiarize the pupils with the notion of proportion by geometrical figures; either appealing to their recollection if they have already done any geometry, or having recourse to direct intuition if they have not, which, moreover, will prepare them to do it. I would add, in conclusion, that after having defined the multiplication of fractions, we must justify this definition by demonstration that it is commutative, associative, and distributive, making it quite clear to the listeners that the verification has been made in order to justify the definition.

We see what part is played in all this by geometrical figures, and this part is justified by the philosophy and the history of the science. If arithmetic had remained free from all intermixture with geometry, it would never have known anything but the whole number. It was in order to adapt itself to the requirements of geometry that it discovered something else.

GEOMETRY

In geometry we meet at once the notion of the straight line. Is it possible to define the straight line? The common definition, the shortest path from one point to another, does not satisfy me at all. I should start simply with the *ruler,* and I should first show the pupil how we can verify a ruler by revolving it. This verification is the true definition of a straight line, for a straight line is an axis of rotation. We should then show him how to verify the ruler by sliding it, and we should have one of the most important properties of a straight line. As for that other property, that of being the shortest path from one point to another, it is a theorem that can be demonstrated apodeictically, but the demonstration is too advanced to find a place in secondary education. It will be better to show that a ruler previously verified can be applied to a taut thread. We must not hesitate, in the presence of difficulties of this kind, to multiply the axioms, justifying them by rough examples.

Some axioms we must admit; and if we admit a few more than is strictly necessary, the harm is not great. The essential thing is to learn to reason exactly with the axioms once admitted. Uncle Sarcey, who loved to repeat himself, often said that the audience at a theatre willingly accepts all the postulates imposed at the start, but that once the curtain has gone up it becomes inexorable on the score of logic. Well, it is just the same in mathematics.

For the circle we can start with the compass. The pupils will readily recognize the curve drawn. We shall then point out to them that the distance of the two points of the instrument remains constant, that one of these points is fixed and the other movable, and we shall thus be led naturally to the logical definition.

The definition of a plane implies an axiom, and we must not attempt to conceal the fact. Take a drawing-board and point out how a movable ruler can be applied constantly to the board, and that while still retaining three degrees of freedom. We should compare this with the cylinder and the cone, surfaces to which a straight line cannot be applied unless we allow it only two degrees of freedom. Then we should take three drawing-boards, and we should show first that they can slide while still remaining in contact with one another, and that with three degrees of freedom. And lastly, in order to distinguish the plane from the sphere, that two of these boards that can be applied to a third can also be applied to one another.

Perhaps you will be surprised at this constant use of movable instruments. It is not a rough artifice, and it is much more philosophical than it would appear at first sight. What is geometry for the philosopher? It is the study of a group. And what group? That of the movements of solid bodies. How are we to define this group, then, without making some solid bodies move?

Are we to preserve the classical definition of parallels, and say that we give this name to two straight lines, situated in the same plane, which, being produced ever so far, never meet? No, because this definition is negative, because it cannot be verified by experience, and cannot consequently be regarded as an immediate datum of intuition, but chiefly because it is totally foreign to the notion of group and to the consideration of the motion of solid bodies, which is, as I have said, the true source of geometry. Would it not be better to define first the rectilineal transposition of an invariable figure as a motion in which all the points of this figure have rectilineal trajectories, and to show that such a transposition is possible, making a square slide on a ruler? From this experimental verification, raised to the form of an axiom, it would be easy to educe the notion of parallel and Euclid's postulate itself.

MECHANICS

I need not go back to the definition of velocity or of acceleration or of the other kinematic notions: they will be more properly connected with ideas of space and time, which alone they involve.

On the contrary, I will dwell on the dynamic notions of force and mass.

There is one thing that strikes me, and that is, how far young people who have received a secondary education are from applying the mechanical laws they have been taught to the real world. It is not only that they are incapable of doing so, but they do not even think of it. For them the world of science and that of reality are shut off in water-tight compartments. It is not uncommon to see a well-dressed man, probably a university man, sitting in a carriage and imagining that he is helping it on by pushing on the dashboard, and that in disregard of the principle of action and reaction.

If we try to analyze the state of mind of our pupils, this will surprise us less. What is for them the true definition of force? Not the one they repeat, but the one that is hidden away in a corner of their intellect, and from thence directs it all. This is their definition: forces are arrows that parallelograms are made of; these arrows are imaginary things that have nothing to do with anything that exists in nature. This would not happen if they were shown forces in reality before having them represented by arrows.

How are we to define force? If we want a logical definition, there is no good one, as I think I have shown satisfactorily elsewhere. There is the anthropomorphic definition, the sensation of muscular effort; but this is really too crude, and we cannot extract anything useful from it.

This is the course we ought to pursue. First, in order to impart a knowledge of the genus force, we must show, one after the other, all the species of this genus. They are very numerous and of great variety. There is the pressure of liquids on the sides of the vessels in which they are contained, the tension of cords, the elasticity of a spring, gravity that acts on all the molecules of a body, friction, the normal mutual action and reaction of two solids in contact.

This is only a qualitative definition; we have to learn to measure a force. For this purpose we shall show first that we can replace one force by another without disturbing the equilibrium, and we shall find the first example of this substitution in the balance and Borda's double scales. Then we shall show that we can replace a weight not only by another weight, but by forces of different nature; for example, Prony's dynamometer break enables us to replace a weight by friction.

From all this arises the notion of the equivalence of two forces.

We must also define the direction of a force. If a force F is equivalent to another force F¹ that is applied to the body we are dealing with through the medium of a taut cord, in such a way that F can be replaced by F¹ without disturbing the equilibrium, then the point of attachment of the cord will be, by definition, the point of application of the force F¹ and that of the equivalent force F, and the direction of the cord will be the direction of the force F¹ and also that of the equivalent force F.

From this we shall pass to the comparison of the magnitude of forces. If one force can replace two others of the same direction, it must be equal to their sum, and we shall show, for instance, that a weight of 20 ounces can replace two weights of 10 ounces.

But this is not all. We know now how to compare the intensity of two forces which have the same direction and the same point of application, but we have to learn to do this when the directions are different. For this purpose we imagine a cord stretched by a weight and passing over a pulley; we say that the tension of the two portions of the cord is the same, and equal to the weight.

Here is our definition. It enables us to compare the tensions of our two portions, and, by using the preceding definitions, to compare two forces of any kind having the same direction as these two portions. We have to justify it by showing that the tension of the last portion remains the same for the same weight, whatever be the number and the disposition of the pulleys. We must then complete it by showing that this is not true unless the pulleys are without friction.

Once we have mastered these definitions we must show that the point of application, the direction, and the intensity are sufficient to determine a force; that two forces for which these three elements are the same are *always* equivalent, and can *always* be replaced one by the other, either in equilibrium or in motion, and that whatever be the other forces coming into play.

We must show that two concurrent forces can always be replaced by a single resultant force, and that *this resultant remains the same* whether the body is in repose or in motion, and whatever be the other forces applied to it.

Lastly, we must show that forces defined as we have defined them satisfy the principle of the equality of action and reaction.

All this we learn by experiment, and by experiment alone.

It will be sufficient to quote some common experiments that the pupils make every day without being aware of it, and to perform before them a small number of simple and well-selected experiments.

It is not until we have passed through all these roundabout ways that we can represent forces by arrows, and even then I think it would be good, from time to time, as the argument develops, to come back from the symbol to the reality. It would not be difficult, for instance, to illustrate the parallelogram of forces with the help of an apparatus composed of three cords passing over pulleys, stretched by weights, and producing equilibrium by pulling on the same point.

Once we know force, it is easy to define mass. This time the definition must be borrowed from dynamics. We cannot do otherwise, since the end in view is to make clear the distinction between mass and weight. Here, again, the definition must be prepared by experiments. There is, indeed, a machine that seems to be made on purpose to show what mass is, and that is Atwood's machine. Besides this we shall recall the laws of falling bodies, and how acceleration of gravity is the same for heavy as for light bodies, and varies according to latitude, etc.

Now if you tell me that all the methods I advocate have long since been applied in schools, I shall be more pleased than surprised to hear it. I know that on the whole our mathematical education is good; I do not wish to upset it, and should even be distressed at this result; I only desire gradual, progressive improvements. This education must not undergo sudden variations at the capricious breath of ephemeral fashions. In such storms its high educative value would soon founder. A good and sound logic must continue to form its foundation. Definition by example is always necessary, but it must prepare the logical definition and not take its place; it must at least make its want felt in cases where the true logical definition cannot be given to any purpose except in higher education.

You will understand that what I have said here in no sense implies the abandonment of what I have written elsewhere. I have often had occasion to criticize definitions which I advocate today. These criticisms hold good in their entirety; the definitions can only be provisional, but it is through them that we must advance.

CHAPTER III

Mathematics and Logic

Introduction

Can mathematics be reduced to logic without having to appeal to principles peculiar to itself? There is a whole school full of ardour and faith who make it their business to establish the possibility. They have their own special language, in which words are used no longer, but only signs. This language can be understood only by the few initiated, so that the vulgar are inclined to bow before the decisive affirmations of the adepts. It will, perhaps, be useful to examine these affirmations somewhat more closely, in order to see whether they justify the peremptory tone in which they are made.

But in order that the nature of the question should be properly understood, it is necessary to enter into some historical details, and more particularly to review the character of Cantor's work.

The notion of infinity had long since been introduced into mathematics, but this infinity was what philosophers call a *becoming*. Mathematical infinity was only a quantity susceptible of growing beyond all limit; it was a variable quantity of which it could not be said that it *had passed*, but only that it *would pass*, all limits.

Cantor undertook to introduce into mathematics an *actual infinity*—that is to say, a quantity which is not only susceptible of passing all limits, but which is regarded as having already done so. He set himself such questions as these: Are there more points in space than there are whole numbers? Are there more points in space than there are points in a plane? etc.

Then the number of whole numbers, that of points in space, etc., constitutes what he terms a *transfinite cardinal number*—that is to say, a cardinal number greater than all the ordinary cardinal numbers. And he amused himself by comparing these transfinite cardinal numbers, by arranging in suitable order the elements of a whole which contains an infinite number of elements; and he also imagined what he terms transfinite ordinal numbers, on which I will not dwell further.

Many mathematicians have followed in his tracks, and have set themselves a series of questions of the same kind. They have become so familiar with transfinite numbers that they have reached the point of making the theory of finite numbers depend on that of Cantor's cardinal numbers. In their opinion, if we wish to teach arithmetic in a truly logical way, we ought to begin by establishing the general properties of the transfinite cardinal numbers, and then distinguish from among them quite a small class, that of the ordinary whole numbers. Thanks to this roundabout proceeding, we might succeed in proving all the propositions relating to this small class (that is to say, our whole arithmetic and algebra) without making use of a single principle foreign to logic.

This method is evidently contrary to all healthy psychology. It is certainly not in this manner that the human mind proceeded to construct mathematics, and I imagine, too, its authors do not dream of introducing it into secondary education. But is it at least logical, or, more properly speaking, is it accurate? We may well doubt it.

Nevertheless, the geometricians who have employed it are very numerous. They have accumulated formulas and imagined that they rid themselves of all that is not pure logic by writing treatises in which the formulas are no longer interspersed with explanatory text, as in the ordinary works on mathematics, but in which the text has disappeared entirely.

Unfortunately, they have arrived at contradictory results, at what are called the *Cantorian antinomies*, to which we shall have occasion to return. These contradictions have not discouraged them, and they have attempted to modify their rules, in order to dispose of those that had already appeared, but without gaining any assurance by so doing that no new ones would appear.

It is time that these exaggerations were treated as they deserve. I have no hope of convincing these logicians, for they have lived too long in this atmosphere. Besides, when we have refuted one of their demonstrations, we are quite sure to find it cropping up again with insignificant changes, and some of them have already risen several times from their ashes. Such in old times was the Lernæan hydra, with its famous heads that always grew again. Hercules was successful because his hydra had only nine heads (unless, indeed, it was eleven), but in this case there are too many, they are in England, in Germany, in Italy, and in France, and he would be forced to abandon the task. And so I appeal only to unprejudiced people of common sense.

I

In these latter years a large number of works have been published on pure mathematics and the philosophy of mathematics, with a view to disengaging and isolating the logical elements of mathematical reasoning. These works have been analyzed and expounded very lucidly by M. Couturat in a work entitled *Les principes des mathématiques.*

In M. Couturat's opinion the new works, and more particularly those of Mr. Russell and Signor Peano, have definitely settled the controversy so long in dispute between Leibnitz and Kant. They have shown that there is no such thing as an *à priori* synthetic judgment (the term employed by Kant to designate the judgments that can neither be demonstrated analytically, nor reduced to identity, nor established experimentally); they have shown that mathematics is entirely reducible to logic, and that intuition plays no part in it whatever.

This is what M. Couturat sets forth in the work I have just quoted. He also stated the same opinions even more explicitly in his speech at Kant's jubilee; so much so that I overheard my neighbour whisper: "It's quite evident that this is the centenary of Kant's *death.*"

Can we subscribe to this decisive condemnation? I do not think so, and I will try to show why.

II

What strikes us first of all in the new mathematics is its purely formal character. "Imagine," says Hilbert, "three kinds of *things,* which we will call points, straight lines, and planes; let us agree that a straight line shall be determined by two points, and that, instead of saying that this straight line is determined by these two points, we may say that it passes through these two points, or that these two points are situated on the straight line." What these *things* are, not only do we not know, but we must not seek to know. It is unnecessary, and anyone who had never seen either a point or a straight line or a plane could do geometry just as well as we can. In order that the words *pass through* or the words *be situated on* should not call up any image in our minds, the former is merely regarded as the synonym of *be determined,* and the latter of *determine.*

Thus it will be readily understood that, in order to demonstrate a theorem, it is not necessary or even useful to know what it means. We might replace geometry by the *reasoning piano* imagined by Stanley Jevons; or, if we prefer, we might imagine a machine where we should put in axioms at one end and take out theorems at the other, like that legendary machine in Chicago where pigs go in alive and come out transformed into hams and sausages. It is no more necessary for the mathematician than it is for these machines to know what he is doing.

I do not blame Hilbert for this formal character of his geometry. He was bound to tend in this direction, given the problem he set himself. He wished to reduce to a minimum the number of the fundamental axioms of geometry, and to make a complete enumeration of them. Now, in the arguments in which our mind remains active, in those in which intuition still plays a part, in the living arguments, so to speak, it is difficult not to introduce an axiom or a postulate that passes unnoticed. Accordingly, it was not till he had reduced all geometrical arguments to a purely mechanical form that he could be certain of having succeeded in his design and accomplished his work.

What Hilbert had done for geometry, others have tried to do for arithmetic and analysis. Even if they had been entirely successful, would the Kantians be finally condemned to silence? Perhaps not, for it is certain that we cannot reduce mathematical thought to an empty form without mutilating it. Even admitting that it has been established that all theorems can be deduced by purely analytical processes, by simple logical combinations of a finite number of axioms, and that these axioms are nothing but conventions, the philosopher would still retain the right to seek the origin of these conventions, and to ask why they were judged preferable to the contrary conventions.

And, further, the logical correctness of the arguments that lead from axioms to theorems is not the only thing we have to attend to. Do the rules of perfect logic constitute the whole of mathematics? As well say that the art of the chess-player reduces itself to the rules for the movement of the pieces. A selection must be made out of all the constructions that can be combined with the materials furnished by logic. The true geometrician makes this selection judiciously, because he is guided by a sure instinct, or by some vague consciousness of I know not what profounder and more hidden geometry, which alone gives a value to the constructed edifice.

To seek the origin of this instinct, and to study the laws of this profound geometry which can be felt but not expressed, would be a noble task for the philosophers who will not allow that logic is all. But this is not the point of view I wish to take, and this is not the way I wish to state the question. This instinct I have been speaking of is necessary to the discoverer, but it seems at first as if we could do without it for the study of the science once created. Well, what I want to find out is, whether it is true that once the principles of logic are admitted we can, I will not say discover, but demonstrate all mathematical truths without making a fresh appeal to intuition.

III

To this question I formerly gave a negative answer. (See *Science and Hypothesis*, Chapter I.) Must our answer be modified by recent works? I said no, because "the principle of complete induction" appeared to me at once necessary to the mathematician, and irreducible to logic. We know the statement of the principle: "If a property is true of the number 1, and if it is established that it is true of $n+1$ provided it is true of n, it will be true of all whole numbers." I recognized in this the typical mathematical argument. I did not mean to say, as has been supposed, that all mathematical arguments can be reduced to an application of this principle. Examining these arguments somewhat closely, we should discover the application of many other similar principles, offering the same essential characteristics. In this category of principles, that of complete induction is only the simplest of all, and it is for that reason that I selected it as a type.

The term *principle of complete induction* which has been adopted is not justifiable. This method of reasoning is nonetheless a true mathematical induction itself, which only differs from the ordinary induction by its certainty.

IV. DEFINITIONS AND AXIOMS

The existence of such principles is a difficulty for the inexorable logicians. How do they attempt to escape it? The principle of complete induction, they say, is not an axiom properly so called, or an *à priori* synthetic judgment; it is simply the definition of the whole number. Accordingly it is a mere convention. In order to discuss this view, it will be necessary to make a close examination of the relations between definitions and axioms.

We will first refer to an article by M. Couturat on mathematical definitions which appeared in *L'Enseignement mathématique,* a review published by Gauthier-Villars and by Georg in Geneva. We find a distinction between *direct definition* and *definition by postulates.*

"Definition by postulates," says M. Couturat, "applies not to a single notion, but to a system of notions; it consists in enumerating the fundamental relations that unite them, which make it possible to demonstrate all their other properties: these relations are postulates..."

If we have previously defined all these notions with one exception, then this last will be by definition the object which verifies these postulates.

Thus certain indemonstrable axioms of mathematics would be nothing but disguised definitions. This point of view is often legitimate, and I have myself admitted it, for instance, in regard to Euclid's postulate.

The other axioms of geometry are not sufficient to define distance completely. Distance, then, will be by definition, the one among all the magnitudes which satisfy the other axioms, that is of such a nature as to make Euclid's postulate true.

Well, the logicians admit for the principle of complete induction what I admit for Euclid's postulate, and they see nothing in it but a disguised definition.

But to give us this right, there are two conditions that must be fulfilled. John Stuart Mill used to say that every definition implies an axiom, that in which we affirm the existence of the object defined. On this score, it would no longer be the axiom that might be a disguised definition, but, on the contrary, the definition that would be a disguised axiom. Mill understood the word *existence* in a material and empirical sense; he meant that in defining a circle we assert that there are round things in nature.

In this form his opinion is inadmissible. Mathematics is independent of the existence of material objects. In mathematics the word *exist* can only have one meaning; it signifies exemption from contradiction. Thus rectified, Mill's thought becomes accurate. In defining an object, we assert that the definition involves no contradiction.

If, then, we have a system of postulates, and if we can demonstrate that these postulates involve no contradiction, we shall have the right to consider them as representing the definition of one of the notions found among them. If we cannot demonstrate this, we must admit it without demonstration, and then it will be an axiom. So that if we wished to find the definition behind the postulate, we should discover the axiom behind the definition.

Generally, for the purpose of showing that a definition does not involve any contradiction, we proceed *by example*, and try to form an example of an object satisfying the definition. Take the case of a definition by postulates. We wish to define a notion A, and we say that, by definition, an A is any object for which certain postulates are true. If we can demonstrate directly that all these postulates are true of a certain object B, the definition will be justified, and the object B will be an *example* of A. We shall be certain that the postulates are not contradictory, since there are cases in which they are all true at once.

But such a direct demonstration by example is not always possible. Then, in order to establish that the postulates do not involve contradiction, we must picture all the propositions that can be deduced from these postulates considered as premises, and show that among these propositions there are no two of which one is the contradiction of the other. If the number of these propositions is finite, a direct verification is possible; but this is a case that is not frequent, and, moreover, of little interest.

If the number of the propositions is infinite, we can no longer make this direct verification. We must then have recourse to processes of demonstration, in which we shall generally be forced to invoke that very principle of complete induction that we are attempting to verify.

I have just explained one of the conditions which the logicians were bound to satisfy, *and we shall see further on that they have not done so.*

V

There is a second condition. When we give a definition, it is for the purpose of using it.

Accordingly, we shall find the word *defined* in the text that follows. Have we the right to assert, of the object represented by this word, the postulate that served as definition? Evidently we have, if the word has preserved its meaning, if we have not assigned it a different meaning by implication. Now this is what sometimes happens, and it is generally difficult to detect it. We must see how the word was introduced into our text, and whether the door through which it came does not really imply a different definition from the one enunciated.

This difficulty is encountered in all applications of mathematics. The mathematical notion has received a highly purified and exact definition, and for the pure mathematician all hesitation has disappeared. But when we come to apply it, to the physical sciences, for instance, we are no longer dealing with this pure notion, but with a concrete object which is often only a rough image of it. To say that this object satisfies the definition, even approximately, is to enunciate a new truth, which has no longer the character of a conventional postulate, and that experience alone can establish beyond a doubt.

But, without departing from pure mathematics, we still meet with the same difficulty. You give a subtle definition of number, and then, once the definition has been given, you think no more about it, because in reality it is not your definition that has taught you what a number is, you knew it long before, and when you come to write the word *number* farther on, you give it the same meaning as anybody else. In order to know what this meaning is, and if it is indeed the same in this phrase and in that, we must see how you have been led to speak of number and to introduce the word into the two phrases. I will not explain my point any further for the moment, for we shall have occasion to return to it.

Thus we have a word to which we have explicitly given a definition A. We then proceed to make use of it in our text in a way which implicitly supposes another definition B. It is possible that these two definitions may designate the same object, but that such is the case is a new truth that must either be demonstrated or else admitted as an independent axiom.

We shall see farther on that the logicians have not fulfilled this second condition any better than the first.

VI

The definitions of number are very numerous and of great variety, and I will not attempt to enumerate even their names and their authors. We must not be surprised that there are so many. If any one of them was satisfactory we should not get any new ones. If each new philosopher who has applied himself to the question has thought it necessary to invent another, it is because he was not satisfied with those of his predecessors; and if he was not satisfied, it was because he thought he detected a *petitio principii.*

I have always experienced a profound sentiment of uneasiness in reading the works devoted to this problem. I constantly expect to run against a *petitio principii,* and when I do not detect it at once I am afraid that I have not looked sufficiently carefully.

The fact is that it is impossible to give a definition without enunciating a phrase, and difficult to enunciate a phrase without putting in a name of number, or at least the word *several,* or at least a word in the plural. Then the slope becomes slippery, and every moment we are in danger of falling into the *petitio principii.*

I will concern myself in what follows with those only of these definitions in which the *petitio principii* is most skilfully concealed.

VII. PASIGRAPHY

The symbolical language created by Signor Peano plays a very large part in these new researches. It is capable of rendering some service, but it appears to me that M. Couturat attaches to it an exaggerated importance that must have astonished Peano himself.

The essential element of this language consists in certain algebraic signs which represent the conjunctions: if, and, or, therefore. That these signs may be convenient is very possible, but that they should be destined to change the face of the whole philosophy is quite another matter. It is difficult to admit that the word *if* acquires, when written ɔ, a virtue it did not possess when written *if*.

This invention of Peano was first called *pasigraphy*, that is to say, the art of writing a treatise on mathematics without using a single word of the ordinary language. This name defined its scope most exactly. Since then it has been elevated to a more exalted dignity, by having conferred upon it the title of *logistic*. The same word is used, it appears, in the École de Guerre to designate the art of the quartermaster, the art of moving and quartering troops.* But no confusion need be feared, and we see at once that the new name implies the design of revolutionizing logic.

We may see the new method at work in a mathematical treatise by Signor Burali-Forti entitled *Una questione sui numeri transfiniti* (An Enquiry Concerning Transfinite Numbers), included in Volume XI of the *Rendiconti del circolo matematico di Palermo* (Reports of the Mathematical Club of Palermo).

I will begin by saying that this treatise is very interesting, and, if I take it here as an example, it is precisely because it is the most important of all that have been written in the new language. Besides, the uninitiated can read it, thanks to an interlined Italian translation.

What gives importance to this treatise is the fact that it presented the first example of those antinomies met with in the study of transfinite numbers, which have become, during the last few years, the despair of mathematicians. The object of this note, says Signor Burali-Forti, is to show that there can be two transfinite (ordinal) numbers, *a* and *b*, such that *a* is neither equal to, greater than, nor smaller than, *b*.

The reader may set his mind at rest. In order to understand the considerations that will follow, he does not require to know what a transfinite ordinal number is.

---

* In the French the confusion is with *"logistique,"* the art of the "maréchal des *logis*," or quartermaster. In English the possibility of confusion does not arise.

Now Cantor had definitely proved that between two transfinite numbers, as between two finite numbers, there can be no relation other than equality or inequality in one direction or the other. But it is not of the matter of this treatise that I desire to speak here; this would take me much too far from my subject. I only wish to concern myself with the form, and I ask definitely whether this form makes it gain much in the way of exactness, and whether it thereby compensates for the efforts it imposes upon the writer and the reader.

To begin with, we find that Signor Burali-Forti defines the number 1 in the following manner:

$$1 = \iota T' \{Ko\_(u,h) \in (u\in \text{One}\},$$

a definition eminently fitted to give an idea of the number 1 to people who had never heard it before.

I do not understand Peanian well enough to venture to risk a criticism, but I am very much afraid that this definition contains a *petitio principii*, seeing that I notice the figure 1 in the first half and the word *One* in the second.

However that may be, Signor Burali-Forti starts with this definition, and, after a short calculation, arrives at the equation

$$(27) \qquad\qquad 1 \in \text{No},$$

which teaches us that One is a number.

And since I am on the subject of these definitions of the first numbers, I may mention that M. Couturat has also defined both 0 and 1.

What is zero? It is the number of elements in the class nil. And what is the class nil? It is the class which contains none.

To define zero as nil and nil as none is really an abuse of the wealth of language, and so M. Couturat has introduced an improvement into his definition by writing

$$0 = \iota \Lambda : \phi x = \Lambda . \supset . \Lambda = (x \in \phi x),$$

which means in English: zero is the number of the objects that satisfy a condition that is never fulfilled. But as never means *in no case*, I do not see that any very great progress has been made.

I hasten to add that the definition M. Couturat gives of the number 1 is more satisfactory.

One, he says in substance, is the number of the elements of a class in which any two elements are identical.

It is more satisfactory, as I said, in this sense, that in order to define 1, he does not use the word *one;* on the other hand, he does use the word *two.* But I am afraid that if we asked M. Couturat what two is, he would be obliged to use the word *one.*

VIII

But let us return to the treatise of Signor Burali-Forti. I said that his conclusions are in direct opposition to those of Cantor. Well, one day I received a visit from M. Hadamard, and the conversation turned upon this antinomy.

"Does not Burali-Forti's reasoning," I said, "seem to you irreproachable?"

"No," he answered; "and, on the contrary, I have no fault to find with Cantor's. Besides, Burali-Forti had no right to speak of the whole of *all* the ordinal numbers."

"Excuse me, he had that right, since he could always make the supposition that

$$\Omega = T' \, (No, \bar{\epsilon} >).$$

I should like to know who could prevent him. And can we say that an object does not exist when we have called it Ω?"

It was quite useless; I could not convince him (besides, it would have been unfortunate if I had, since he was right). Was it only because I did not speak Peanian with sufficient eloquence? Possibly, but, between ourselves, I do not think so.

Thus, in spite of all this pasigraphical apparatus, the question is not solved. What does this prove? So long as it is merely a question of demonstrating that one is a number, pasigraphy is equal to the task; but if a difficulty presents itself, if there is an antinomy to be resolved, pasigraphy becomes powerless.

CHAPTER IV

THE NEW LOGICS

I. RUSSELL'S LOGIC

In order to justify its pretensions, logic has had to transform itself. We have seen new logics spring up, and the most interesting of these is Mr. Bertrand Russell's. It seems as if there could be nothing new written about formal logic, and as if Aristotle had gone to the very bottom of the subject. But the field that Mr. Russell assigns to logic is infinitely more extensive than that of the classical logic, and he has succeeded in expressing views on this subject that are original and sometimes true.

To begin with, while Aristotle's logic was, above all, the logic of classes, and took as its starting-point the relation of subject and predicate, Mr. Russell subordinates the logic of classes to that of propositions. The classical syllogism, "Socrates is a man," etc., gives place to the hypothetical syllogism, "If A is true, B is true; now if B is true, C is true, etc." This is, in my opinion, one of the happiest of ideas, for the classical syllogism is easily reduced to the hypothetical syllogism, while the inverse transformation cannot be made without considerable difficulty.

But this is not all. Mr. Russell's logic of propositions is the study of the laws in accordance with which combinations are formed with the conjunctions *if, and, or,* and the negative *not.* This is a considerable extension of the ancient logic. The properties of the classical syllogism can be extended without any difficulty to the hypothetical syllogism, and in the forms of this latter we can easily recognize the scholastic forms; we recover what is essential in the classical logic. But the theory of the syllogism is still only the syntax of the conjunction *if* and, perhaps, of the negative.

By adding two other conjunctions, *and* and *or,* Mr. Russell opens up a new domain to logic. The signs *and* and *or* follow the same laws as the two signs $\times$ and $+$, that is to say, the commutative, associative, and distributive laws. Thus *and* represents logical multiplication, while *or* represents logical addition. This, again, is most interesting.

Mr. Russell arrives at the conclusion that a false proposition of any kind involves all the other propositions, whether true or false. M. Couturat says that this conclusion will appear paradoxical at first sight. However, one has only to correct a bad mathematical paper to recognize how true Mr. Russell's view is. The candidate often takes an immense amount of trouble to find the first false equation; but as soon as he has obtained it, it is no more than child's play for him to accumulate the most surprising results, some of which may actually be correct.

II

We see how much richer this new logic is than the classical logic. The symbols have been multiplied and admit of varied combinations, *which are no longer of limited number.* Have we any right to give this extension of meaning to the word *logic?* It would be idle to examine this question, and to quarrel with Mr. Russell merely on the score of words. We will grant him what he asks; but we must not be surprised if we find that certain truths which had been declared to be irreducible to logic, in the old sense of the word, have become reducible to logic, in its new sense, which is quite different.

We have introduced a large number of new notions, and they are not mere combinations of the old. Moreover, Mr. Russell is not deceived on this point, and not only at the beginning of his first chapter—that is to say, his logic of propositions—but at the beginning of his second and third chapters also—that is to say, his logic of classes and relations—he introduces new words which he declares to be undefinable.

And that is not all. He similarly introduces principles which he declares to be undemonstrable. But these undemonstrable principles are appeals to intuition, *à priori* synthetic judgments. We regarded them as intuitive when we met them more or less explicitly enunciated in treatises on mathematics. Have they altered in character because the meaning of the word *logic* has been extended, and we find them now in a book entitled *Treatise on Logic? They have not changed in nature, but only in position.*

III

Could these principles be considered as disguised definitions? That they should be so, we should require to be able to demonstrate that they involve no contradiction. We should have to establish that, however far we pursue the series of deductions, we shall never be in danger of contradicting ourselves.

We might attempt to argue as follows. We can verify the fact that the operations of the new logic, applied to premisses free from contradiction, can only give consequences equally free from contradiction. If then, after n operations, we have not met with contradiction, we shall not meet it anymore after $n + 1$. Accordingly, it is impossible that there can be a moment when contradiction will *begin,* which shows that we shall never meet it. Have we the right to argue in this way? No, for it would be making complete induction, and we must not forget that *we do not yet know the principle of complete induction.*

Therefore we have no right to regard these axioms as disguised definitions, and we have only one course left. Each one of them, we admit, is a new act of intuition. This is, moreover, as I believe, the thought of Mr. Russell and M. Couturat.

Thus each of the nine undefinable notions and twenty undemonstrable propositions (I feel sure that, if I had made the count, I should have found one or two more) which form the groundwork of the new logic—of the logic in the broad sense—presupposes a new and independent act of our intuition, and why should we not term it a true *à priori* synthetic judgment? On this point everybody seems to be agreed; but what Mr. Russell claims, *and what appears to me doubtful, is that after these appeals to intuition we shall have finished: we shall have no more to make, and we shall be able to construct the whole of mathematics without bringing in a single new element.*

IV

M. Couturat is fond of repeating that this new logic is quite independent of the idea of number. I will not amuse myself by counting how many instances his statement contains of adjectives of number, cardinal as well as ordinal, or of indefinite adjectives such as *several.* However, I will quote a few examples:

"The logical product of *two* or of *several* propositions is…"

"All propositions are susceptible of *two* values only, truth or falsehood."

"The relative product of *two* relations is a relation."

"A relation is established between *two* terms."

Sometimes this difficulty would not be impossible to avoid, but sometimes it is essential. A relation is incomprehensible without two terms. It is impossible to have the intuition of a relation, without having at the same time the intuition of its two terms, and without remarking that they are two, since, for a relation to be conceivable, they must be two and two only.

V. ARITHMETIC

I come now to what M. Couturat calls the *ordinal theory,* which is the groundwork of arithmetic properly so called. M. Couturat begins by enunciating Peano's five axioms, which are independent, as Signor Peano and Signor Padoa have demonstrated.

1. Zero is a whole number.
2. Zero is not the sequent of any whole number.
3. The sequent of a whole number is a whole number. To which it would be good to add: every whole number has a sequent.
4. Two whole numbers are equal if their sequents are equal.

The fifth axiom is the principle of complete induction.

M. Couturat considers these axioms as disguised definitions; they constitute the definition by postulates of zero, of the "sequent," and of the whole number.

But we have seen that, in order to allow of a definition by postulates being accepted, we must be able to establish that it implies no contradiction.

Is this the case here? Not in the very least.

The demonstration cannot be made *by example*. We cannot select a portion of whole numbers—for instance, the three first—and demonstrate that they satisfy the definition.

If I take the series 0, 1, 2, I can readily see that it satisfies axioms 1, 2, 4, and 5; but in order that it should satisfy axiom 3, it is further necessary that 3 should be a whole number, and consequently that the series 0, 1, 2, 3 should satisfy the axioms. We could verify that it satisfies axioms 1, 2, 4, and 5, but axiom 3 requires besides that 4 should be a whole number, and that the series 0, 1, 2, 3, 4 should satisfy the axioms, and so on indefinitely.

It is, therefore, impossible to demonstrate the axioms for some whole numbers without demonstrating them for all, and so we must give up the demonstration by example.

It is necessary, then, to take all the consequences of our axioms and see whether they contain any contradiction. If the number of these consequences were finite, this would be easy; but their number is infinite—they are the whole of mathematics, or at least the whole of arithmetic.

What are we to do, then? Perhaps, if driven to it, we might repeat the reasoning of Section III. But, as I have said, *this reasoning is complete induction,* and it is precisely the principle of complete induction that we are engaged in justifying.

VI. HILBERT'S LOGIC

I come now to Mr. Hilbert's important work, addressed to the Mathematical Congress at Heidelberg, a French translation of which, by M. Pierre Boutroux, appeared in *L'Enseignement mathématique*, while an English translation by Mr. Halsted appeared in *The Monist*. In this work, in which we find the most profound thought, the author pursues an aim similar to Mr. Russell's, but he diverges on many points from his predecessor.

"However," he says, "if we look closely, we recognize that in logical principles, as they are commonly presented, certain arithmetical notions are found already implied; for instance, the notion of whole, and, to a certain extent, the notion of number. Thus we find ourselves caught in a circle, and that is why it seems to me necessary, if we wish to avoid all paradox, to develop the principles of logic and of arithmetic simultaneously."

We have seen above that what Mr. Hilbert says of the principles of logic, *as they are commonly presented*, applies equally to Mr. Russell's logic. For Mr. Russell logic is anterior to arithmetic, and for Mr. Hilbert they are "simultaneous." Farther on we shall find other and yet deeper differences; but we will note them as they occur. I prefer to follow the development of Hilbert's thought step by step, quoting the more important passages verbatim.

"Let us first take into consideration the object 1." We notice that in acting thus we do not in any way imply the notion of number, for it is clearly understood that 1 here is nothing but a symbol, and that we do not in any way concern ourselves with knowing its signification. "The groups formed with this object, two, three, or several times repeated..." This time the case is quite altered, for if we introduce the words *two, three,* and, above all, *several,* we introduce the notion of number; and then the definition of the finite whole number that we find later on comes a trifle late. The author was much too wary not to perceive this *petitio principii.* And so, at the end of his work, he seeks to effect a real *patching-up.*

Hilbert then introduces two simple objects, 1 and =, and pictures all the combinations of these two objects, all the combinations of their combinations, and so on. It goes without saying that we must forget the ordinary signification of these two signs, and not attribute any to them. He then divides these combinations into two classes, that of entities and that of nonentities, and, until further orders, this partition is entirely arbitrary. Every affirmative proposition teaches us that a combination belongs to the class of entities, and every negative proposition teaches us that a certain combination belongs to the class of nonentities.

<h1 style="text-align:center">VII</h1>

We must now note a difference that is of the highest importance. For Mr. Russell a chance object, which he designates by x, is an absolutely indeterminate object, about which he assumes nothing. For Hilbert it is one of those combinations formed with the symbols 1 and $=$; he will not allow the introduction of anything but combinations of objects already defined. Moreover, Hilbert formulates his thought in the most concise manner, and I think I ought to reproduce his statement *in extenso:* "The indeterminates which figure in the axioms (in place of the 'some' or the 'all' of ordinary logic) represent exclusively the whole of the objects and combinations that we have already acquired in the actual state of the theory, or that we are in course of introducing. Therefore, when we deduce propositions from the axioms under consideration, it is these objects and these combinations alone that we have the right to substitute for the indeterminates. Neither must we forget that when we increase the number of the fundamental objects, the axioms at the same time acquire a new extension, and must, in consequence, be put to the proof afresh and, if necessary, modified."

The contrast with Mr. Russell's point of view is complete. According to this latter philosopher, we may substitute in place of x not only objects already known, but anything whatsoever. Russell is faithful to his point of view, which is that of comprehension. He starts with the general idea of entity, and enriches it more and more, even while he restricts it, by adding to it new qualities. Hilbert, on the contrary, only recognizes as possible entities combinations of objects already known; so that (looking only at one side of his thought) we might say that he takes the point of view of extension.

VIII

Let us proceed with the exposition of Hilbert's ideas. He introduces two axioms which he enunciates in his symbolical language, but which signify, in the language of the uninitiated like us, that every quantity is equal to itself, and that every operation upon two identical quantities gives identical results. So stated they are evident, but such a presentation of them does not faithfully represent Hilbert's thought. For him mathematics has to combine only pure symbols, and a true mathematician must base his reasoning upon them without concerning himself with their meaning. Accordingly, his axioms are not for him what they are for the ordinary man.

He considers them as representing the definition by postulates of the symbol =, up to this time devoid of all signification. But in order to justify this definition, it is necessary to show that these two axioms do not lead to any contradiction.

For this purpose Hilbert makes use of the reasoning of Section III, without apparently perceiving that he is making complete induction.

IX

The end of Mr. Hilbert's treatise is altogether enigmatical, and I will not dwell upon it. It is full of contradictions, and one feels that the author is vaguely conscious of the *petitio principii* he has been guilty of, and that he is vainly trying to plaster up the cracks in his reasoning.

What does this mean? It means that *when he comes to demonstrate that the definition of the whole number by the axiom of complete induction does not involve contradiction, Mr. Hilbert breaks down, just as Mr. Russell and M. Couturat broke down, because the difficulty is too great.*

X. GEOMETRY

Geometry, M. Couturat says, is a vast body of doctrine upon which complete induction does not intrude. This is true to a certain extent: we cannot say that it does not intrude at all, but that it intrudes very little. If we refer to Mr. Halsted's *Rational Geometry* (New York: John Wiley and Sons, 1904), founded on Hilbert's principles, we find the principle of induction intruding for the first time at page 114 (unless, indeed, I have not searched carefully enough, which is quite possible).

Thus geometry, which seemed, only a few years ago, the domain in which intuition held undisputed sway, is today the field in which the logisticians appear to triumph. Nothing could give a better measure of the importance of Hilbert's geometrical works, and of the profound impression they have left upon our conceptions.

But we must not deceive ourselves. *What is, in fact, the fundamental theorem of geometry? It is that the axioms of geometry do not involve contradiction, and this cannot be demonstrated without the principle of induction.*

How does Hilbert demonstrate this essential point? He does it by relying upon analysis, and, through it, upon arithmetic, and, through it, upon the principle of induction.

If another demonstration is ever discovered, it will still be necessary to rely on this principle, since the number of the possible consequences of the axioms which we have to show are not contradictory is infinite.

XI. CONCLUSION

Our conclusion is, first of all, that *the principle of induction cannot be regarded as the disguised definition of the whole number.*

Here are three truths:

The principle of complete induction;
Euclid's postulate;
The physical law by which phosphorus melts at 44° centigrade (quoted by M. Le Roy).

We say: these are three disguised definitions—the first that of the whole number, the second that of the straight line, and the third that of phosphorus.

I admit it for the second, but I do not admit it for the two others, and I must explain the reason of this apparent inconsistency.

In the first place, we have seen that a definition is only acceptable if it is established that it does not involve contradiction. We have also shown that, in the case of the first definition, this demonstration is impossible; while in the case of the second, on the contrary, we have just recalled the fact that Hilbert has given a complete demonstration.

So far as the third is concerned, it is clear that it does not involve contradiction. But does this mean that this definition guarantees, as it should, the existence of the object defined? We are here no longer concerned with the mathematical sciences, but with the physical sciences, and the word *existence* has no longer the same meaning; it no longer signifies absence of contradiction, but objective existence.

This is one reason already for the distinction I make between the three cases, but there is a second. In the applications we have to make of these three notions, do they present themselves as defined by these three postulates?

The possible applications of the principle of induction are innumerable. Take, for instance, one of those we have expounded above, in which it is sought to establish that a collection of axioms cannot lead to a contradiction. For this purpose we consider one of the series of syllogisms that can be followed out, starting with these axioms as premises.

When we have completed the nth syllogism, we see that we can form still another, which will be the $(n+1)^{th}$: thus the number n serves for counting a series of successive operations; it is a number that can be obtained by successive additions. Accordingly, it is a number from which we can return to unity by *successive subtractions*. It is evident that we could not do so if we had $n = n-1$, for then subtraction would always give us the same number. Thus, then, the way in which we have been brought to consider this number n involves a definition of the finite whole number, and this definition is as follows: *a finite whole number is that which can be obtained by successive additions, and which is such that n is not equal to $n-1$*.

This being established, what do we proceed to do? We show that if no contradiction has occurred up to the nth syllogism, it will not occur any the more at the $(n+1)^{th}$, and we conclude that it will never occur. You say I have the right to conclude thus, because whole numbers are, by definition, those for which such reasoning is legitimate. But that involves another definition of the whole number, which is as follows: *a whole number is that about which we can reason by recurrence*. In the species it is that of which we can state that, if absence of contradiction at the moment of occurrence of a syllogism whose number is a whole number carries with it the absence of contradiction at the moment of occurrence of the syllogism whose number is the following whole number, then we need not fear any contradiction for any of the syllogisms whose numbers are whole numbers.

The two definitions are not identical. They are equivalent, no doubt, but they are so by virtue of an *à priori* synthetic judgment; we cannot pass from one to the other by purely logical processes. Consequently, we have no right to adopt the second after having introduced the whole number by a road which presupposes the first.

On the contrary, what happens in the case of the straight line? I have already explained this so often that I feel some hesitation about repeating myself once more. I will content myself with a brief summary of my thought.

We have not, as in the previous case, two equivalent definitions logically irreducible one to the other. We have only one expressible in words. It may be said that there is another that we feel without being able to enunciate it, because we have the intuition of a straight line, or because we can picture a straight line. But, in the first place, we cannot picture it in geometric space, but only in representative space; and then we can equally well picture objects which possess the other properties of a straight line, and not that of satisfying Euclid's postulate. These objects are "non-Euclidian straight lines," which, from a certain point of view, are not entities destitute of meaning, but circles (true circles of true space) orthogonal to a certain sphere. If, among these objects equally susceptible of being pictured, it is the former (the Euclidian straight lines) that we call straight lines, and not the latter (the non-Euclidian straight lines), it is certainly so by definition.

And if we come at last to the third example, the definition of phosphorus, we see that the true definition would be: phosphorus is this piece of matter that I see before me in this bottle.

XII

Since I am on the subject, let me say one word more. Concerning the example of phosphorus, I said: "This proposition is a true physical law that can be verified, for it means: all bodies which possess all the properties of phosphorus except its melting-point, melt, as it does, at 44° centigrade." It has been objected that this law is not verifiable, for if we came to verify that two bodies resembling phosphorus melt one at 44° and the other at 50° centigrade, we could always say that there is, no doubt, besides the melting-point, some other property in which they differ.

This was not exactly what I meant to say, and I should have written: "all bodies which possess such and such properties in finite number (namely, the properties of phosphorus given in chemistry books, with the exception of its melting-point) melt at 44° centigrade."

In order to make still clearer the difference between the case of the straight line and that of phosphorus, I will make one more remark. The straight line has several more or less imperfect images in nature, the chief of which are rays of light and the axis of rotation of a solid body. Assuming that we ascertain that the ray of light does not satisfy Euclid's postulate (by showing, for instance, that a star has a negative parallax), what shall we do? Shall we conclude that, as a straight line is by definition the trajectory of light, it does not satisfy the definition, or, on the contrary, that, as a straight line by definition satisfies the postulate, the ray of light is not rectilineal?

Certainly we are free to adopt either definition, and, consequently, either conclusion. But it would be foolish to adopt the former, because the ray of light probably satisfies in a most imperfect way not only Euclid's postulate but the other properties of the straight line; because, while it deviates from the Euclidian straight, it deviates nonetheless from the axis of rotation of solid bodies, which is another imperfect image of the straight line; and lastly, because it is, no doubt, subject to change, so that such and such a line which was straight yesterday will no longer be so tomorrow if some physical circumstance has altered.

Assume, now, that we succeed in discovering that phosphorus melts not at 44° but at 43.9° centigrade. Shall we conclude that, as phosphorus is by definition that which melts at 44°, this substance that we called phosphorus is not true phosphorus, or, on the contrary, that phosphorus melts at 43.9°? Here, again, we are free to adopt either definition, and, consequently, either conclusion; but it would be foolish to adopt the former, because we cannot change the name of a substance every time we add a fresh decimal to its melting-point.

XIII

To sum up, Mr. Russell and Mr. Hilbert have both made a great effort, and have both of them written a book full of views that are original, profound, and often very true. These two books furnish us with subject for much thought, and there is much that we can learn from them. Not a few of their results are substantial and destined to survive.

But to say that they have definitely settled the controversy between Kant and Leibnitz and destroyed the Kantian theory of mathematics is evidently untrue. I do not know whether they actually imagined they had done it, but if they did they were mistaken.

CHAPTER V

The Last Efforts of the Logisticians

I

The logisticians have attempted to answer the foregoing considerations. For this purpose they have been obliged to transform logistic, and Mr. Russell in particular has modified his original views on certain points. Without entering into the details of the controversy, I should like to return to what are, in my opinion, the two most important questions. Have the rules of logistic given any proof of fruitfulness and of infallibility? Is it true that they make it possible to demonstrate the principle of complete induction without any appeal to intuition?

II. The Infallibility of Logistic

As regards fruitfulness, it seems that M. Couturat has most childish illusions. Logistic, according to him, lends "stilts and wings" to discovery, and on the following page he says, "*It is ten years* since Signor Peano published the first edition of his "Formulaire."

What! You have had wings for ten years, and you haven't flown yet!

I have the greatest esteem for Signor Peano, who has done some very fine things (for instance, his curve which fills a whole area); but, after all, he has not gone any further, or higher, or faster than the majority of wingless mathematicians, and he could have done everything just as well on his feet.

On the contrary, I find nothing in logistic for the discoverer but shackles. It does not help us at all in the direction of conciseness, far from it; and if it requires 27 equations to establish that 1 is a number, how many will it require to demonstrate a real theorem? If we distinguish, as Mr. Whitehead does, the individual x, the class whose only member is x, which we call ιx, then the class whose only member is the class whose only member is x, which we call $\iota\iota x$, do we imagine that these distinctions, however useful they may be, will greatly expedite our progress?

Logistic forces us to say all that we commonly assume, it forces us to advance step by step; it is perhaps surer, but it is not more expeditious.

It is not wings you have given us, but leading-strings. But we have the right to demand that these leading-strings should keep us from falling; this is their only excuse. When an investment does not pay a high rate of interest, it must at least be a gilt-edged security.

Must we follow your rules blindly? Certainly, for otherwise it would be intuition alone that would enable us to distinguish between them. But in that case they must be infallible, for it is only in an infallible authority that we can have blind confidence. Accordingly, this is a necessity for you: you must be infallible or cease to exist.

You have no right to say to us: "We make mistakes, it is true, but you make mistakes too." For us, making mistakes is a misfortune, a very great misfortune, but for you it is death.

Neither must you say, "Does the infallibility of arithmetic prevent errors of addition?" The rules of calculation are infallible, and yet we find people making mistakes *through not applying these rules*. But a revision of their calculation will show at once just where they went astray. Here the case is quite different. The logisticians *have applied* their rules, and yet they have fallen into contradiction. So true is this, that they are preparing to alter these rules and "sacrifice the notion of class." Why alter them if they were infallible?

"We are not obliged," you say, "to solve *hic et nunc* all possible problems." Oh, we do not ask as much as that. If, in face of a problem, you gave *no* solution, we should have nothing to say; but, on the contrary, you give *two*, and these two are contradictory, and consequently one at least of them is false, and it is this that constitutes a failure.

Mr. Russell attempts to reconcile these contradictions, which can only be done, according to him, "by restricting or even sacrificing the notion of class." And M. Couturat, discounting the success of this attempt, adds: "If logisticians succeed where others have failed, M. Poincaré will surely recollect this sentence, and give logistic the credit of the solution."

Certainly not. Logistic exists; it has its code, which has already gone through four editions; or, rather, it is this code which is logistic itself. Is Mr. Russell preparing to show that one at least of the two contradictory arguments has transgressed the code? Not in the very least; he is preparing to alter these laws and to revoke a certain number of them. If he succeeds, I shall give credit to Mr. Russell's intuition, and not to Peanian Logistic, which he will have destroyed.

III. Liberty of Contradiction

I offered two principal objections to the definition of the whole number adopted by the logisticians. What is M. Couturat's answer to the first of these objections?

What is the meaning in mathematics of the words *to exist*? It means, I said, to be free from contradiction. This is what M. Couturat disputes. "Logical existence," he says, "is quite a different thing from absence of contradiction. It consists in the fact that a class is not empty. To say that some *a*'s exist is, by definition, to assert that the class *a* is not void." And, no doubt, to assert that the class *a* is not void is, by definition, to assert that some *a*'s exist. But one of these assertions is just as destitute of meaning as the other if they do not both signify either that we can see or touch *a*, which is the meaning given them by physicists or naturalists, or else that we can conceive of an *a* without being involved in contradictions, which is the meaning given them by logicians and mathematicians.

In M. Couturat's opinion it is not non-contradiction that proves existence, but existence that proves non-contradiction. In order to establish the existence of a class, we must accordingly establish, by an *example*, that there is an individual belonging to that class. "But it will be said, How do we demonstrate the existence of this individual? Is it not necessary that this existence should be established, to enable us to deduce the existence of the class of which it forms part? It is not so. Paradoxical as the assertion may appear, we never demonstrate the existence of an individual. Individuals, from the very fact that they are individuals, are always considered as existing. We have never to declare that an individual exists, absolutely speaking, but only that it exists in a class." M. Couturat finds his own assertion paradoxical, and he will certainly not be alone in so finding it. Nevertheless it must have some sense, and it means, no doubt, that the existence of an individual alone in the world, of which nothing is asserted, cannot involve contradiction. As long as it is quite alone, it is evident that it cannot interfere with anyone. Well, be it so; we will admit the existence of the individual, "absolutely speaking," but with it we have nothing to do. It still remains to demonstrate the existence of the individual "in a class," and, in order to do this, you will still have to prove that the assertion that such an individual belongs to such a class is neither contradictory in itself nor with the other postulates adopted.

"Accordingly," M. Couturat continues, "to assert that a definition is not valid unless it is first proved that it is not contradictory, is to impose an arbitrary and improper condition." The claim for the liberty of contradiction could not be stated in more emphatic or haughtier terms. "In any case, the *onus probandi* rests with those who think these principles are contradictory." Postulates are presumed to be compatible, just as a prisoner is presumed to be innocent, until the contrary is proved.

It is unnecessary to add that I do not acquiesce in this claim. But, you say, the demonstration you demand of us is impossible, and you cannot require us to "aim at the moon." Excuse me; it is impossible for you, but not for us who admit the principle of induction as an *à priori* synthetic judgment. This would be necessary for you as it is for us.

In order to demonstrate that a system of postulates does not involve contradiction, it is necessary to apply the principle of complete induction. Not only is there nothing "extraordinary" in this method of reasoning, but it is the only correct one. It is not "inconceivable" that anyone should ever have used it, and it is not difficult to find "examples and precedents." In my article I have quoted two, and they were borrowed from Hilbert's pamphlet. He is not alone in having made use of it, and those who have not done so have been wrong. What I reproach Hilbert with, is not that he has had recourse to it (a born mathematician such as he could not but see that a demonstration is required, and that this is the only possible one), but that he has had recourse to it without recognizing the reasoning by recurrence.

IV. THE SECOND OBJECTION

I had noted a second error of the logisticians in Hilbert's article. Today Hilbert is excommunicated, and M. Couturat no longer considers him as a logistician. He will, therefore, ask me if I have found the same mistake in the orthodox logisticians. I have not seen it in the pages I have read, but I do not know whether I should find it in the three hundred pages they have written that I have no wish to read.

Only, they will have to commit the error as soon as they attempt to make any sort of an application of mathematical science. The eternal contemplation of its own navel is not the sole object of this science. It touches nature, and one day or other it will come into contact with it. Then it will be necessary to shake off purely verbal definitions and no longer to content ourselves with words.

Let us return to Mr. Hilbert's example. It is still a question of reasoning by recurrence and of knowing whether a system of postulates is not contradictory. M. Couturat will no doubt tell me that in that case it does not concern him, but it may perhaps interest those who do not claim, as he does, the liberty of contradiction.

We wish to establish, as above, that we shall not meet with contradiction after some particular number of arguments, a number which may be as large as you please, provided it is finite. For this purpose we must apply the principle of induction. Are we to understand here by finite number every number to which the principle of induction applies? Evidently not, for otherwise we should be involved in the most awkward consequences.

To have the right to lay down a system of postulates, we must be assured that they are not contradictory. This is a truth that is admitted by *the majority* of scientists; I should have said *all* before reading M. Couturat's last article. But what does it signify? Does it mean that we must be sure of not meeting with contradiction after a *finite* number of propositions, the *finite* number being, by definition, that which possesses all the properties of a recurrent nature in such a way that if one of these properties were found wanting—if, for instance, we came upon a contradiction—we should *agree* to say that the number in question was not finite?

In other words, do we mean that we must be sure of not meeting a contradiction, with this condition, that we agree to stop just at the moment when we are on the point of meeting one? The mere statement of such a proposition is its sufficient condemnation.

Thus not only does Mr. Hilbert's reasoning assume the principle of induction, but he assumes that this principle is given us, not as a simple definition, but as an *à priori* synthetic judgment.

I would sum up as follows:

—

A demonstration is necessary.

The only possible demonstration is the demonstration by recurrence.

This demonstration is legitimate only if the principle of induction is admitted, and if it is regarded not as a definition but as a synthetic judgment.

V. THE CANTORIAN ANTINOMIES

I will now take up the examination of Mr. Russell's new treatise. This treatise was written with the object of overcoming the difficulties raised by those *Cantorian antinomies* to which I have already made frequent allusion. Cantor thought it possible to construct a Science of the Infinite. Others have advanced further along the path he had opened, but they very soon ran against strange contradictions. These antinomies are already numerous, but the most celebrated are:

1. Burali-Forti's antinomy.
2. The Zermelo-König antinomy.
3. Richard's antinomy.

Cantor had demonstrated that ordinal numbers (it is a question of transfinite ordinal numbers, a new notion introduced by him) can be arranged in a lineal series; that is to say, that of two unequal ordinal numbers, there is always one that is smaller than the other. Burali-Forti demonstrates the contrary; and indeed, as he says in substance, if we could arrange *all* the ordinal numbers in a lineal series, this series would define an ordinal number that would be greater than *all* the others, to which we could then add 1 and so obtain yet another ordinal number which would be still greater. And this is contradictory.

We will return later to the Zermelo-König antinomy, which is of a somewhat different nature. Richard's antinomy is as follows (*Revue générale des sciences,* June 30, 1905). Let us consider all the decimal numbers that can be defined with the help of a finite number of words. These decimal numbers form an aggregate E, and it is easy to see that this aggregate is denumerable—that is to say, that it is possible to *number* the decimal numbers of this aggregate from one to infinity. Suppose the numeration effected, and let us define a number N in the following manner. If the nth decimal of the nth number of the aggregate E is

$$0, 1, 2, 3, 4, 5, 6, 7, 8, \text{ or } 9,$$

the nth decimal of N will be

$$1, 2, 3, 4, 5, 6, 7, 8, 1, \text{ or } 1.$$

As we see, N is not equal to the nth number of E, and since n is any chance number, N does not belong to E, and yet N should belong to this aggregate, since we have defined it in a finite number of words.

We shall see farther on that M. Richard himself has, with much acuteness, given the explanation of his paradox, and that his explanation can be extended, *mutatis mutandis,* to the other paradoxes of like nature. Mr. Russell quotes another rather amusing antinomy:

What is the smallest whole number that cannot be defined in a sentence formed of less than a hundred English words?

This number exists, and, indeed, the number of numbers capable of being defined by such a sentence is evidently finite, since the number of words in the English language is not infinite. Therefore among them there will be one that is smaller than all the others.

On the other hand the number does not exist, for its definition involves contradiction. The number, in fact, is found to be defined by the sentence in italics, which is formed of less than a hundred English words, and, by definition, the number must not be capable of being defined by such a sentence.

VI. Zigzag Theory and No Classes Theory

What is Mr. Russell's attitude in face of these contradictions? After analysing those I have just spoken of, and quoting others, after putting them in a form that recalls Epimenides, he does not hesitate to conclude as follows:

"A propositional function of one variable does not always determine a class."* A "propositional function" (that is to say, a definition) or "norm" can be "non-predicative." And this does not mean that these non-predicative propositions determine a class that is empty or void; it does not mean that there is no value of x that satisfies the definition and can be one of the elements of the class. The

---

* This and the following quotations are from Mr. Russell's paper, "On some difficulties in the theory of transfinite numbers and order types," *Proceedings of the London Mathematical Society,* Ser. 2, Vol. 4, Part 1.

elements exist, but they have no right to be grouped together to form a class.

But this is only the beginning, and we must know how to recognize whether a definition is or is not predicative. For the purpose of solving this problem, Mr. Russell hesitates between three theories, which he calls

A. The zigzag theory.
B. The theory of limitation of size.
C. The no classes theory.

According to the zigzag theory, "definitions (propositional functions) determine a class when they are fairly simple, and only fail to do so when they are complicated and recondite." Now who is to decide whether a definition can be regarded as sufficiently simple to be acceptable? To this question we get no answer except a candid confession of powerlessness. "The axioms as to what functions are predicative have to be exceedingly complicated, and cannot be recommended by any intrinsic plausibility. This is a defect which might be remedied by greater ingenuity, or by the help of some hitherto unnoticed distinction. But hitherto, in attempting to set up axioms for this theory, I have found no guiding principle except the avoidance of contradictions."

This theory therefore remains very obscure. In the darkness there is a single glimmer, and that is the word *zigzag*. What Mr. Russell calls *zigzagginess* is no doubt this special character which distinguishes the argument of Epimenides.

According to the theory of limitation of size, a class must not be too extensive. It may, perhaps, be infinite, but it must not be too infinite.

But we still come to the same difficulty. At what precise moment will it begin to be too extensive? Of course this difficulty is not solved, and Mr. Russell passes to the third theory.

In the no classes theory all mention of the word *class* is prohibited, and the word has to be replaced by various paraphrases. What a change for the logisticians who speak of nothing but class and classes of classes! The whole of logistic will have to be refashioned. Can we imagine the appearance of a page of logistic when all propositions dealing with class have been suppressed? There will be nothing left but a few scattered survivors in the midst of a blank page. *Apparent rari nantes in gurgite vasto.*

However that may be, we understand Mr. Russell's hesitation at the modifications to which he is about to submit the fundamental principles he has hitherto adopted. Criteria will be necessary to decide whether a definition is too complicated or too extensive, and these criteria cannot be justified except by an appeal to intuition.

It is towards the no classes theory that Mr. Russell eventually inclines.

However it is, logistic must be refashioned, and it is not yet known how much of it can be saved. It is unnecessary to add that it is Cantorism and logistic alone that are in question. The true mathematics, the mathematics that is of some use, may continue to develop according to its own principles, taking no heed of the tempests that rage without, and step by step it will pursue its wonted conquests, which are decisive and have never to be abandoned.

VII. The True Solution

How are we to choose between these different theories? It seems to me that the solution is contained in M. Richard's letter mentioned above, which will be found in the *Revue générale des sciences* of June 30, 1905. After stating the antinomy that I have called Richard's antinomy, he gives the explanation.

Let us refer to what was said of this antinomy in Section V. E is the aggregate of *all* the numbers that can be defined by a finite number of words, *without introducing the notion of the aggregate E itself,* otherwise the definition of E would contain a vicious circle, for we cannot define E by the aggregate E itself.

Now we have defined N by a finite number of words, it is true, but only with the help of the notion of the aggregate E, and that is the reason why N does not form a part of E.

In the example chosen by M. Richard, the conclusion is presented with complete evidence, and the evidence becomes the more apparent on a reference to the actual text of the letter. But the same explanation serves for the other antinomies, as may be easily verified.

Thus *the definitions that must be regarded as non-predicative are those which contain a vicious circle.* The above examples show sufficiently clearly what I mean by this. Is this what Mr. Russell calls "zigzaggi-ness"? I merely ask the question without answering it.

VIII. The Demonstrations of the Principle of Induction

We will now examine the so-called demonstrations of the principle of induction, and more particularly those of Mr. Whitehead and Signor Burali-Forti.

And first we will speak of Whitehead's, availing ourselves of some new denominations happily introduced by Mr. Russell in his recent treatise.

We will call *recurrent class* every class of numbers that includes zero, and also includes $n+1$ if it includes n.

We will call *inductive number* every number which forms a part of *all* recurrent classes.

Upon what condition will this latter definition, which plays an essential part in Whitehead's demonstration, be "predicative" and consequently acceptable?

Following upon what has been said above, we must understand by *all* recurrent classes all those whose definition does not contain the notion of inductive number; otherwise we shall be involved in the vicious circle which engendered the antinomies.

Now, *Whitehead has not taken this precaution.*

Whitehead's argument is therefore vicious; it is the same that led to the antinomies. It was illegitimate when it gave untrue results, and it remains illegitimate when it leads by chance to a true result.

A definition which contains a vicious circle defines nothing. It is of no use to say we are sure, whatever be the meaning given to our definition, that there is at least zero which belongs to the class of inductive numbers. It is not a question of knowing whether this class is empty, but whether it can be rigidly delimited. A "non-predicative class" is not an empty class, but a class with uncertain boundaries.

It is unnecessary to add that this particular objection does not invalidate the general objections that apply to all the demonstrations.

IX

Signor Burali-Forti has given another demonstration in his article "Le classi finite" (*Atti di Torino*, Vol. xxxii). But he is obliged to admit two postulates:

The first is that there exists always at least one infinite class.

The second is stated thus:

$$u \in K\,(K - \iota\, \Lambda).\, \mathfrak{O}.\, u < v'\, u.$$

The first postulate is no more evident than the principle to be demonstrated. The second is not only not evident, but it is untrue, as Mr. Whitehead has shown, as, moreover, the veriest schoolboy could have seen at the first glance if the axiom had been stated in intelligible language, since it means: the number of combinations that can be formed with several objects is smaller than the number of those objects.

X. ZERMELO'S AXIOM

In a celebrated demonstration, Signor Zermelo relies on the following axiom:

In an aggregate of any kind (or even in each of the aggregates of an aggregate of aggregates) we can always select one element *at random* (even if the aggregate of aggregates contains an infinity of aggregates).

This axiom had been applied a thousand times without being stated, but as soon as it was stated, it raised doubts. Some mathematicians, like M. Borel, rejected it resolutely, while others admitted it. Let us see what Mr. Russell thinks of it according to his last article.

He pronounces no opinion, but the considerations which he gives are most suggestive.

To begin with a picturesque example, suppose that we have as many pairs of boots as there are whole numbers, so that we can number *the pairs* from 1 to infinity, how many boots shall we have? Will the number of boots be equal to the number of pairs? It will be so if, in each pair, the right boot is distinguishable from the left; it will be sufficient in fact to give the number $2n-1$ to the right boot of the nth pair, and the number $2n$ to the left boot of the nth pair. But it will not be so if the right boot is similar to the left, because such an operation then becomes impossible; unless we admit Zermelo's axiom, since in that case we can select *at random* from each pair the boot we regard as the right.

XI. CONCLUSIONS

A demonstration really based upon the principles of Analytical logic will be composed of a succession of propositions; some, which will serve as premises, will be identities or definitions; others will be deduced from the former step by step; but although the connexion between each proposition and the succeeding proposition can be grasped immediately, it is not obvious at a glance how it has been possible to pass from the first to the last, which we may be tempted to look upon as a new truth. But if we replace successively the various expressions that are used by their definitions, and if we pursue this operation to the furthest possible limit, there will be nothing left at the end but identities, so that all will be reduced to one immense tautology. Logic therefore remains barren, unless it is fertilized by intuition.

This is what I wrote formerly. The logisticians assert the contrary, and imagine that they have proved it by effectively demonstrating new truths. But what mechanism have they used?

Why is it that by applying to their arguments the procedure

I have just described, that is, by replacing the terms defined by their definitions, we do not see them melt into identities like the ordinary arguments? It is because the procedure is not applicable to them. And why is this? Because their definitions are non-predicative and present that kind of hidden vicious circle I have pointed out above, and non-predicative definitions cannot be substituted for the term *defined.* Under these conditions, *Logistic is no longer barren, it engenders antinomies.*

It is the belief in the existence of actual infinity that has given birth to these non-predicative definitions. I must explain myself. In these definitions we find the word *all,* as we saw in the examples quoted above. The word *all* has a very precise meaning when it is a question of a finite* number of objects; but for it still to have a precise meaning when the number of the objects is infinite, it is necessary that there should exist an actual infinity. Otherwise *all* these objects cannot be conceived as existing prior to their definition, and then, if the definition of a notion N depends on *all* the objects A, it may be tainted with the vicious circle, if among the objects A there is one that cannot be defined without bringing in the notion N itself.

The rules of formal logic simply express the properties of all the possible classifications. But in order that they should be applicable, it is necessary that these classifications should be immutable and not require to be modified in the course of the argument. If we have only to classify a finite number of objects, it is easy to preserve these classifications without change. If the number of the objects is indefinite, that is to say if we are constantly liable to find new and unforeseen objects springing up, it may happen that the appearance of a new object will oblige us to modify the classification, and it is thus that we are exposed to the antinomies.

There is no actual infinity. The Cantorians forgot this, and so fell into contradiction. It is true that Cantorism has been useful, but that was when it was applied to a real problem, whose terms were clearly defined, and then it was possible to advance without danger.

---

* The original has "infinite," obviously a slip.

Like the Cantorians, the logisticians have forgotten the fact, and they have met with the same difficulties. But it is a question whether they took this path by accident or whether it was a necessity for them.

In my view, there is no doubt about the matter; belief in an actual infinity is essential in the Russellian logistic, and this is exactly what distinguishes it from the Hilbertian logistic. Hilbert takes the point of view of extension precisely in order to avoid the Cantorian antinomies. Russell takes the point of view of comprehension, and consequently for him the genus is prior to the species, and the *summum genus* prior to all. This would involve no difficulty if the *summum genus* were finite; but if it is infinite, it is necessary to place the infinite before the finite—that is to say, to regard the infinite as actual.

And we have not only infinite classes; when we pass from the genus to the species by restricting the concept by new conditions, the number of these conditions is still infinite, for they generally express that the object under consideration is in such and such a relation with all the objects of an infinite class.

But all this is ancient history. Mr. Russell has realized the danger and is going to reconsider the matter. He is going to change everything, and we must understand clearly that he is preparing not only to introduce new principles which permit of operations formerly prohibited, but also to prohibit operations which he formerly considered legitimate. He is not content with adoring what he once burnt, but he is going to burn what he once adored, which is more serious. He is not adding a new wing to the building, but sapping its foundations.

The old Logistic is dead, and so true is this, that the zigzag theory and the no classes theory are already disputing the succession. We will wait until the new exists before we attempt to judge it.

PART III

THE
NEW
MECHANICS

CHAPTER I

MECHANICS AND RADIUM

I. INTRODUCTION

Are the general principles of Dynamics, which have served since Newton's day as the foundation of Physical Science, and appear immutable, on the point of being abandoned, or, at the very least, profoundly modified? This is the question many people have been asking for the last few years. According to them the discovery of radium has upset what were considered the most firmly rooted scientific doctrines, the impossibility of the transmutation of metals on the one hand, and, on the other, the fundamental postulates of Mechanics. Perhaps they have been in too great haste to consider these novelties as definitely established, and to shatter our idols of yesterday; perhaps it would be good to await more numerous and more convincing experiments. It is nonetheless necessary that we should at once acquire a knowledge of the new doctrines and of the arguments, already most weighty, upon which they rely.

I will first recall in a few words what these principles are.

A. The motion of a material point, isolated and unaffected by any exterior force, is rectilineal and uniform. This is the principle of inertia; no acceleration without force.

B. The acceleration of a moving point has the same direction as the resultant of all the forces to which the point is subjected; it is equal to the quotient of this resultant by a co-efficient called the *mass* of the moving point.

The mass of a moving point, thus defined, is constant; it does not depend upon the velocity acquired by the point, it is the same whether the force is parallel to this velocity and only tends to accelerate or retard the motion of the point, or whether it is, on the contrary, perpendicular to that velocity and tends to cause the motion to deviate to right or left, that is to say, to *curve* the trajectory.

C. All the forces to which a material point is subjected arise from the action of other material points; they depend only upon the *relative* positions and velocities of these different material points.

By combining the two principles B and C we arrive at the *principle of relative motion,* by virtue of which the laws of motion of a system are the same whether we refer the system to fixed axes, or whether we refer it to moving axes animated with a rectilineal and uniform forward motion, so that it is impossible to distinguish absolute motion from a relative motion referred to such moving axes.

D. If a material point A acts upon another material point B, the body B reacts upon A, and these two actions are two forces that are equal and directly opposite to one another. This is *the principle of the equality of action and reaction,* or more briefly, *the principle of reaction.*

Astronomical observations, and the commonest physical phenomena, seem to have afforded the most complete, unvarying, and precise confirmation of these principles. That is true, they tell us now, but only because we have never dealt with any but low velocities. Mercury, for instance, which moves faster than any of the other planets, scarcely travels sixty miles a second—Would it behave in the same way if it travelled a thousand times as fast? It is clear that we have still no cause for anxiety; whatever may be the progress of automobilism, it will be some time yet before we have to give up applying the classical principles of Dynamics to our machines.

How is it then that we have succeeded in realizing velocities a thousand times greater than that of Mercury, equal, for instance, to a tenth or a third of the velocity of light, or coming nearer to it even than that? It is by the help of the cathode rays and the rays of radium.

We know that radium emits three kinds of rays, which are designated by the three Greek letters α, β, γ. In what follows, unless I specifically state the contrary, I shall always speak of the β rays, which are analogous to the cathode rays.

After the discovery of the cathode rays, two opposite theories were propounded. Crookes attributed the phenomena to an actual molecular bombardment, Hertz to peculiar undulations of the ether. It was a repetition of the controversy that had divided physicists a century before with regard to light. Crookes returned to the emission theory, abandoned in the case of light, while Hertz held to the undulatory theory. The facts seemed to be in favour of Crookes.

It was recognized in the first place that the cathode rays carry with them a negative electric charge: they are deviated by a magnetic and by an electric field, and these deviations are precisely what would be produced by these same fields upon projectiles animated with a very great velocity, and highly charged with negative electricity. These two deviations depend upon two quantities; the velocity on the one hand, and the proportion of the projectile's electric charge to its mass on the other. We cannot know the absolute value of this mass, nor that of the charge, but only their proportion. It is clear, in fact, that if we double both the charge and the mass, without changing the velocity, we shall double the force that tends to deviate the projectile; but as its mass is similarly doubled, the observable acceleration and deviation will not be changed. Observation of the two deviations will accordingly furnish us with two equations for determining these two unknown quantities. We find a velocity of 6,000 to 20,000 miles a second. As for the proportion of the charge to the mass, it is very great; it may be compared with the corresponding proportion in the case of a hydrogen ion in electrolysis, and we find then that a cathode projectile carries with it about a thousand times as much electricity as an equal mass of hydrogen in an electrolyte.

In order to confirm these views, we should require a direct measure of this velocity, that could then be compared with the velocity so calculated. Some old experiments of Sir J. J. Thomson's had given results more than a hundred times too low, but they were subject to certain causes of error. The question has been taken up again by Wiechert, with the help of an arrangement by which he makes use of the Hertzian oscillations, and this has given results in accordance with the theory, at least in the matter of magnitude, and it would be most interesting to take up these experiments again. However it be, the theory of undulations seems to be incapable of accounting for this body of facts.

The same calculations made upon the β rays of radium have yielded still higher velocities—60,000, 120,000 miles a second, and even more. These velocities greatly surpass any that we know. It is true that light, as we have long known, travels 186,000 miles a second, but it is not a transportation of matter, while, if we adopt the emission theory for the cathode rays, we have material molecules actually animated with the velocities in question, and we have to enquire whether the ordinary laws of Mechanics are still applicable to them.

II. Longitudinal and Transversal Mass

We know that electric currents give rise to phenomena of induction, in particular to *self-induction*. When a current increases it develops an electro-motive force of self-induction which tends to oppose the current. On the contrary, when the current decreases, the electro-motive force of self-induction tends to maintain the current. Self-induction then opposes all variation in the intensity of a current, just as in Mechanics, the inertia of a body opposes all variation in its velocity. *Self-induction is an actual inertia.* Everything takes place as if the current could not be set up without setting the surrounding ether in motion, and as if the inertia of this ether consequently tended to keep the intensity of the current constant. The inertia must be overcome to set up the current, and it must be overcome again to make it cease.

A cathode ray, which is a rain of projectiles charged with negative electricity, can be likened to a current. No doubt this current differs, at first sight at any rate, from the ordinary conduction currents, where the matter is motionless and the electricity circulates through the matter. It is a *convection current,* where the electricity is attached to a material vehicle and carried by the movement of that vehicle. But Rowland has proved that convection currents produce the same magnetic effects as conduction currents. They must also produce the same effects of induction. Firstly, if it were not so, the principle of the conservation of energy would be violated; and secondly, Crémien and Pender have employed a method in which these effects of induction are *directly* demonstrated.

If the velocity of a cathode corpuscle happens to vary, the intensity of the corresponding current will vary equally, and there will be developed effects of self-induction which tend to oppose this variation. These corpuscles must therefore possess a double inertia, first their actual inertia, and then an apparent inertia due to self-induction, which produces the same effects. They will therefore have a total apparent mass, composed of their real mass and of a fictitious mass of electro-magnetic origin. Calculation shows that this fictitious mass varies with the velocity (when this is comparable with the velocity of light), and that the force of the inertia of self-induction is not the same when the velocity of the projectile is increased or diminished, as when its direction is changed, and accordingly the same holds good of the apparent total force of inertia.

The total apparent mass is therefore not the same when the actual force applied to the corpuscle is parallel with its velocity and tends to accelerate its movement, as when it is perpendicular to the velocity and tends to alter its direction. Accordingly we must distinguish between the *total longitudinal mass* and the *total transversal mass,* and, moreover, these two total masses depend upon the velocity. Such are the results of Abraham's theoretical work.

In the measurements spoken of in the last section, what was it that was determined by measuring the two deviations? The velocity on the one hand, and on the other the proportion of the charge to the *total transversal mass.* Under these conditions, how are we to determine what are the proportions, in this total mass, of the actual mass and of the fictitious electro-magnetic mass? If we had only the cathode rays properly so called, we could not dream of doing so, but fortunately we have the rays of radium, whose velocity, as we have seen, is considerably higher. These rays are not all identical, and do not behave in the same way under the action of an electric and a magnetic field. We find that the electric deviation is a function of the magnetic deviation, and by receiving upon a sensitive plate rays of radium that have been subjected to the action of the two fields, we can photograph the curve which represents the relation between these two deviations. This is what Kaufmann has done, and he has deduced the relation between the velocity and the proportion of the charge to the total apparent mass, a proportion that we call ϵ.

We might suppose that there exist several kinds of rays, each characterized by a particular velocity, by a particular charge, and by a particular mass; but this hypothesis is most improbable. What reason indeed could there be why all the corpuscles of the same mass should always have the same velocity? It is more natural to suppose that the charge and the *actual* mass are the same for all the projectiles, and that they differ only in velocity. If the proportion ϵ is a function of the velocity, it is not because the actual mass varies with the velocity, but, as the fictitious electro-magnetic mass depends upon that velocity, the total apparent mass, which is alone observable, must depend upon it also, even though the actual mass does not depend upon it but is constant.

Abraham's calculations make us acquainted with the law in accordance with which the *fictitious* mass varies as a function of the velocity, and Kaufmann's experiment makes us acquainted with the law of variation of the *total* mass. A comparison of these two laws will therefore enable us to determine the proportion of the *actual* mass to the total mass.

Such is the method employed by Kaufmann to determine this proportion. The result is most surprising: *the actual mass is nil.*

We have thus been led to quite unexpected conceptions. What had been proved only in the case of the cathode corpuscles has been extended to all bodies. What we call mass would seem to be nothing but an appearance, and all inertia to be of electro-magnetic origin. But if this be true, mass is no longer constant; it increases with the velocity: while apparently constant for velocities up to as much as 600 miles a second, it grows thenceforward and becomes infinite for the velocity of light. Transversal mass is no longer equal to longitudinal mass, but only about equal if the velocity is not too great. Principle B of Mechanics is no longer true.

III. CANAL-RAYS

At the point we have reached, this conclusion may seem premature. Can we apply to the whole of matter what has only been established for these very light corpuscles which are only an emanation of matter and perhaps not true matter? But before broaching this question, we must say a word about other kinds of rays—I mean the *canal-rays,* Goldstein's *Kanalstrahlen.* Simultaneously with the cathode rays charged with negative electricity, the cathode emits canal-rays charged with positive electricity. In general these canal-rays, not being repelled by the cathode, remain confined in the immediate neighbourhood of that cathode, where they form the "buff stratum" that is not very easy to detect. But if the cathode is pierced with holes and blocks the tube almost completely, the canal-rays will be generated *behind* the cathode, in the opposite direction from that of the cathode rays, and it will become possible to study them. It is thus that we have been enabled to demonstrate their positive charge and to show that the magnetic and electric deviations still exist, as in the case of the cathode rays, though they are much weaker.

Radium likewise emits rays similar to the canal-rays, and relatively very absorbable, which are called α rays.

As in the case of the cathode rays, we can measure the two deviations and deduce the velocity and the proportion ϵ. The results are less constant than in the case of the cathode rays, but the velocity is lower, as is also the proportion ϵ. The positive corpuscles are less highly charged than the negative corpuscles; or if, as is more natural, we suppose that the charges are equal and of opposite sign, the positive corpuscles are much larger. These corpuscles, charged some positively and others negatively, have been given the name of *electrons.**

IV. Lorentz's Theory

But the electrons do not only give evidence of their existence in these rays in which they appear to us animated with enormous velocities. We shall see them in very different parts, and it is they that explain for us the principal phenomena of optics and of electricity. The brilliant synthesis about which I am going to say a few words is due to Lorentz.

Matter is entirely formed of electrons bearing enormous charges, and if it appears to us neutral, it is because the electrons' charges of opposite sign balance. For instance, we can picture a kind of solar system consisting of one great positive electron, about which gravitate numerous small planets which are negative electrons, attracted by the electricity of opposite sign with which the central electron is charged. The negative charges of these planets balance the positive charge of the sun, so that the algebraic sum of all these charges is nil.

All these electrons are immersed in ether. The ether is everywhere identical with itself, and perturbations are produced in it,

---

*The name is now applied only to the negative corpuscles, which seem to possess no actual mass and only a fictitious electro-magnetic mass, and not to the canal-rays, which appear to consist of ordinary chemical atoms positively charged, owing to the fact that they have lost one or more of the electrons they possess in their ordinary neutral state.

following the same laws as light or the Hertzian oscillations in empty space. Beyond the electrons and the ether there is nothing. When a luminous wave penetrates a part of the ether where the electrons are numerous, these electrons are set in motion under the influence of the perturbation of the ether, and then react upon the ether. This accounts for refraction, dispersion, double refraction, and absorption. In the same way, if an electron was set in motion for any reason, it would disturb the ether about it and give birth to luminous waves, and this explains the emission of light by incandescent bodies.

In certain bodies—metals, for instance—we have motionless electrons, about which circulate movable electrons, enjoying complete liberty, except of leaving the metallic body and crossing the surface that separates it from exterior space, or from the air, or from any other non-metallic body. These movable electrons behave then inside the metallic body as do the molecules of a gas, according to the kinetic theory of gases, inside the vessel in which the gas is contained. But under the influence of a difference of potential the negative movable electrons would all tend to go to one side and the positive movable electrons to the other. This is what produces electric currents, *and it is for this reason that such bodies act as conductors.* Moreover, the velocities of our electrons will become greater as the temperature rises, if we accept the analogy of the kinetic theory of gases. When one of these movable electrons meets the surface of the metallic body, a surface it cannot cross, it is deflected like a billiard ball that has touched the cushion, and its velocity undergoes a sudden change of direction. But when an electron changes its direction, as we shall see farther on, it becomes the source of a luminous wave, and it is for this reason that hot metals are incandescent.

In other bodies, such as dielectric and transparent bodies, the movable electrons enjoy much less liberty. They remain, as it were, attached to fixed electrons which attract them. The farther they stray, the greater becomes the attraction that tends to bring them back. Accordingly they can only suffer slight displacements; they cannot circulate throughout the body, but only oscillate about their mean position. It is for this reason that these bodies are non-conductors; they are, moreover, generally transparent, and they are refractive because the luminous vibrations are communicated to the movable electrons which are susceptible of oscillation, and a refraction of the original beam of light results.

I cannot here give the details of the calculations. I will content myself with saying that this theory accounts for all the known facts, and has enabled us to foresee new ones, such as Zeeman's phenomenon.

V. MECHANICAL CONSEQUENCES

Now we can form two hypotheses in explanation of the above facts.

1. The positive electrons possess an actual mass, much greater than their fictitious electro-magnetic mass, and the negative electrons alone are devoid of actual mass. We may even suppose that, besides the electrons of both signs, there are neutral atoms which have no other mass than their actual mass. In this case Mechanics is not affected, we have no need to touch its laws, actual mass is constant, only the movements are disturbed by the effects of self-induction, as has always been known. These perturbations are, moreover, almost negligible, except in the case of the negative electrons which, having no actual mass, are not true matter.

2. But there is another point of view. We may suppose that the neutral atom does not exist, and that the positive electrons are devoid of actual mass just as much as the negative electrons. But if this be so, actual mass disappears, and either the word *mass* will have no further meaning, or else it must designate the fictitious electro-magnetic mass; in that case, mass will no longer be constant, transversal mass will no longer be equal to longitudinal mass, and the principles of Mechanics will be upset.

And first a word by way of explanation. I said that, for the same charge, the *total* mass of a positive electron is much greater than that of a negative electron. Then it is natural to suppose that this difference is explained by the fact that the positive electron has, in addition to its fictitious mass, a considerable actual mass, which would bring us back to the first hypothesis. But we may equally well admit that the actual mass is nil for the one as for the other, but that the fictitious mass of the positive electron is much greater, because this electron is much smaller. I say advisedly, much smaller. And indeed, in this hypothesis, inertia is of exclusively electromagnetic origin, and is reduced to the inertia of the ether; the electrons are no longer anything in themselves, they are only holes in the ether, around which the ether is agitated; the smaller these holes are, the more ether there will be, and the greater, consequently, will be its inertia.

How are we to decide between these two hypotheses? By working upon the canal-rays, as Kaufmann has done upon the β rays? This is impossible, for the velocity of these rays is much too low. So each must decide according to his temperament, the conservatives taking one side and the lovers of novelty the other. But perhaps, to gain a complete understanding of the innovators' arguments, we must turn to other considerations.

CHAPTER II

MECHANICS AND OPTICS

We know the nature of the phenomenon of aberration discovered by Bradley. The light emanating from a star takes a certain time to traverse the telescope. During this time the telescope is displaced by the earth's motion. If, therefore, the telescope were pointed in the *true* direction of the star, the image would be formed at the point occupied by the crossed threads of the reticule when the light reached the object-glass. When the light reached the plane of the reticule the crossed threads would no longer be in the same spot, owing to the earth's motion. We are therefore obliged to alter the direction of the telescope to bring the image back to the crossed threads. It follows that the astronomer will not point his telescope exactly in the direction of the absolute velocity of the light from the star—that is to say, upon the true position of the star—but in the direction of the relative velocity of the light in relation to the earth—that is to say, upon what is called the apparent position of the star.

The velocity of light is known, and accordingly we might imagine that we have the means of calculating the *absolute* velocity of the earth. (I shall explain the meaning of this word *absolute* later.) But it is not so at all. We certainly know the apparent position of the star we are observing, but we do not know its true position. We know the velocity of light only in terms of magnitude and not of direction.

If, therefore, the earth's velocity were rectilineal and uniform, we should never have suspected the phenomenon of aberration. But it is variable: it is composed of two parts—the velocity of the solar system, which is, as far as we know, rectilineal and uniform; and the velocity of the earth in relation to the sun, which is variable. If the velocity of the solar system—that is to say, the constant part—alone existed, the observed direction would be invariable. The position we should thus observe is called the *mean* apparent position of the star.

Now if we take into account at once both parts of the earth's velocity, we shall get the actual apparent position, which describes a small ellipse about the mean apparent position, and it is this ellipse that is observed.

Neglecting very small quantities, we shall see that the dimensions of this ellipse depend only upon the relation between the earth's velocity in relation to the sun and the velocity of light, so that the *relative* velocity of the earth in relation to the sun is alone in question.

We must pause, however. This result is not exact, but only approximate. Let us push the approximation a step further. The dimensions of the ellipse will then depend upon the absolute velocity of the earth. If we compare the great axes of ellipse for the different stars, we shall have, theoretically at least, the means of determining this absolute velocity.

This is perhaps less startling than it seems at first. It is not a question, indeed, of the velocity in relation to absolute space, but of the velocity in relation to the ethics, which is regarded, *by definition,* as being in absolute repose.

Moreover, this method is purely theoretical. In fact the aberration is very small, and the possible variations of the ellipse of aberration are much smaller still, and, accordingly, if we regard the aberration as of the first order, the variations must be regarded as of the second order, about a thousandth of a second of arc, and ab-

516 · *Science and Method*

solutely inappreciable by our instruments. Lastly, we shall see further on why the foregoing theory must be rejected, and why we could not determine this absolute velocity even though our instruments were ten thousand times as accurate.

Another method may be devised, and, indeed, has been devised. The velocity of light is not the same in the water as in the air: could we not compare the two apparent positions of a star seen through a telescope filled first with air and then with water? The results have been negative; the apparent laws of reflection and of refraction are not altered by the earth's motion. This phenomenon admits of two explanations.

1. We may suppose that the ether is not in repose, but that it is displaced by bodies in motion. It would not then be astonishing that the phenomenon of refraction should not be altered by the earth's motion, since everything—lenses, telescopes, and ether—would be carried along together by the same motion. As for aberration itself, it would be explained by a kind of refraction produced at the surface of separation of the ether in repose in the interstellar spaces and the ether carried along by the earth's movement. It is upon this hypothesis (the total translation of the ether) that *Hertz's theory* of the electro-dynamics of bodies in motion is founded.

2. Fresnel, on the contrary, supposes that the ether is in absolute repose in space, and almost in absolute repose in the air, whatever be the velocity of that air, and that it is partially displaced by refringent mediums. Lorentz has given this theory a more satisfactory form. In his view the ether is in repose and the electrons alone are in motion. In space, where the ether alone comes into play, and in the air, where it comes almost alone into play, the displacement is nil or almost nil. In refringent mediums, where the perturbation is produced both by the vibrations of the ether and by those of the electrons set in motion by the agitation of the ether, the undulations are *partially* carried along.

To help us to decide between these two hypotheses, we have the experiment of Fizeau, who compared, by measurements of fringes of interference, the velocity of light in the air in repose and in motion as well as in water in repose and in motion. These experiments have confirmed Fresnel's hypothesis of partial displacement, and they have been repeated with the same result by Michelson. *Hertz's theory, therefore, must be rejected.*

II. The Principle of Relativity

But if the ether is not displaced by the earth's motion, is it possible by means of optical phenomena to demonstrate the absolute velocity of the earth, or rather its velocity in relation to the motionless ether? Experience has given a negative reply, and yet the experimental processes have been varied in every possible way. Whatever be the method employed, we shall never succeed in disclosing any but relative velocities; I mean the velocities of certain material bodies in relation to other material bodies. Indeed, when the source of the light and the apparatus for observation are both on the earth and participate in its motion, the experimental results have always been the same, whatever be the direction of the apparatus in relation to the direction of the earth's orbital motion. That astronomical aberration takes place is due to the fact that the source, which is a star, is in motion in relation to the observer.

The hypotheses formed up to now account perfectly for this general result, *if we neglect very small quantities on the order of the square of aberration.* The explanation relies on the notion of *local time* introduced by Lorentz, which I will try to make clear. Imagine two observers placed, one at a point A and the other at a point B, wishing to set their watches by means of optical signals. They agree that B shall send a signal to A at a given hour by his watch, and A sets his watch to that hour as soon as he sees the signal. If the operation were performed in this way only, there would be a systematic error; for, since light takes a certain time, t, to travel from B to A, A's watch would always be slower than B's to the extent of t. This error is easily corrected, for it is sufficient to interchange the signals. A in its turn must send signals to B, and after this new setting it will be B's watch that will be slower than A's to the extent of t. Then it will only be necessary to take the arithmetic mean between the two settings.

But this method of operating assumes that light takes the same time to travel from A to B and to return from B to A. This is true if the observers are motionless, but it is no longer true if they are involved in a common transposition, because in that case A, for instance, will be meeting the light that comes from B, while B is retreating from the light that comes from A. Accordingly, if the observers are involved in a common transposition without suspecting it, their setting will be defective; their watches will not show the same time, but each of them will mark the *local time* proper to the place where it is.

The two observers will have no means of detecting this, if the motionless ether can only transmit luminous signals all travelling at the same velocity, and if the other signals they can send are transmitted to them by mediums involved with them in their transposition. The phenomenon each of them observes will be either early or late—it will not occur at the moment it would have if there were no transposition; but since their observations are made with a watch defectively set, they will not detect it, and the appearances will not be altered.

It follows from this that the compensation is easy to explain so long as we neglect the square of aberration, and for a long time experiments were not sufficiently accurate to make it necessary to take this into account. But one day Michelson thought out a much more delicate process. He introduced rays that had traversed different distances after being reflected by mirrors. Each of the distances being about a yard, and the fringes of interference making it possible to detect differences of a fraction of a millionth of a millimeter ($\frac{1}{25000000}$ th of an inch), the square of aberration could no longer be neglected, and yet *the results were still negative*. Accordingly, the theory required to be completed, and this has been done by *the hypothesis of Lorentz and Fitz-Gerald*.

These two physicists assume that all bodies involved in a transposition undergo a contraction in the direction of this transposition, while their dimensions perpendicular to the transposition remain invariable. *This contraction is the same for all bodies.* It is, moreover, very slight, about one part in two hundred million for a velocity such as that of the earth. Moreover, our measuring instruments could not disclose it, even though they were very much more accurate, since indeed the yard-measures with which we measure undergo the same contraction as the objects to be measured. If a body fits exactly to a measure when the body, and consequently the measure, are turned in the direction of the earth's motion, it will not cease to fit exactly to the measure when turned in another direction, in spite of the fact that the body and the measure have changed their length in changing their direction, precisely because the change is the same for both. But it is not so if we measure a distance, no longer with a yard-measure, but by the time light takes to traverse it, and this is exactly what Michelson has done.

A body that is spherical when in repose will thus assume the form of a flattened ellipsoid of revolution when it is in motion. But the observer will always believe it to be spherical, because he has himself undergone an analogous deformation, as well as all the objects that serve him as points of reference. On the contrary, the surfaces of the waves of light, which have remained exactly spherical, will appear to him as elongated ellipsoids.

What will happen then? Imagine an observer and a source involved together in the transposition. The wave surfaces emanating from the source will be spheres, having as the centre the successive positions of the source. The distance of this centre from the actual position of the source will be proportional to the time elapsed since the emission—that is to say, to the radius of the sphere. All these spheres are accordingly homothetic one to the other, in relation to the actual position S of the source. But for our observer, on account of the contraction, all these spheres will appear as elongated ellipsoids, and all these ellipsoids will still be homothetic in relation to the point S; the excentricity of all the ellipsoids is the same, and depends solely upon the earth's velocity. *We shall select our law of contraction in such a way that S will be the focus of the meridian section of the ellipsoid.*

This time the compensation is *exact*, and this is explained by Michelson's experiments.

I said above that, according to the ordinary theories, observations of astronomical aberration could make us acquainted with the absolute velocity of the earth, if our instruments were a thousand times as accurate, but this conclusion must be modified. It is true that the angles observed would be modified by the effect of this absolute velocity, but the graduated circles we use for measuring the angles would be deformed by the motion; they would become ellipses, the result would be an error in the angle measured, *and this second error would exactly compensate the former.*

This hypothesis of Lorentz and Fitz-Gerald will appear most extraordinary at first sight. All that can be said in its favour for the moment is that it is merely the immediate interpretation of Michelson's experimental result, if we *define* distances by the time taken by light to traverse them.

However that be, it is impossible to escape the impression that the *principle of relativity* is a general law of Nature, and that we shall never succeed, by any imaginable method, in demonstrating any but relative velocities; and by this I mean not merely the velocities of bodies in relation to the ether, but the velocities of bodies in relation to each other. So many different experiments have given similar results that we cannot but feel tempted to attribute to this principle of relativity a value comparable, for instance, to that of the principle of equivalence. It is good in any case to see what are the consequences to which this point of view would lead, and then to submit these consequences to the test of experiment.

III. The Principle of Reaction

Let us see what becomes, under Lorentz's theory, of the principle of the equality of action and reaction. Take an electron, A, which is set in motion by some means. It produces a disturbance in the ether, and after a certain time this disturbance reaches another electron, B, which will be thrown out of its position of equilibrium. Under these conditions there can be no equality between the action and the reaction, at least if we do not consider the ether, but only the electrons *which are alone observable,* since our matter is composed of electrons.

It is indeed the electron A that has disturbed the electron B; but even if the electron B reacts upon A, this reaction, though possibly equal to the action, cannot in any case be simultaneous, since the electron B cannot be set in motion until after a certain length of time necessary for the effect to travel through the ether. If we submit the problem to a more precise calculation, we arrive at the following result. Imagine a Hertz excitator placed at the focus of a parabolic mirror to which it is attached mechanically; this excitator emits electro-magnetic waves, and the mirror drives all these waves in the same direction: the excitator will accordingly radiate energy in a particular direction. Well, calculations show that *the excitator will recoil* like a cannon that has fired a projectile. In the case of the cannon, the recoil is the natural result of the equality of action and reaction. The cannon recoils because the projectile on which it has acted reacts upon it.

But here the case is not the same. What we have fired away is no longer a material projectile; it is energy, and energy has no mass—there is no counterpart. Instead of an excitator, we might have considered simply a lamp with a reflector concentrating its rays in a single direction.

It is true that if the energy emanating from the excitator or the lamp happens to reach a material object, this object will experience a mechanical thrust as if it had been struck by an actual projectile, and this thrust will be equal to the recoil of the excitator or the lamp, if no energy has been lost on the way, and if the object absorbs the energy in its entirety. We should then be tempted to say that there is still compensation between the action and the reaction. But this compensation, even though it is complete, is always late. It never occurs at all if the light, after leaving the source, strays in the interstellar spaces without ever meeting a material body, and it is incomplete if the body it strikes is not perfectly absorbent.

Are these mechanical actions too small to be measured, or are they appreciable by experiment? They are none other than the actions due to the *Maxwell-Bartholi* pressures. Maxwell had predicted these pressures by calculations relating to Electro-statics and Magnetism, and Bartholi had arrived at the same results on thermodynamic grounds.

It is in this way that *tails of comets* are explained. Small particles are detached from the head of the comet, they are struck by the light of the sun, which repels them just as would a shower of projectiles coming from the sun. The mass of these particles is so small that this repulsion overcomes the Newtonian gravitation, and accordingly they form the tail as they retreat from the sun.

Direct experimental verification of this pressure of radiation was not easy to obtain. The first attempt led to the construction of the *radiometer*. But this apparatus *turns the wrong way*, the reverse of the theoretical direction, and the explanation of its rotation, which has since been discovered, is entirely different. Success has been attained at last by creating a more perfect vacuum on the one hand; and on the other, by not blackening one of the faces of the plates, and by directing a luminous beam upon one of these faces. The radiometric effects and other disturbing causes are eliminated by a series of minute precautions, and a deviation is obtained which is extremely small, but is, it appears, in conformity with the theory.

The same effects of the Maxwell-Bartholi pressure are similarly predicted by Hertz's theory, of which I spoke above, and by that of Lorentz, but there is a difference. Suppose the energy, in the form of light, for instance, travels from a luminous source to any body through a transparent medium. The Maxwell-Bartholi pressure will act not only upon the source at its start and upon the body lighted at its arrival, but also upon the matter of the transparent medium it traverses. At the moment the luminous wave reaches a new portion of this medium, the pressure will drive forward the matter there distributed, and will drive it back again when the wave leaves that portion. So that the recoil of the source has for its counterpart the forward motion of the transparent matter that is in contact with the source; a little later the recoil of this same matter has for its counterpart the forward motion of the transparent matter a little further off, and so on.

Only, is the compensation perfect? Is the action of the Maxwell-Bartholi pressure upon the matter of the transparent medium equal to its reaction upon the source, and that, whatever that matter may be? Or rather, is the action less in proportion as the medium is less refringent and more rarefied, becoming nil in a vacuum? If we admit Hertz's theory, which regards the ether as mechanically attached to matter, so that the ether is completely carried along by matter, we must answer the first and not the second question in the affirmative.

There would then be perfect compensation, such as the principle of the equality of action and reaction demands, even in the least refringent media, even in the air, even in the interplanetary space, where it would be sufficient to imagine a bare remnant of matter, however attenuated. If we admit Lorentz's theory, on the contrary, the compensation, always imperfect, is inappreciable in the air, and becomes nil in space.

But we have seen above that Fizeau's experiment does not permit of our retaining Hertz's theory. We must accordingly adopt Lorentz's theory, and consequently *give up the principle of reaction*.

IV. Consequences of the Principle of Relativity

We have seen above the reasons that incline us to regard the principle of relativity as a general law of nature. Let us see what consequences the principle will lead us to if we regard it as definitely proved.

First of all, it compels us to generalize the hypothesis of Lorentz and Fitz-Gerald on the contraction of all bodies in the direction of their transposition. More particularly, we must extend the hypothesis to the electrons themselves. Abraham considered these electrons as spherical and undeformable, but we shall have to admit that the electrons, while spherical when in repose, undergo Lorentz's contraction when they are in motion, and then take the form of flattened ellipsoids.

This deformation of the electrons will have an influence upon their mechanical properties. In fact, I have said that the displacement of these charged electrons is an actual convection current, and that their apparent inertia is due to the self-induction of this current, exclusively so in the case of the negative electrons, but whether exclusively or not in the case of the positive electrons we do not yet know.

On these terms the compensation will be perfect, and in conformity with the requirements of the principle of relativity, but only upon two conditions:

1. That the positive electrons have no real mass, but only a fictitious electro-magnetic mass; or at least that their real mass, if it exists, is not constant, but varies with the velocity, following the same laws as their fictitious mass.

2. That all forces are of electro-magnetic origin, or at least that they vary with the velocity, following the same laws as forces of electro-magnetic origin.

It is Lorentz again who has made this remarkable synthesis. Let us pause a moment to consider what results from it. In the first place, there is no more matter, since the positive electrons have no longer any real mass, or at least no constant real mass. The actual principles of our Mechanics, based upon the constancy of mass, must accordingly be modified.

Secondly, we must seek an electro-magnetic explanation of all known forces, and especially of gravitation, or at least modify the law of gravitation in the sense that this force must be altered by velocity in the same way as electro-magnetic forces. We shall return to this point.

All this appears somewhat artificial at first sight, and more particularly the deformation of the electrons seems extremely hypothetical. But the matter can be presented differently, so as to avoid taking this hypothesis of deformation as the basis of the argument. Let us imagine the electrons as material points, and enquire how their mass ought to vary as a function of the velocity so as not to violate the principle of relativity. Or rather let us further enquire what should be their acceleration under the influence of an electric or magnetic field, so that the principle should not be violated and that we should return to the ordinary laws when we imagine the velocity very low. We shall find that the variations of this mass or of these accelerations must occur *as if* the electron underwent Lorentz's deformation.

V. Kaufmann's Experiment

Two theories are thus presented to us: one in which the electrons are undeformable, which is Abraham's; the other, in which they undergo Lorentz's deformation. In either case their mass grows with their velocity, becoming infinite when that velocity becomes equal to that of light; but the law of the variation is not the same. The method employed by Kaufmann to demonstrate the law of variation of the mass would accordingly seem to give us the means of deciding experimentally between the two theories.

Unfortunately his first experiments were not sufficiently accurate for this purpose, so much so that he has thought it necessary to repeat them with more precautions, and measuring the intensity of the fields with greater care. In their new form *they have shown Abraham's theory to be right.* Accordingly, it would seem that the principle of relativity has not the exact value we have been tempted to give it, and that we have no longer any reason for supposing that the positive electrons are devoid of real mass like the negative electrons.

Nevertheless, before adopting this conclusion some reflexion is necessary. The question is one of such importance that one would wish to see Kaufmann's experiment repeated by another experimenter.[*]

Unfortunately, the experiment is a very delicate one, and cannot be performed successfully, except by a physicist as skilful as Kaufmann. All suitable precautions have been taken, and one cannot well see what objection can be brought.

[*] At the moment of going to press we learnt that M. Bucherer has repeated the experiment, surrounding it with new precautions, and that, unlike Kaufmann, he has obtained results confirming Lorentz's views.

There is, nevertheless, one point to which I should wish to call attention, and that is the measurement of the electro-static field, the measurement upon which everything depends. This field was produced between the two armatures of a condenser, and between these two armatures an extremely perfect vacuum had to be created in order to obtain complete isolation. The difference in the potential of the two armatures was then measured, and the field was obtained by dividing this difference by the distance between the armatures. This assumes that the field is uniform; but is this certain? May it not be that there is a sudden drop in the potential in the neighbourhood of one of the armatures, of the negative armature, for instance? There may be a difference in potential at the point of contact between the metal and the vacuum, and it may be that this difference is not the same on the positive as on the negative side. What leads me to think this is the electric valve effect between mercury and vacuum? It would seem that we must at least take into account the possibility of this occurring, however slight the probability may be.

VI. The Principle of Inertia

In the new dynamics the principle of inertia is still true—that is to say, that an *isolated* electron will have a rectilineal and uniform motion. At least it is generally agreed to admit it, though Lindemann has raised objections to the assumption. I do not wish to take sides in the discussion, which I cannot set out here on account of its extremely difficult nature. In any case, the theory would only require slight modifications to escape Lindemann's objections.

We know that a body immersed in a fluid meets with considerable resistance when it is in motion; but that is because our fluids are viscous. In an ideal fluid, absolutely devoid of viscidity, the body would excite behind it a liquid stern-wave, a kind of wake. At the start, it would require a great effort to set it in motion, since it would be necessary to disturb not only the body itself but the liquid of its wake. But once the motion was acquired, it would continue without resistance, since the body, as it advanced, would simply carry with it the disturbance of the liquid, without any increase in the total *vis viva* of the liquid. Everything would take place, therefore, as if its inertia had been increased. An electron advancing through the ether will behave in the same way. About it the ether will be disturbed, but this disturbance will accompany the body in its motion, so that, to an observer moving with the electron, the electric and magnetic fields which accompany the electron would appear invariable, and could only change if the velocity of the electron happened to vary. An effort is therefore required to set the electron in motion, since it is necessary to create the energy of these fields. On the other hand, once the motion is acquired, no effort is necessary to maintain it, since the energy created has only to follow the electron like a wake. This energy, therefore, can only increase the inertia of the electron, as the agitation of the liquid increases that of the body immersed in a perfect fluid. And actually the electrons, at any rate the negative electrons, have no other inertia but this.

In Lorentz's hypothesis, the *vis viva,* which is nothing but the energy of the ether, is not proportional to v^2. No doubt if v is very small, the *vis viva* is apparently proportional to v^2, the amount of momentum apparently proportional to v, and the two masses apparently constant and equal to one another. But *when the velocity approaches the velocity of light, the vis viva, the amount of momentum, and the two masses increase beyond all limit.*

In Abraham's hypothesis the expressions are somewhat more complicated, but what has just been said holds good in its essential features.

Thus the mass, the amount of momentum, and the *vis viva* become infinite when the velocity is equal to that of light. Hence it follows that *no body can, by any possibility, attain a velocity higher than that of light.* And, indeed, as its velocity increases its mass increases, so that its inertia opposes a more and more serious obstacle to any fresh increase in its velocity.

A question then presents itself. Admitting the principle of relativity, an observer in motion can have no means of perceiving his own motion. If, therefore, no body in its actual motion can exceed the velocity of light, but can come as near it as we like, it must be the same with regard to its relative motion in relation to our observer. Then we might be tempted to reason as follows: the observer can attain a velocity of 120,000 miles a second, the body in its relative motion in relation to the observer can attain the same velocity; its absolute velocity will then be 240,000 miles, which is impossible, since this is a figure higher than that of the velocity of light. But this is only an appearance which vanishes when we take into account Lorentz's method of valuing local times.

VII. The Wave of Acceleration

When an electron is in motion it produces a disturbance in the ether which surrounds it. If its motion is rectilineal and uniform, this disturbance is reduced to the wake I spoke of in the last section. But it is not so if the motion is in a curve or not uniform. The disturbance may then be regarded as the superposition of two others, to which Langevin has given the names of *wave of velocity* and *wave of acceleration*.

The wave of velocity is nothing else than the wake produced by the uniform motion.

As for the wave of acceleration, it is a disturbance absolutely similar to light waves, which starts from the electron the moment it undergoes an acceleration, and is then transmitted in successive spherical waves with the velocity of light.

Hence it follows that in a rectilineal and uniform motion there is complete conservation of energy, but as soon as there is acceleration there is loss of energy, which is dissipated in the form of light waves and disappears into infinite space through the ether.

Nevertheless, the effects of this wave of acceleration, and more particularly the corresponding loss of energy, are negligible in the majority of cases—that is to say, not only in the ordinary Mechanics and in the motions of the celestial bodies, but even in the case of the radium rays, where the velocity, but not the acceleration, is very great. We may then content ourselves with the application of the laws of Mechanics, stating that the force is equal to the product of the acceleration and the mass, this mass, however, varying with the velocity according to the laws set forth above. The motion is then said to be *quasi-stationary*.

It is not so in all the cases where the acceleration is great, the chief of which are as follows. (1.) In incandescent gases certain electrons take on an oscillatory motion of very high frequency; the displacements are very small, the velocities finite, and the accelerations very great; the energy is then communicated to the ether, and it is for this reason that these gases radiate light of the same periodicity as the oscillations of the electron. (2.) Inversely, when a gas receives light, these same electrons are set in motion with violent accelerations, and they absorb light. (3.) In Hertz's excitator, the electrons which circulate in the metallic mass undergo a sudden acceleration at the moment of the discharge, and then take on an oscillatory motion of high frequency. It follows that a part of the energy is radiated in the form of Hertzian waves. (4.) In an incandescent metal, the electrons enclosed in the metal are animated with great velocities. On arriving at the surface of the metal, which they cannot cross, they are deflected, and so undergo a considerable acceleration, and it is for this reason that the metal emits light. This I have already explained in Book III, Chap. I, Sec. 4. The details of the laws of the emission of light by dark bodies are perfectly explained by this hypothesis. (5.) Lastly, when the cathode rays strike the anti-cathode, the negative electrons constituting these rays, which are animated with very great velocities, are suddenly stopped. In consequence of the acceleration they thus undergo, they produce undulations in the ether. This, according to certain physicists, is the origin of the Röntgen rays, which are nothing else than light rays of very short wave length.

CHAPTER III

THE NEW MECHANICS AND ASTRONOMY

I. GRAVITATION

Mass may be defined in two ways—firstly, as the quotient of the force by the acceleration, the true definition of mass, which is the measure of the body's inertia; and secondly, as the attraction exercised by the body upon a foreign body, by virtue of Newton's law. We have therefore to distinguish between mass, the co-efficient of inertia, and mass, the co-efficient of attraction. According to Newton's law, there is a rigorous proportion between these two co-efficients, but this is only demonstrated in the case of velocities to which the general principles of dynamics are applicable. Now we have seen that the mass co-efficient of inertia increases with the velocity; must we conclude that the mass co-efficient of attraction increases similarly with the velocity, and remains proportional to the co-efficient of inertia, or rather that the co-efficient of attraction remains constant? This is a question that we have no means of deciding.

On the other hand, if the co-efficient of attraction depends upon the velocity, as the velocities of bodies mutually attracting each other are generally not the same, how can this co-efficient depend upon these two velocities?

Upon this subject we can but form hypotheses, but we are naturally led to enquire which of these hypotheses will be compatible with the principle of relativity. There are a great number, but the only one I will mention here is Lorentz's hypothesis, which I will state briefly.

Imagine first of all electrons in repose. Two electrons of similar sign repel one another, and two electrons of opposite sign attract one another. According to the ordinary theory, their mutual actions are proportional to their electric charges. If, therefore, we have four electrons, two positive, A and A', and two negative, B and B', and the charges of these four electrons are the same in absolute value, the repulsion of A upon A' will be, at the same distance, equal to the repulsion of B upon B', and also equal to the attraction of A upon B' or of A' upon B. Then if A and B are very close to each other, as also A' and B', and we examine the action of the system A+B upon the system A'+B', we shall have two repulsions and two attractions that are exactly compensated, and the resultant action will be nil.

Now material molecules must precisely be regarded as kinds of solar systems in which the electrons circulate, some positive and others negative, *in such a way that the algebraic sum of all the charges is nil.* A material molecule is thus in all points comparable to the system A+B I have just spoken of, so that the total electric action of two molecules upon each other should be nil.

But experience shows us that these molecules attract one another in accordance with Newtonian gravitation, and that being so we can form two hypotheses. We may suppose that gravitation has no connexion with electro-static attraction, that it is due to an entirely different cause, and that it is merely superimposed upon it; or else we may admit that there is no proportion between the attractions and the charges, and that the attraction exercised by a charge +1 upon a charge −1 is greater than the mutual repulsion of two charges +1 or of two charges −1.

In other words, the electric field produced by the positive electrons and that produced by the negative electrons are superimposed and remain distinct. The positive electrons are more sensitive to the field produced by the negative electrons than to the field produced by the positive electrons, and contrariwise for the negative electrons. It is clear that this hypothesis somewhat complicates electro-statics, but makes it include gravitation. It was, in the main, Franklin's hypothesis.

Now, what happens if the electrons are in motion? The positive electrons will create a disturbance in the ether, and will give rise in it to an electric field and a magnetic field. The same will be true of the negative electrons. The electrons, whether positive or negative, then receive a mechanical impulse by the action of these different fields. In the ordinary theory, the electro-magnetic field due to the motion of the positive electrons exercises, upon two electrons of opposite sign and of the same absolute charge, actions that are equal and of opposite sign. We may, then, without impropriety make no distinction between the field due to the motion of the positive electrons and the field due to the motion of the negative electrons, and consider only the algebraic sum of these two fields—that is to say, the resultant field.

In the new theory, on the contrary, the action upon the positive electrons of the electro-magnetic field due to the positive electrons takes place in accordance with the ordinary laws, and the same is true of the action upon the negative electrons of the field due to the negative electrons. Let us now consider the action of the field due to the positive electrons upon the negative electrons, or *vice versa*. It will still follow the same laws, but *with a different co-efficient*. Each electron is more sensitive to the field created by the electrons of opposite denomination than to the field created by the electrons of the same denomination.

Such is Lorentz's hypothesis, which is reduced to Franklin's hypothesis for low velocities. It agrees with Newton's law in the case of these low velocities. More than that, as gravitation is brought down to forces of electro-dynamic origin, Lorentz's general theory will be applicable to it, and consequently the principle of relativity will not be violated.

We see that Newton's law is no longer applicable to great velocities, and that it must be modified, for bodies in motion, precisely in the same way as the laws of electro-statics have to be for electricity in motion.

We know that electro-magnetic disturbances are transmitted with the velocity of light. We shall therefore be tempted to reject the foregoing theory, remembering that gravitation is transmitted, according to Laplace's calculations, at least ten million times as quickly as light, and that consequently it cannot be of electro-magnetic origin. Laplace's result is well known, but its significance is generally lost sight of. Laplace assumed that, if the transmission of gravitation is not instantaneous, its velocity of transmission combines with that of the attracted body, as happens in the case of light in the phenomenon of astronomical aberration, in such a way that the effective force is not directed along the straight line joining the two bodies, but makes a small angle with that straight line. This is quite an individual hypothesis, not very well substantiated, and in any case entirely different from that of Lorentz. Laplace's result proves nothing against Lorentz's theory.

II. Comparison with Astronomical Observations

Are the foregoing theories reconcilable with astronomical observations? To begin with, if we adopt them, the energy of the planetary motions will be constantly dissipated by the effect of the *wave of acceleration*. It would follow from this that there would be a constant acceleration of the mean motions of the planets, as if these planets were moving in a resisting medium. But this effect is exceedingly slight, much too slight to be disclosed by the most minute observations. The acceleration of the celestial bodies is relatively small, so that the effects of the wave of acceleration are negligible, and the motion may be regarded as *quasi-stationary*. It is true that the effects of the wave of acceleration are constantly accumulating, but this accumulation itself is so slow that it would certainly require thousands of years of observation before it became perceptible.

Let us therefore make the calculation, taking the motion as quasi-stationary, and that under the three following hypotheses:

A. Admitting Abraham's hypothesis (undeformable electrons), and retaining Newton's law in its ordinary form.
B. Admitting Lorentz's hypothesis concerning the deformation of the electrons, and retaining Newton's ordinary law.
C. Admitting Lorentz's hypothesis concerning the electrons, and modifying Newton's law, as in the foregoing section, so as to make it compatible with the principle of relativity.

It is in the motion of Mercury that the effect will be most perceptible, because it is the planet that has the highest velocity. Tisserand formerly made a similar calculation, admitting Weber's law. I would remind the reader that Weber attempted to explain both the electro-static and the electro-dynamic phenomena, assuming that the electrons (whose name had not yet been invented) exercise upon each other attractions and repulsions in the direction of the straight line joining them, and depending not only upon their distances, but also upon the first and second derivatives of these distances, that is consequently upon their velocities and their accelerations. This law of Weber's, different as it is from those that tend to gain acceptance today, presents nonetheless a certain analogy with them.

Tisserand found that if the Newtonian attraction took place in conformity with Weber's law, there would result, in the perihelion of Mercury, a secular variation of 14″, *in the same direction as that which has been observed and not explained,* but smaller, since the latter is 38″.

Let us return to the hypotheses A, B, and C, and study first the motion of a planet attracted by a fixed centre. In this case there will be no distinction between hypotheses B and C, since, if the attracting point is fixed, the field it produces is a purely electro-static field, in which the attraction varies in the inverse ratio of the square of the distance, in conformity with Coulomb's electro-static law, which is identical with Newton's.

The *vis viva* equation holds good if we accept the new definition of *vis viva*. In the same way the equation of the areas is replaced by another equivalent. The moment of the quantity of motion is a constant, but the quantity of motion must be defined in the new way.

The only observable effect will be a secular motion of the perihelion. For this motion we shall get, with Lorentz's theory, a half, and with Abraham's theory two-fifths, of what was given by Weber's law.

If we now imagine two moving bodies gravitating about their common centre of gravity, the effects are but very slightly different, although the calculations are somewhat more complicated. The motion of Mercury's perihelion will then be 7″ in Lorentz's theory, and 5.6″ in Abraham's.

The effect is, moreover, proportional to $n^3 a^2$, n being the mean motion of the planet, and a the radius of its orbit. Accordingly for the planets, by virtue of Kepler's law, the effect varies in the inverse ratio of $\sqrt{a^5}$, and it is therefore imperceptible except in the case of Mercury.

It is equally imperceptible in the case of the moon, because, though n is large, a is extremely small. In short, it is five times as small for Venus, and six hundred times as small for the moon, as it is for Mercury. I would add that as regards Venus and the earth, the motion of the perihelion (for the same angular velocity of this motion) would be much more difficult to detect by astronomical observations, because the excentricity of their orbits is much slighter than in the case of Mercury.

To sum up, *the only appreciable effect upon astronomical observations would be a motion of Mercury's perihelion, in the same direction as that which has been observed without being explained, but considerably smaller.*

This cannot be regarded as an argument in favour of the new dynamics, since we still have to seek another explanation of the greater part of the anomaly connected with Mercury; but still less can it be regarded as an argument against it.

III. Lesage's Theory

It would be good to set these considerations beside a theory put forward long ago to explain universal gravitation. Imagine the interplanetary spaces full of very tiny corpuscles, travelling in all directions at very high velocities. An isolated body in space will not be affected apparently by the collisions with these corpuscles, since the collisions are distributed equally in all directions. But if two bodies, A and B, are in proximity, the body B will act as a screen, and intercept a portion of the corpuscles, which, but for it, would have struck A. Then the collisions received by A from the side away from B will have no counterpart, or will be only imperfectly compensated, and will drive A towards B.

Such is Lesage's theory, and we will discuss it first from the point of view of ordinary Mechanics. To begin with, how must the collisions required by this theory occur? Must it be in accordance with the laws of perfectly elastic bodies, or of bodies devoid of elasticity, or in accordance with some intermediate law? Lesage's corpuscles cannot behave like perfectly elastic bodies, for in that case the effect would be nil, because the corpuscles intercepted by the body B would be replaced by others which would have rebounded from B, and calculation proves that the compensation would be perfect.

The collision must therefore cause a loss of energy to the corpuscles, and this energy should reappear in the form of heat. But what would be the amount of heat so produced? We notice that the attraction passes through the body, and we must accordingly picture the earth, for instance, not as a complete screen, but as composed of a very large number of extremely small spherical molecules, acting individually as little screens, but allowing Lesage's corpuscles to travel freely between them. Thus, not only is the earth not a complete screen, but it is not even a strainer, since the unoccupied spaces are much larger than the occupied. To realize this, we must remember that Laplace demonstrated that the attraction, in passing through the earth, suffers a loss, at the very most, of a ten-millionth part, and his demonstration is perfectly satisfactory.

Indeed, if the attraction were absorbed by the bodies it passes through, it would no longer be proportional to their masses; it would be *relatively* weaker for large than for small bodies, since it would have a greater thickness to traverse. The attraction of the sun for the earth would therefore be *relatively* weaker than that of the sun for the moon, and a very appreciable inequality in the moon's motion would result. We must therefore conclude, if we adopt Lesage's theory, that the total surface of the spherical molecules of which the earth is composed is, at the most, the ten-millionth part of the total surface of the earth.

Darwin proved that Lesage's theory can only lead exactly to Newton's law if we assume the corpuscles to be totally devoid of elasticity. The attraction exercised by the earth upon a mass 1 at a distance 1 will then be proportional both to S, the total surface of the spherical molecules of which it is composed, to v, the velocity of the corpuscles, and to the square root of p, the density of the medium formed by the corpuscles. The heat produced will be proportional to S, to the density p, and to the cube of the velocity v.

But we must take account of the resistance experienced by a body moving in such a medium. It cannot move, in fact, without advancing towards certain collisions, and on the other hand retreating before those that come from the opposite direction, so that the compensation realized in a state of repose no longer exists. The calculated resistance is proportional to S, to p, and to v. Now we know that the heavenly bodies move as if they met with no resistance, and the precision of the observations enables us to assign a limit to the resistance.

This resistance varying as Spv, while the attraction varies as $S\sqrt{pv}$, we see that the relation of the resistance to the square of the attraction is in inverse ratio of the product Sv.

We get thus an inferior limit for the product Sv. We had already a superior limit for S (by the absorption of the attraction by the bodies it traverses). We thus get an inferior limit for the velocity v, which must be at least equal to 24.10^{17} times the velocity of light.

From this we can deduce p and the amount of heat produced.

This would suffice to elevate the temperature 10^{26} degrees a second. In any given time the earth would receive 10^{20} as much heat as the sun emits in the same time, and I am not speaking of the heat that reaches the earth from the sun, but of the heat radiated in all directions. It is clear that the earth could not long resist such conditions.

We shall be led to results no less fantastic if, in opposition to Darwin's views, we endow Lesage's corpuscles with an elasticity that is imperfect but not nil. It is true that the *vis viva* of the corpuscles will not then be entirely converted into heat, but the attraction produced will equally be less, so that it will only be that portion of the *vis viva* converted into heat that will contribute towards the production of attraction, and so we shall get the same result. A judicious use of the theorem of virial will enable us to realize this.

We may transform Lesage's theory by suppressing the corpuscles and imagining the ether traversed in all directions by luminous waves coming from all points of space. When a material object receives a luminous wave, this wave exercises upon it a mechanical action due to the Maxwell-Bartholi pressure, just as if it had received a blow from a material projectile. The waves in question may accordingly play the part of Lesage's corpuscles. This is admitted, for instance, by M. Tommasina.

This does not get over the difficulties. The velocity of transmission cannot be greater than that of light, and we are thus brought to an inadmissible figure for the resistance of the medium. Moreover, if the light is wholly reflected, the effect is nil, just as in the hypothesis of the perfectly elastic corpuscles. In order to create attraction, the light must be partially absorbed, but in that case heat will be produced. The calculations do not differ essentially from those made in regard to Lesage's ordinary theory, and the result retains the same fantastic character.

On the other hand, attraction is not absorbed, or but very slightly absorbed, by the bodies it traverses, while this is not true of the light we know. Light that would produce Newtonian attraction would require to be very different from ordinary light, and to be, for instance, of very short wave length. This makes no allowance for the fact that, if our eyes were sensible to this light, the whole sky would appear much brighter than the sun, so that the sun would be seen to stand out in black, as otherwise it would repel instead of attract us. For all these reasons, the light that would enable us to explain attraction would require to be much more akin to Röntgen's X rays than to ordinary light.

Furthermore, the X rays will not do. However penetrating they may appear to us, they cannot pass through the whole earth, and we must accordingly imagine X′ rays much more penetrating than the ordinary X rays. Then a portion of the energy of these X′ rays must be destroyed, as otherwise there would be no attraction. If we do not wish it to be transformed into heat, which would lead to the production of an enormous heat, we must admit that it is radiated in all directions in the form of secondary rays, which we may call X″ rays, which must be much more penetrating even than the X′ rays, failing which they would in their turn disturb the phenomena of attraction.

Such are the complicated hypotheses to which we are led when we seek to make Lesage's theory tenable.

But all that has been said assumes the ordinary laws of Mechanics. Will the case be stronger if we admit the new dynamics? And in the first place, can we preserve the principle of relativity? First let us give Lesage's theory its original form, and imagine space furrowed by material corpuscles. If these corpuscles were perfectly elastic, the laws of their collision would be in conformity with this principle of relativity, but we know that in that case their effect would be nil. We must therefore suppose that these corpuscles are not elastic; and then it is difficult to imagine a law of collision compatible with the principle of relativity. Besides, we should still get a considerable production of heat, and, notwithstanding that, a very appreciable resistance of the medium.

If we suppress the corpuscles and return to the hypothesis of the Maxwell-Bartholi pressure, the difficulties are no smaller. It is this that tempted Lorentz himself in his *Mémoire* to the Academy of Sciences of Amsterdam of the 25th of April 1900.

Let us consider a system of electrons immersed in an ether traversed in all directions by luminous waves. One of these electrons struck by one of these waves will be set in vibration. Its vibration will be synchronous with that of the light, but there may be a difference of phase, if the electron absorbs a part of the incident energy. If indeed it absorbs energy, it means that it is the vibration of the ether that keeps the electron in vibration, and the electron must accordingly be behind the ether. An electron in motion may be likened to a convection current, therefore every magnetic field, and particularly that due to the luminous disturbance itself, must exercise a mechanical action upon the electron. This action is very slight, and more than that, it changes its sign in the course of the period; nevertheless the mean action is not nil if there is a difference of phase between the vibrations of the electron and those of the ether. The mean action is proportional to this difference, and consequently to the energy absorbed by the electron.

I cannot here enter into the details of the calculations. I will merely state that the final result is an attraction between any two electrons varying in the inverse ratio of the square of the distance, and proportional to the energy absorbed by the two electrons.

There cannot, therefore, be attraction without absorption of light, and consequently without production of heat, and it is this that determined Lorentz to abandon this theory, which does not differ fundamentally from the Lesage-Maxwell-Bartholi theory. He would have been still more alarmed if he had pushed the calculations to the end, for he would have found that the earth's temperature must increase 10^{13} degrees a second.

IV. Conclusions

I have attempted to give in a few words as complete an idea as possible of these new doctrines; I have tried to explain how they took birth, as otherwise the reader would have had cause to be alarmed by their boldness. The new theories are not yet demonstrated—they are still far from it, and rest merely upon an aggregation of probabilities sufficiently imposing to forbid our treating them with contempt. Further experiments will no doubt teach us what we must finally think of them. The root of the question is in Kaufmann's experiment and such as may be attempted in verification of it.

In conclusion, may I be permitted to express a wish? Suppose that in a few years from now these theories are subjected to new tests and come out triumphant, our secondary education will then run a great risk. Some teachers will no doubt wish to make room for the new theories. Novelties are so attractive, and it is so hard not to appear sufficiently advanced! At least they will wish to open up prospects to the children, who will be warned, before they are taught the ordinary Mechanics, that it has had its day, and that at most it was only good for such an old fogey as Laplace. Then they will never become familiar with the ordinary mechanics.

Is it good to warn them that it is only approximate? Certainly, but not till later on; when they are steeped to the marrow in the old laws, when they have got into the way of thinking in them, and are no longer in danger of unlearning them, then they may safely be shown their limitations.

It is with the ordinary mechanics that they have to live; it is the only kind they will ever have to apply. Whatever be the progress of motoring, our cars will never attain the velocities at which its laws cease to be true. The other is only a luxury, and we must not think of luxury until there is no longer any risk of its being detrimental to what is necessary.

PART IV

ASTRONOMICAL

SCIENCE

CHAPTER I

The Milky Way and the Theory of Gases

The considerations I wish to develop here have so far attracted but little attention from astronomers. I have merely to quote an ingenious idea of Lord Kelvin's, which has opened to us a new field of research, but still remains to be followed up. Neither have I any original results to make known, and all that I can do is to give an idea of the problems that are presented, but that no one, up to this time, has made it his business to solve.

Everyone knows how a great number of modern physicists represent the constitution of gases. Gases are composed of an innumerable multitude of molecules which are animated with great velocities, and cross and re-cross each other in all directions. These molecules probably act at a distance one upon another, but this action decreases very rapidly with the distance, so that their trajectories remain apparently rectilineal, and only cease to be so when two molecules happen to pass sufficiently close to one another, in which case their mutual attraction or repulsion causes them to deviate to right or left. This is what is sometimes called a collision, but we must not understand this word *collision* in its ordinary sense; it is not necessary that the two molecules should come into contact, but only that they should come near enough to each other for their mutual attraction to become perceptible. The laws of the deviation they undergo are the same as if there had been an actual collision.

It seems at first that the orderless collisions of this innumerable dust can only engender an inextricable chaos before which the analyst must retire. But the law of great numbers, that supreme law of chance, comes to our assistance. In face of a semi-disorder we should be forced to despair, but in extreme disorder this statistical law re-establishes a kind of average or mean order in which the mind can find itself again. It is the study of this mean order that constitutes the kinetic theory of gases; it shows us that the velocities of the molecules are equally distributed in all directions, that the amount of these velocities varies for the different molecules, but that this very variation is subject to a law called Maxwell's law. This law teaches us how many molecules there are animated with such and such a velocity. As soon as a gas departs from this law, the mutual collisions of the molecules tend to bring it back promptly, by modifying the amount and direction of their velocities. Physicists have attempted, and not without success, to explain in this manner the experimental properties of gases—for instance, Mariotte's (or Boyle's) law.

Consider now the Milky Way. Here also we see an innumerable dust, only the grains of this dust are no longer atoms but stars; these grains also move with great velocities, they act at a distance one upon another, but this action is so slight at great distances that their trajectories are rectilineal; nevertheless, from time to time, two of them may come near enough together to be deviated from their course, like a comet that passed too close to Jupiter. In a word, in the eyes of a giant, to whom our suns were what our atoms are to us, the Milky Way would only look like a bubble of gas.

Such was Lord Kelvin's leading idea. What can we draw from this comparison, and to what extent is it accurate? This is what we are going to enquire into together; but before arriving at a definite conclusion, and without wishing to prejudice the question, we anticipate that the kinetic theory of gases will be, for the astronomer, a model which must not be followed blindly, but may afford him useful inspiration. So far celestial mechanics has attacked only the solar system, or a few systems of double stars. It retired before the aggregations presented by the Milky Way, or clusters of stars, or resoluble nebulæ, because it saw in them only chaos. But the Milky Way is no more complicated than a gas; the statistical methods based upon the calculation of probabilities applicable to the one are also applicable to the other. Above all, it is important to realize the resemblance and also the difference between the two cases.

Lord Kelvin attempted to determine by this means the dimensions of the Milky Way. For this purpose we are reduced to counting the stars visible in our telescopes, but we cannot be sure that, behind the stars we see, there are not others which we do not see; so that what we should measure in this manner would not be the size of the Milky Way, but the scope of our instruments. The new theory will offer us other resources. We know, indeed, the motions of the stars nearest to us, and we can form an idea of the amount and direction of their velocities. If the ideas expounded above are correct, these velocities must follow Maxwell's law, and their mean value will teach us, so to speak, what corresponds with the temperature of our fictitious gas. But this temperature itself depends upon the dimensions of our gaseous bubble. How, in fact, will a gaseous mass, left undisturbed in space, behave, if its elements are attracted in accordance with Newton's law? It will assume a spherical shape; further, in consequence of gravitation, the density will be greater at the centre, and the pressure will also increase from the surface to the centre on account of the weight of the exterior parts attracted towards the centre; lastly, the temperature will increase towards the centre, the temperature and the pressure being connected by what is called the adiabatic law, as is the case in the successive layers of our atmosphere. At the surface itself the pressure will be nil, and the same will be true of the absolute temperature, that is to say, of the velocity of the molecules.

Here a question presents itself. I have spoken of the adiabatic law, but this law is not the same for all gases, since it depends upon the proportion of their two specific heats. For air and similar gases this proportion is 1.41; but is it to air that the Milky Way should be compared? Evidently not. It should be regarded as a monatomic gas, such as mercury vapour, argon, or helium—that is to say, the proportion of the specific heats should be taken as equal to 1.66. And, indeed, one of our molecules would be, for instance, the solar system; but the planets are very unimportant personages and the sun alone counts, so that our molecule is clearly monatomic. And even if we take a double star, it is probable that the action of a foreign star that happened to approach would become sufficiently appreciable to deflect the general motion of the system long before it was capable of disturbing the relative orbits of the two components. In a word, the double star would behave like an indivisible atom.

However this may be, the pressure, and consequently the temperature, at the centre of the gaseous sphere are proportional to the size of the sphere, since the pressure is increased by the weight of all the overlying strata. We may suppose that we are about at the centre of the Milky Way, and, by observing the actual mean velocity of the stars, we shall know what corresponds to the central temperature of our gaseous sphere and be able to determine its radius.

We may form an idea of the result by the following considerations. Let us make a simple hypothesis. The Milky Way is spherical, and its masses are distributed homogeneously: it follows that the stars describe ellipses having the same centre. If we suppose that the velocity drops to nothing at the surface, we can calculate this velocity at the centre by the equation of *vis viva*. We thus find that this velocity is proportional to the radius of the sphere and the square root of its density. If the mass of this sphere were that of the sun, and its radius that of the terrestrial orbit, this velocity, as is easily seen, would be that of the earth upon its orbit. But in the case we have supposed, the sun's mass would have to be distributed throughout a sphere with a radius 1,000,000 times as great, this radius being the distance of the nearest stars. The density is accordingly 10^{18} times as small; now the velocities are upon the same scale, and therefore the radius must be 10^9 as great, or 1,000 times the distance of the nearest stars, which would give about a thousand million stars in the Milky Way.

But you will tell me that these hypotheses are very far removed from reality. Firstly, the Milky Way is not spherical (we shall soon return to this point); and secondly, the kinetic theory of gases is not compatible with the hypothesis of a homogeneous sphere. But if we made an exact calculation in conformity with this theory, though we should no doubt obtain a different result, it would still be of the same order of magnitude: now in such a problem the data are so uncertain that the order of magnitude is the only end we can aim at.

And here a first observation suggests itself. Lord Kelvin's result, which I have just obtained again by an approximate calculation, is in marked accordance with the estimates that observers have succeeded in making with their telescopes, so that we must conclude that we are on the point of piercing the Milky Way. But this enables us to solve another question. There are the stars we see because they shine, but might there not be dark stars travelling in the interstellar spaces, whose existence might long remain unknown? But in that case, what Lord Kelvin's method gives us would be the total number of stars, including the dark stars, and as his figure compares with that given by the telescope, there is not any dark matter, or at least not as much dark as there is brilliant matter.

Before going further we must consider the problem under another aspect. Is the Milky Way, thus constituted, really the image of a gas properly so called? We know that Crookes introduced the notion of a fourth state of matter, in which gases, becoming too rarefied, are no longer true gases, but become what he calls radiant matter. In view of the slightness of its density, is the Milky Way the image of gaseous or of radiant matter? It is the consideration of what is called the *free path* of the molecules that will supply the answer.

A gaseous molecule's trajectory may be regarded as composed of rectilineal segments connected by very small arcs corresponding with the successive collisions. The length of each of these segments is what is called the free path. This length is obviously not the same for all the segments and for all the molecules; but we may take an average, and this is called the *mean free path*, and its length is in inverse proportion to the density of the gas. Matter will be radiant when the mean path is greater than the dimensions of the vessel in which it is enclosed, so that a molecule is likely to traverse the whole vessel in which the gas is enclosed, without experiencing a collision, and it remains gaseous when the contrary is true. It follows that the same fluid may be radiant in a small vessel and gaseous in a large one, and this is perhaps the reason why, in the case of Crookes' tubes, a more perfect vacuum is required for a larger tube.

What, then, is the case of the Milky Way? It is a mass of gas of very low density, but of very great dimensions. Is it likely that a star will traverse it without meeting with any collision—that is to say, without passing near enough to another star to be appreciably diverted from its course? What do we mean by *near enough*? This is necessarily somewhat arbitrary, but let us assume that it is the distance from the sun to Neptune, which represents a deviation of about ten degrees. Supposing, now, that each of our stars is surrounded by a danger sphere of this radius, will a straight line be able to pass between these spheres? At the mean distance of the stars of the Milky Way, the radius of these spheres will subtend an angle of about a tenth of a second, and we have a thousand million stars. If we place upon the celestial sphere a thousand million little circles with radius of a tenth of a second, will these circles cover the celestial sphere many times over? Far from it. They will only cover a sixteen-thousandth part. Thus the Milky Way is not the image of gaseous matter, but of Crookes' radiant matter. Nevertheless, as there was very little precision in our previous conclusions, we do not require to modify them to any appreciable extent.

But there is another difficulty. The Milky Way is not spherical, and up to now we have reasoned as though it were so, since that is the form of equilibrium that would be assumed by a gas isolated in space. On the other hand, there are clusters of stars whose form is globular, to which what we have said up to this point would apply better. Herschel had already applied himself to the explanation of their remarkable appearance. He assumed that the stars of these clusters are uniformly distributed in such a way that a cluster is a homogeneous sphere. Each star would then describe an ellipse, and all these orbits would be accomplished in the same time, so that at the end of a certain period the cluster would return to its original configuration, and that configuration would be stable. Unfortunately the clusters do not appear homogeneous. We observe a condensation at the centre, and we should still observe it even though the sphere were homogeneous, since it is thicker at the centre, but it would not be so marked. A cluster may, therefore, better be compared to a gas in adiabatic equilibrium which assumes a spherical form, because that is the figure of equilibrium of a gaseous mass.

But, you will say, these clusters are much smaller than the Milky Way, of which it is even probable that they form a part, and although they are denser, they give us rather something analogous to radiant matter. Now, gases only arrive at their adiabatic equilibrium in consequence of innumerable collisions of the molecules. We might perhaps find a method of reconciling these facts. Suppose the stars of the cluster have just sufficient energy for their velocity to become nil when they reach the surface. Then they may traverse the cluster without a collision, but on reaching the surface they turn back and traverse it again. After traversing it a great number of times, they end by being deflected by a collision. Under these conditions we should still have a matter that might be regarded as gaseous. If by chance there were stars in the cluster with greater velocities, they have long since emerged from it, and have left it never to return. For all these reasons it would be interesting to examine the known clusters and try to get an idea of the law of their densities and see if it is the adiabatic law of gases.

But to return to the Milky Way. It is not spherical, and would be more properly represented as a flattened disc. It is clear, then, that a mass starting without velocity from the surface will arrive at the centre with varying velocities, according as it has started from the surface in the neighbourhood of the middle of the disc or from the edge of the disc. In the latter case the velocity will be considerably greater.

Now up to the present we have assumed that the individual velocities of the stars, the velocities we observe, must be comparable to those that would be attained by such masses. This involves a certain difficulty. I have given above a value for the dimensions of the Milky Way, and I deduced it from the observed individual velocities, which are of the same order of magnitude as that of the earth upon its orbit; but what is the dimension I have thus measured? Is it the thickness or the radius of the disc? It is, no doubt, something between the two, but in that case what can be said of the thickness itself, or of the radius of the disc? Data for making the calculation are wanting, and I content myself with foreshadowing the possibility of basing at least an approximate estimate upon a profound study of the individual motions.

Now, we find ourselves confronted by two hypotheses. Either the stars of the Milky Way are animated with velocities which are in the main parallel with the galactic plane, but otherwise distributed uniformly in all directions parallel with this plane. If so, observation of the individual motions should reveal a preponderance of components parallel with the Milky Way. This remains to be ascertained, for I do not know that any systematic study has been made from this point of view. On the other hand, such an equilibrium could only be provisional, for, in consequence of collisions, the molecules—I mean the stars—will acquire considerable velocities in a direction perpendicular to the Milky Way, and will end by emerging from its plane, so that the system will tend towards the spherical form, the only figure of equilibrium of an isolated gaseous mass.

Or else the whole system is animated with a common rotation, and it is for this reason that it is flattened, like the earth, like Jupiter, and like all rotating bodies. Only, as the flattening is considerable, the rotation must be rapid. Rapid, no doubt, but we must understand the meaning of the word. The density of the Milky Way is 10^{25} times as low as the sun's; a velocity of revolution $\sqrt{10^{25}}$ times smaller than the sun's would therefore be equivalent in its case from the point of view of the flattening. A velocity 10^{12} times as slow as the earth's, or the thirtieth of a second of arc in a century, will be a very rapid revolution, almost too rapid for stable equilibrium to be possible.

In this hypothesis, the observable individual motions will appear to us uniformly distributed, and there will be no more preponderance of the components parallel with the galactic plane. They will teach us nothing with respect to the rotation itself, since we form part of the rotating system. If the spiral nebulæ are other Milky Ways foreign to ours, they are not involved in this rotation, and we might study their individual motions. It is true that they are very remote, for if a nebula has the dimensions of the Milky Way, and if its apparent radius is, for instance, 20″, its distance is 10,000 times the radius of the Milky Way.

But this does not matter, since it is not about the rectilinear motion of our system that we ask them for information, but about its rotation. The fixed stars, by their apparent motion, disclose the diurnal rotation of the earth, although their distance is immense. Unfortunately, the possible rotation of the Milky Way, rapid as it is, relatively speaking, is very slow from the absolute point of view, and, moreover, bearings upon nebulæ cannot be very exact. It would accordingly require thousands of years of observation to learn anything.

However it be, in this second hypothesis, the figure of the Milky Way would be a figure of ultimate equilibrium.

I will not discuss the relative value of these two hypotheses at any greater length, because there is a third which is perhaps more probable. We know that among the irresoluble nebulæ several families can be distinguished, the irregular nebulæ such as that in Orion, the planetary and annular nebulæ, and the spiral nebulæ. The spectra of the first two families have been determined, and prove to be discontinuous. These nebulæ are accordingly not composed of stars. Moreover, their distribution in the sky appears to depend upon the Milky Way, whether they show a tendency to be removed from it, or on the contrary to approach it, and therefore they form part of the system. On the contrary, the spiral nebulæ are generally considered as independent of the Milky Way: it is assumed that they are, like it, composed of a multitude of stars; that they are, in a word, other Milky Ways very remote from ours. The work recently done by Stratonoff tends to make us look upon the Milky Way itself as a spiral nebula, and this is the third hypothesis of which I wished to speak.

How are we to explain the very singular appearances presented by the spiral nebulæ, which are too regular and too constant to be due to chance? To begin with, it is sufficient to cast one's eyes upon one of these figures to see that the mass is in rotation, and we can even see the direction of the rotation: all the spiral radii are curved in the same direction, and it is evident that it is the *advancing wing* hanging back upon the *pivot,* and that determines the direction of the rotation. But that is not all. It is clear that these nebulæ cannot be likened to a gas in repose, nor even to a gas in relative equilibrium under the domination of a uniform rotation; they must be compared to a gas in permanent motion in which internal currents rule.

Suppose, for example, that the rotation of the central nucleus is rapid (you know what I mean by this word), too rapid for stable equilibrium. Then at the equator the centrifugal force will prevail over the attraction, and the stars will tend to escape from the equator, and will form divergent currents. But as they recede, since their momentum of rotation remains constant and the radius vector increases, their angular velocity will diminish, and it is for this reason that the advancing wing appears to hang back.

Under this aspect of the case there would not be a true permanent motion, for the central nucleus would constantly lose matter which would go out never to return, and would be gradually exhausted. But we may modify the hypothesis. As it recedes, the star loses its velocity and finally stops. At that moment the attraction takes possession of it again and brings it back towards the nucleus, and accordingly there will be centripetal currents. We must assume that the centripetal currents are in the first rank and the centrifugal currents in the second rank, if we take as a comparison a company in battle executing a turning movement. Indeed the centrifugal force must be compensated by the attraction exercised by the central layers of the swarm upon the exterior layers.

Moreover, at the end of a certain length of time, a permanent status is established. As the swarm becomes curved, the attraction exercised by the advancing wing upon the pivot tends to retard the pivot, and that of the pivot upon the advancing wing tends to accelerate the advance of this wing, whose retrograde motion increases no further, so that finally all the radii end by revolving at a uniform velocity. We may nevertheless assume that the rotation of the nucleus is more rapid than that of the radii.

One question remains. Why do these centripetal and centrifugal swarms tend to concentrate into radii instead of being dispersed more or less throughout, and why are these radii regularly distributed? The reason for the concentration of the swarms is the attraction exercised by the swarms already existing upon the stars that emerge from the nucleus in their neighbourhood. As soon as an inequality is produced, it tends to be accentuated by this cause.

Why are the radii regularly distributed? This is a more delicate matter. Suppose there is no rotation, and that all the stars are in two rectangular planes in such a way that their distribution is symmetrical in relation to the two planes. By symmetry, there would be no reason for their emerging from the planes nor for the symmetry to be altered. This configuration would accordingly give equilibrium, but *it would be an unstable equilibrium.*

If there is rotation on the contrary, we shall get an analogous configuration of equilibrium with four curved radii, equal to one another, and intersecting at an angle of 90°, and if the rotation is sufficiently rapid, this equilibrium may be stable.

I am not in a position to speak more precisely. It is enough for me to foreshadow the possibility that these spiral forms may, perhaps, some day be explained by the help only of the law of gravitation and statistical considerations, recalling those of the theory of gases.

What I have just said about internal currents shows that there might be some interest in a systematic study of the aggregate of the individual motions. This might be undertaken a hundred years hence, when the second edition of the astrographic chart of the heavens is brought out and compared with the first, the one that is being prepared at present.

But I should wish, in conclusion, to call your attention to the question of the age of the Milky Way and the nebulæ. We might form an idea of this age if we obtained confirmation of what we have imagined to be the case. This kind of statistical equilibrium of which gases supply the model, cannot be established except as a consequence of a great number of collisions. If these collisions are rare, it can only be produced after a very long time. If actually the Milky Way (or at least the clusters that form part of it), and if the nebulæ have obtained this equilibrium, it is because they are very ancient, and we shall get an inferior limit for their age. We shall likewise obtain a superior limit, for this equilibrium is not ultimate and cannot last forever. Our spiral nebulæ would be comparable to gases animated with permanent motions. But gases in motion are viscous and their velocities are finally expended. What corresponds in this case to viscidity (and depends upon the chances of collision of the molecules) is exceedingly slight, so that the actual status may continue for a very long time, but not forever, so that our Milky Ways cannot be everlasting nor become infinitely ancient.

But this is not all. Consider our atmosphere. At the surface an infinitely low temperature must prevail, and the velocity of the molecules is in the neighbourhood of zero. But this applies only to the mean velocity. In consequence of collisions, one of these molecules may acquire (rarely, it is true) an enormous velocity, and then it will leave the atmosphere, and once it has left it, it will never return. Accordingly our atmosphere is being exhausted exceedingly slowly. By the same mechanism the Milky Way will also lose a star from time to time, and this likewise limits its duration.

Well, it is certain that if we calculate the age of the Milky Way by this method, we shall arrive at enormous figures. But here a difficulty presents itself. Certain physicists, basing their calculations on other considerations, estimate that suns can have but an ephemeral existence of about fifty million years, while our minimum would be much greater than that. Must we believe that the evolution of the Milky Way began while matter was still dark? But how have all the stars that compose it arrived at the same time at the adult period, a period which lasts for so short a time? Or do they all reach it successively, and are those that we see only a small minority as compared with those that are extinct or will become luminous some day? But how can we reconcile this with what has been said above about the absence of dark matter in any considerable proportion? Must we abandon one of the two hypotheses, and, if so, which? I content myself with noting the difficulty, without pretending to solve it, and so I end with a great mark of interrogation. Still, it is interesting to state problems even though their solution seems very remote.

CHAPTER II

FRENCH GEODESY[*]

Everyone understands what an interest we have in knowing the shape and the dimensions of our globe, but some people would perhaps be astonished at the precision that is sought for. Is this a useless luxury? What is the use of the efforts geodesists devote to it?

If a Member of Parliament were asked this question, I imagine he would answer: "I am led to think that Geodesy is one of the most useful of sciences, for it is one of those that cost us most money." I shall attempt to give a somewhat more precise answer.

The great works of art, those of peace as well as those of war, cannot be undertaken without long studies, which save many gropings, miscalculations, and useless expense. These studies cannot be made without a good map. But a map is nothing but a fanciful picture, of no value whatever if we try to construct it without basing it upon a solid framework. As well might we try to make a human body stand upright with the skeleton removed.

---

[*] Throughout this chapter the author is speaking of the work of his own countrymen. In the translation words such as *we* and *our* have been avoided, as far as possible; but where they occur, they must be understood to refer to France and not to England.

Now this framework is obtained by geodetic measurements. Therefore without Geodesy we can have no good map, and without a good map no great public works.

These reasons would no doubt be sufficient to justify much expense, but they are reasons calculated to convince practical men. It is not upon these that we should insist here; there are higher and, upon the whole, more important reasons.

We will therefore state the question differently: Can Geodesy make us better acquainted with nature? Does it make us understand its unity and harmony? An isolated fact indeed is but of little worth, and the conquests of science have a value only if they prepare new ones.

Accordingly, if we happened to discover a little hump upon the terrestrial ellipsoid, this discovery would be of no great interest in itself. It would become precious on the contrary if, in seeking for the cause of the hump, we had the hope of penetrating new secrets.

So when Maupertuis and La Condamine in the eighteenth century braved such diverse climates, it was not only for the sake of knowing the shape of our planet, it was a question of the system of the whole world. If the earth was flattened, Newton was victorious, and with him the doctrine of gravitation and the whole of the modern celestial mechanics. And today, a century and a half since the victory of the Newtonians, are we to suppose that Geodesy has nothing more to teach us? We do not know what there is in the interior of the globe. Mine shafts and borings have given us some knowledge of a stratum one or two miles deep—that is to say, the thousandth part of the total mass; but what is there below that?

Of all the extraordinary voyages dreamt of by Jules Verne, it was perhaps the voyage to the centre of the earth that led us to the most unexplored regions.

But those deep sunk rocks that we cannot reach, exercise at a distance the attraction that acts upon the pendulum and deforms the terrestrial spheroid. Geodesy can therefore weigh them at a distance, so to speak, and give us information about their disposition. It will thus enable us really to see those mysterious regions which Jules Verne showed us only in imagination.

This is not an empty dream. By comparing all the measurements, M. Faye has reached a result well calculated to cause surprise. In the depths beneath the oceans, there are rocks of very great density, while, on the contrary, beneath the continents there seem to be empty spaces.

New observations will perhaps modify these conclusions in their details, but our revered master has, at any rate, shown us in what direction we must push our researches, and what it is that the geodesist can teach the geologist who is curious about the interior constitution of the earth, and what material he can supply to the thinker who wishes to reflect upon the past and the origin of this planet.

Now why have I headed this chapter "French Geodesy"? It is because, in different countries, this science has assumed, more perhaps than any other, a national character; and it is easy so see the reason for this.

There must certainly be rivalries. Scientific rivalries are always courteous, or, at least, almost always. In any case they are necessary, because they are always fruitful.

Well, in these enterprises that demand such long efforts and so many collaborators, the individual is effaced, in spite of himself of course. None has the right to say, this is my work. So the rivalry is not between individuals, but between nations. Thus we are led to ask what share France has taken in the work, and I think we have a right to be proud of what she has done.

At the beginning of the eighteenth century there arose long discussions between the Newtonians, who believed the earth to be flattened as the theory of gravitation demands, and Cassini, who was misled by inaccurate measurements, and believed the globe to be elongated. Direct observation alone could settle the question. It was the French Academy of Sciences that undertook this task, a gigantic one for that period.

While Maupertuis and Clairaut were measuring a degree of longitude within the Arctic Circle, Bouguer and La Condamine turned their faces towards the mountains of the Andes, in regions that were then subject to Spain, and today form the Republic of

Ecuador. Our emissaries were exposed to great fatigues, for journeys then were not so easy as they are today.

It is true that the country in which Maupertuis' operations were conducted was not a desert, and it is even said that he enjoyed among the Lapps those soft creature comforts that are unknown to the true Arctic navigator. It was more or less in the neighbourhood of places to which, in our day, comfortable steamers carry, every summer, crowds of tourists and young English ladies. But at that date Cook's Agency did not exist, and Maupertuis honestly thought that he had made a Polar expedition.

Perhaps he was not altogether wrong. Russians and Swedes are today making similar measurements at Spitzbergen, in a country where there are real ice-packs. But their resources are far greater, and the difference of date fully compensates for the difference of latitude.

Maupertuis' name has come down to us considerably mauled by the claws of Dr. Akakia, for Maupertuis had the misfortune to displease Voltaire, who was then king of the mind. At first he was extravagantly praised by Voltaire; but the flattery of kings is as much to be dreaded as their disfavour, for it is followed by a terrible day of reckoning. Voltaire himself learnt something of this.

Voltaire called Maupertuis "my kind master of thought," "Marquess of the Arctic Circle," "dear flattener of the world and of Cassini," and even, as supreme flattery, "Sir Isaac Maupertuis"; and he wrote, "There is none but the King of Prussia that I place on a level with you; his sole defect is that he is not a geometrician." But very soon the scene changes; he no longer speaks of deifying him, like the Argonauts of old, or of bringing down the council of the gods from Olympus to contemplate his work, but of shutting him up in a mad-house. He speaks no more of his sublime mind, but of his despotic pride, backed by very little science and much absurdity.

I do not wish to tell the tale of these mock-heroic conflicts, but I should like to make a few reflections upon two lines of Voltaire's. In his *Discours sur la modération* (there is no question of moderation in praise or blame), the poet wrote:

> Vous avez confirmé dans des lieux pleins d'ennui
> Ce que Newton connut sans sortir de chez lui.
>
> (You have confirmed, in dreary far-off lands,
> What Newton knew without e'er leaving home.)

These two lines, which take the place of the hyperbolical praises of earlier date, are most unjust, and without any doubt, Voltaire was too well informed not to realize it.

At that time men valued only the discoveries that can be made without leaving home. Today it is theory rather that is held in low esteem. But this implies a misconception of the aim of science.

Is nature governed by caprice, or is harmony the reigning influence? That is the question. It is when science reveals this harmony that it becomes beautiful, and for that reason worthy of being cultivated. But whence can this revelation come if not from the accordance of a theory with experience? Our aim then is to find out whether or not this accordance exists. From that moment, these two terms, which must be compared with each other, become one as indispensable as the other. To neglect one for the other would be folly. Isolated, theory is empty and experience blind; and both are useless and of no interest alone.

Maupertuis is therefore entitled to his share of the fame. Certainly it is not equal to that of Newton, who had received the divine spark, or even of his collaborator Clairaut. It is not to be despised, however, because his work was necessary; and if France, after being outstripped by England in the seventeenth century, took such full revenge in the following century, it was not only to the genius of the Clairauts, the d'Alemberts, and the Laplaces that she owed it, but also to the long patience of such men as Maupertuis and La Condamine.

We come now to what may be called the second heroic period of Geodesy. France was torn with internal strife, and the whole of Europe was in arms against her. One would suppose that these tremendous struggles must have absorbed all her energies. Far from that, however, she had still some left for the service of science. The men of that day shrank before no enterprise—they were men of faith.

Delambre and Méchain were commissioned to measure an arc running from Dunkirk to Barcelona. This time there is no journey to Lapland or Peru; the enemy's squadrons would close the roads. But if the expeditions are less distant, the times are so troublous that the obstacles and even the dangers are quite as great.

In France Delambre had to fight against the ill-will of suspicious municipalities. One knows that steeples, which can be seen a long way off, and observed with precision, often serve as signals for geodesists. But in the country Delambre was working through, there were no steeples left. I forget now what proconsul it was who had passed through it and boasted that he had brought down all the steeples that raised their heads arrogantly above the humble dwellings of the common people.

So they erected pyramids of planks covered with white linen to make them more conspicuous. This was taken to mean something quite different. White linen! Who was the foolhardy man who ventured to set up, on our heights so recently liberated, the odious standard of the counter-revolution? The white linen must needs be edged with blue and red stripes.

Méchain, operating in Spain, met with other but no less serious difficulties. The Spanish country folk were hostile. There was no lack of steeples, but was it not sacrilege to take possession of them with instruments that were mysterious and perhaps diabolical? The revolutionaries were the allies of Spain, but they were allies who smelt a little of the stake.

"We are constantly threatened," writes Méchain, "with having our throats cut." Happily, thanks to the exhortations of the priests, and to the pastoral letters from the bishops, the fiery Spaniards contented themselves with threats.

Some years later, Méchain made a second expedition to Spain. He proposed to extend the meridian from Barcelona to the Balearic Isles. This was the first time that an attempt had been made to cross a large arm of the sea by triangulation, by taking observations of signals erected upon some high mountain in a distant island. The enterprise was well conceived and well planned, but it failed nevertheless. The French scientist met with all kinds of difficulties, of which he complains bitterly in his correspondence. "Hell," he writes, perhaps with some exaggeration, "hell, and all the scourges it vomits upon the earth—storms, war, pestilence, and dark intrigues—are let loose against me!"

The fact is that he found among his collaborators more headstrong arrogance than good-will, and that a thousand incidents delayed his work. The plague was nothing; fear of the plague was much more formidable. All the islands mistrusted the neighbouring islands, and were afraid of receiving the scourge from them. It was only after long weeks that Méchain obtained permission to land, on condition of having all his papers vinegared—such were the antiseptics of those days. Disheartened and ill, he had just applied for his recall, when he died.

It was Arago and Biot who had the honour of taking up the unfinished work and bringing it to a happy conclusion. Thanks to the support of the Spanish Government and the protection of several bishops, and especially of a celebrated brigand chief, the operations progressed rapidly enough. They were happily terminated, and Biot had returned to France, when the storm burst.

It was the moment when the whole of Spain was taking up arms to defend her independence against France. Why was this stranger climbing mountains to make signals? It was evidently to call the French army. Arago only succeeded in escaping from the populace by giving himself up as a prisoner. In his prison his only distraction was reading the account of his own execution in the Spanish newspapers. The newspapers of those days sometimes gave premature news. He had at least the consolation of learning that he had died a courageous and a Christian death.

Prison itself was not safe, and he had to make his escape and reach Algiers. Thence he sailed for Marseilles on an Algerian ship. This ship was captured by a Spanish privateer, and so Arago was brought back to Spain, and dragged from dungeon to dungeon in the midst of vermin and in the most horrible misery.

If it had only been a question of his subjects and his guests, the Dey would have said nothing. But there were two lions on board, a present the African sovereign was sending to Napoleon. The Dey threatened war.

The vessel and the prisoners were released. The point should have been correctly made, since there was an astronomer on board; but the astronomer was seasick, and the Algerian sailors, who wished to go to Marseilles, put in at Bougie. Thence Arago travelled to Algiers, crossing Kabylia on foot through a thousand dangers. He was detained for a long time in Africa and threatened with penal servitude. At last he was able to return to France. His observations, which he had preserved under his shirt, and more extraordinary still, his instruments, had come through these terrible adventures without damage.

Up to this point, France not only occupied the first place, but she held the field almost alone. In the years that followed she did not remain inactive, and the French ordnance map is a model. Yet the new methods of observation and of calculation came principally from Germany and England. It is only during the last forty years that France has regained her position.

She owes it to a scientific officer, General Perrier, who carried out successfully a truly audacious enterprise, the junction of Spain and Africa. Stations were established upon four peaks on the two shores of the Mediterranean. There were long months of waiting for a calm and clear atmosphere. At last there was seen the slender thread of light that had travelled two hundred miles over the sea, and the operation had succeeded.

Today still more daring projects have been conceived. From a mountain in the vicinity of Nice signals are to be sent to Corsica, no longer with a view to the determination of geodetic questions, but in order to measure the velocity of light. The distance is only one hundred and twenty-five miles, but the ray of light is to make the return journey, after being reflected from a mirror in Corsica.

And it must not go astray on the journey, but must return to the exact spot from which it started.

Latterly the activity of French Geodesy has not slackened. We have no more such astonishing adventures to relate, but the scientific work accomplished is enormous. The territory of France beyond the seas, just as that of the mother country, is being covered with triangles measured with precision.

We have become more and more exacting, and what was admired by our fathers does not satisfy us today. But as we seek greater exactness, the difficulties increase considerably. We are surrounded by traps, and have to beware of a thousand unsuspected causes of error. It becomes necessary to make more and more infallible instruments.

Here again France has not allowed herself to be outdone. Her apparatus for the measurement of bases and of angles leaves nothing to be desired, and I would also mention Colonel Defforges' pendulum, which makes it possible to determine gravity with a precision unknown till now.

The future of French Geodesy is now in the hands of the geographical department of the army, which has been directed successively by General Bassot and General Berthaut. This has advantages that can hardly be overestimated. For good geodetic work, scientific aptitude alone is not sufficient. A man must be able to endure long fatigues in all climates. The chief must know how to command the obedience of his collaborators and to enforce it upon his native helpers. These are military qualities, and, moreover, it is known that science has always gone hand in hand with courage in the French army.

I would add that a military organization assures the indispensable unity of action. It would be more difficult to reconcile the pretensions of rival scientists, jealous of their independence and anxious about what they call their honour, who would nevertheless have to operate in concert, though separated by great distances. There arose frequent discussions between geodesists of former times, some of which started echoes that were heard long after. The Academy long rang with the quarrel between Bouguer and La

Condamine. I do not mean to say that soldiers are free from passions, but discipline imposes silence upon over-sensitive vanity.

Several foreign governments have appealed to French officers to organize their geodetic departments. This is a proof that the scientific influence of France abroad has not been weakened.

Her hydrographic engineers also supply a famous contingent to the common work. The chart of her coasts and of her colonies, and the study of tides, offer them a vast field for research. Finally, I would mention the general levelling of France, which is being carried out by M. Lallemand's ingenious and accurate methods.

With such men, we are sure of the future. Work for them to do will not be wanting. The French colonial empire offers them immense tracts imperfectly explored. And that is not all. The International Geodetic Association has recognized the necessity of a new measurement of the arc of Quito, formerly determined by La Condamine. It is the French who have been entrusted with the operation. They had every right, as it was their ancestors who achieved, so to speak, the scientific conquest of the Cordilleras. Moreover, these rights were not contested, and the French Government determined to exercise them.

Captains Maurain and Lacombe made a preliminary survey, and the rapidity with which they accomplished their mission, travelling through difficult countries, and climbing the most precipitous peaks, deserves the highest praise. It excited the admiration of General Alfaro, President of the Republic of Ecuador, who surnamed them *los hombres de hierro,* the men of iron.

The definitive mission started forthwith, under the command of Lieutenant-Colonel (then Commandant) Bourgeois. The results obtained justified the hopes that had been entertained. But the officers met with unexpected difficulties due to the climate. More than once one of them had to remain for several months at an altitude of 13,000 feet, in clouds and snow, without seeing anything of the signals he had to observe, which refused to show themselves. But thanks to their perseverance and courage, the only result was a delay, and an increase in the expenses, and the accuracy of the measurements did not suffer.

General Conclusions

What I have attempted to explain in the foregoing pages is how the scientist is to set about making a selection of the innumerable facts that are offered to his curiosity, since he is compelled to make a selection, if only by the natural infirmity of his mind, though a selection is always a sacrifice. To begin with, I explained it by general considerations, recalling, on the one hand, the nature of the problem to be solved, and on the other, seeking a better understanding of the nature of the human mind, the principal instrument in the solution. Then I explained it by examples, but not an infinity of examples, for I too had to make a selection, and I naturally selected the questions I had studied most carefully. Others would no doubt have made a different selection, but this matters little, for I think they would have reached the same conclusions.

There is a hierarchy of facts. Some are without any positive bearing, and teach us nothing but themselves. The scientist who ascertains them learns nothing but facts, and becomes no better able to foresee new facts. Such facts, it seems, occur but once, and are not destined to be repeated.

There are, on the other hand, facts that give a large return, each of which teaches us a new law. And since he is obliged to make a selection, it is to these latter facts that the scientist must devote himself.

No doubt this classification is relative, and arises from the frailty of our mind. The facts that give but a small return are the complex facts, upon which a multiplicity of circumstances exercise an appreciable influence—circumstances so numerous and so diverse that we cannot distinguish them all. But I should say, rather, that they are the facts that we consider complex, because the entanglement of these circumstances exceeds the compass of our mind. No doubt a vaster and a keener mind than ours would judge otherwise. But that matters little; it is not this superior mind that we have to use, but our own.

The facts that give a large return are those that we consider simple, whether they are so in reality, because they are only influenced by a small number of well-defined circumstances, or whether they take on an appearance of simplicity, because the multiplicity of circumstances upon which they depend obey the laws of chance, and so arrive at a mutual compensation. This is most frequently the case, and is what compelled us to enquire somewhat closely into the nature of chance. The facts to which the laws of chance apply become accessible to the scientist, who would lose heart in face of the extraordinary complication of the problems to which these laws are not applicable.

We have seen how these considerations apply not only to the physical but also to the mathematical sciences. The method of demonstration is not the same for the physicist as for the mathematician. But their methods of discovery are very similar. In the case of both they consist in rising from the fact to the law, and in seeking the facts that are capable of leading up to a law.

In order to elucidate this point, I have exhibited the mathematician's mind at work, and that under three forms: the mind of the inventive and creative mathematician; the mind of the unconscious geometrician who, in the days of our far-off ancestors or in the hazy years of our infancy, constructed for us our instinctive notion of space; and the mind of the youth in a secondary school for whom the master unfolds the first principles of the science, and seeks to make him understand its fundamental definitions. Throughout we have seen the part played by intuition and the spirit of generalization, without which these three grades of mathematicians, if I may venture so to express myself, would be reduced to equal impotence.

And in demonstration itself logic is not all. The true mathematical reasoning is a real induction, differing in many respects from physical induction, but, like it, proceeding from the particular to the universal. All the efforts that have been made to upset this order, and to reduce mathematical induction to the rules of logic, have ended in failure, but poorly disguised by the use of a language inaccessible to the uninitiated.

The examples I have drawn from the physical sciences have shown us a good variety of instances of facts that give a large return. A single experiment of Kaufmann's upon radium rays revolutionizes at once Mechanics, Optics, and Astronomy. Why is this? It is because, as these sciences developed, we have recognized more clearly the links which unite them, and at last we have perceived a kind of general design of the map of universal science. There are facts common to several sciences, like the common fountain head of streams diverging in all directions, which may be compared to that nodal point of the St. Gothard from which there flow waters that feed four different basins.

Then we can make our selection of facts with more discernment than our predecessors, who regarded these basins as distinct and separated by impassable barriers.

It is always simple facts that we must select, but among these simple facts we should prefer those that are situated in these kinds of nodal points of which I have just spoken.

And when sciences have no direct link, they can still be elucidated mutually by analogy. When the laws that regulate gases were being studied, it was realized that the fact in hand was one that would give a great return, and yet this return was still estimated below its true value, since gases are, from a certain point of view, the image of the Milky Way; and these facts, which seemed to be of interest only to the physicist, will soon open up new horizons to the astronomer, who little expected it.

Lastly, when the geodesist finds that he has to turn his glass a few seconds of arc in order to point it upon a signal that he has erected with much difficulty, it is a very small fact, but it is a fact giving a great return, not only because it reveals the existence of a little hump upon the terrestrial geoid, for the little hump would of itself be of small interest, but because this hump gives him indications as to the distribution of matter in the interior of the globe, and, through that, as to the past of our planet, its future, and the laws of its development.

INDEX

A NOTE ON THE TYPE

The principal text of this Modern Library edition was set in a digitized version of Janson, a typeface that dates from about 1690 and was cut by Nicholas Kis, a Hungarian working in Amsterdam. The original matrices have survived and are held by the Stempel foundry in Germany. Hermann Zapf redesigned some of the weights and sizes for Stempel, basing his revisions on the original design.